T0392514

VOLUME ONE HUNDRED AND SIXTY FIVE

INTERNATIONAL REVIEW OF NEUROBIOLOGY

Covid-19 and Parkinsonism

INTERNATIONAL REVIEW OF NEUROBIOLOGY

VOLUME 165

SERIES EDITOR

PATRICIA JANAK

Janak Lab, Department of Neuroscience
Dunning Hall, John Hopkins University
Baltimore, MD, USA

PETER JENNER

Division of Pharmacology and Therapeutics
GKT School of Biomedical Sciences
King's College, London, UK

EDITORIAL BOARD

RICHARD L. BELL
SHAFIQUR RAHMAN
MARIA STAMELOU
K RAY CHAUDHURI
NATALIYA TITOVA
BAI-YUN ZENG
TODD E. THIELE
HARI SHANKER SHARMA
ARUNA SHARMA
GRAŻYNA SÖDERBOM
NATALIE WITEK
NINA SMYTH
ERIN CALIPARI
LIISA GALEA

VOLUME ONE HUNDRED AND SIXTY FIVE

INTERNATIONAL REVIEW OF NEUROBIOLOGY

Covid-19 and Parkinsonism

Edited by

K RAY CHAUDHURI
Department of Basic and Clinical Neuroscience; Department of Neurosciences, Institute of Psychiatry, Psychology & Neuroscience, King's College London; Parkinson's Foundation Centre of Excellence, King's College Hospital NHS Foundation Trust; Institute of Psychiatry, Psychology & Neuroscience at King's College London and King's College Hospital NHS Foundation Trust, London, United Kingdom

MAYELA RODRÍGUEZ-VIOLANTE
Clinical Neurodegenerative Diseases Research Unit, Movement Disorders Clinic at the National Institute of Neurology and Neurosurgery, American British Cowdray Medical Center, Ciudad de México, Mexico

ANGELO ANTONINI
Parkinson and Movement Disorders Unit, Department of Neuroscience, Centre for Rare Neurological Diseases (ERN-RND); Center for Neurodegenerative Disease Research (CESNE), University of Padova, Padova, Italy

IRO BOURA
Department of Neurology, University Hospital of Heraklion; Medical School, University of Crete, Heraklion, Crete, Greece; Department of Neurosciences; Department of Basic and Clinical Neuroscience, Institute of Psychiatry, Psychology & Neuroscience, King's College London; Parkinson's Foundation Centre of Excellence, King's College Hospital NHS Foundation Trust, London, United Kingdom

Academic Press is an imprint of Elsevier
50 Hampshire Street, 5th Floor, Cambridge, MA 02139, United States
525 B Street, Suite 1650, San Diego, CA 92101, United States
The Boulevard, Langford Lane, Kidlington, Oxford OX5 1GB, United Kingdom
125 London Wall, London, EC2Y 5AS, United Kingdom

First edition 2022

Copyright © 2022 Elsevier Inc. All rights reserved.

No part of this publication may be reproduced or transmitted in any form or by any means, electronic or mechanical, including photocopying, recording, or any information storage and retrieval system, without permission in writing from the publisher. Details on how to seek permission, further information about the Publisher's permissions policies and our arrangements with organizations such as the Copyright Clearance Center and the Copyright Licensing Agency, can be found at our website: www.elsevier.com/permissions.

This book and the individual contributions contained in it are protected under copyright by the Publisher (other than as may be noted herein).

Notices
Knowledge and best practice in this field are constantly changing. As new research and experience broaden our understanding, changes in research methods, professional practices, or medical treatment may become necessary.

Practitioners and researchers must always rely on their own experience and knowledge in evaluating and using any information, methods, compounds, or experiments described herein. In using such information or methods they should be mindful of their own safety and the safety of others, including parties for whom they have a professional responsibility.

To the fullest extent of the law, neither the Publisher nor the authors, contributors, or editors, assume any liability for any injury and/or damage to persons or property as a matter of products liability, negligence or otherwise, or from any use or operation of any methods, products, instructions, or ideas contained in the material herein.

ISBN: 978-0-323-99173-5
ISSN: 0074-7742

For information on all Academic Press publications
visit our website at https://www.elsevier.com/books-and-journals

Publisher: Zoe Kruze
Acquisitions Editor: Sam Mahfoudh
Developmental Editor: Federico Paulo Mendoza
Production Project Manager: Abdulla Sait
Cover Designer: Matthew Limbert

Typeset by STRAIVE, India

Contents

Contributors

Angelo Antonini
Parkinson and Movement Disorders Unit, Department of Neuroscience, Centre for Rare Neurological Diseases (ERN-RND); Center for Neurodegenerative Disease Research (CESNE), University of Padova, Padova, Italy

Lucia Batzu
Department of Basic and Clinical Neuroscience, Institute of Psychiatry, Psychology & Neuroscience, King's College London; Parkinson's Foundation Centre of Excellence, King's College Hospital NHS Foundation Trust, London, United Kingdom

Iro Boura
Department of Neurology, University Hospital of Heraklion; Medical School, University of Crete, Heraklion, Crete, Greece; Department of Neurosciences; Department of Basic and Clinical Neuroscience, Institute of Psychiatry, Psychology & Neuroscience, King's College London; Parkinson's Foundation Centre of Excellence, King's College Hospital NHS Foundation Trust, London, United Kingdom

Kallol Ray Chaudhuri
Department of Basic and Clinical Neuroscience; Department of Neurosciences, Institute of Psychiatry, Psychology & Neuroscience, King's College London; Parkinson's Foundation Centre of Excellence, King's College Hospital NHS Foundation Trust; Institute of Psychiatry, Psychology & Neuroscience at King's College London and King's College Hospital NHS Foundation Trust, London, United Kingdom

Ştefania Diaconu
County Clinic Hospital; Faculty of Medicine, Transilvania University, Braşov, Romania

Espen Dietrichs
Department of Neurology, Oslo University Hospital and University of Oslo, Oslo, Norway

Aron Emmi
Institute of Human Anatomy, Department of Neuroscience; Parkinson and Movement Disorders Unit, Department of Neuroscience, Centre for Rare Neurological Diseases (ERN-RND); Center for Neurodegenerative Disease Research (CESNE), University of Padova, Padova, Italy

Cristian Falup-Pecurariu
County Clinic Hospital; Faculty of Medicine, Transilvania University, Braşov, Romania

Alfonso Fasano
Edmond J. Safra Program in Parkinson's Disease, Morton and Gloria Shulman Movement Disorders Clinic, Toronto Western Hospital–UHN, Division of Neurology, University of Toronto; Krembil Research Institute; Center for Advancing Neurotechnological Innovation to Application (CRANIA), Toronto, ON, Canada

Conor Fearon
Edmond J. Safra Program in Parkinson's Disease, Morton and Gloria Shulman Movement Disorders Clinic, Toronto Western Hospital–UHN, Division of Neurology, University of Toronto, Toronto, ON, Canada

James E. Goldman
Department of Pathology and Cell Biology, and the Taub Institute for Research on Alzheimer's Disease and the Aging Brain, Vagelos College of Physicians and Surgeons, Columbia University and the New York Presbyterian Hospital, New York, NY, United States

Annette Hand
Newcastle upon Tyne Hospitals NHS Foundation Trust and Northumbria University, Newcastle upon Tyne, United Kingdom

Laura Irincu
County Clinic Hospital, Braşov, Romania

Irina Ivan
County Clinic Hospital, Braşov, Romania

Prashanth Lingappa Kukkle
Parkinson's Disease and Movement Disorders Clinic; Center for Parkinson's Disease and Movement Disorders, Manipal Hospital, Miller's Road, Bangalore, India

Yue Hui Lau
Department of Neurosciences; Department of Basic and Clinical Neuroscience, Institute of Psychiatry, Psychology & Neuroscience, King's College London; Parkinson's Foundation Centre of Excellence, King's College Hospital NHS Foundation Trust, London, United Kingdom

Claudia Lazcano-Ocampo
Parkinson's Foundation Centre of Excellence, King's College Hospital NHS Foundation Trust, London, United Kingdom; Department of Neurology, Movement Disorders Unit, Hospital Sotero del Rio; Department of Neurology, Clínica INDISA, Santiago, Chile

Valentina Leta
Parkinson's Foundation Centre of Excellence, King's College Hospital NHS Foundation Trust; Department of Basic and Clinical Neuroscience; Department of Neurosciences, Institute of Psychiatry, Psychology & Neuroscience, King's College London, London, United Kingdom

Donna Mathew
Neuroscience and Mental Health Research Institute, School of Medicine, Cardiff University, Cardiff, United Kingdom

Victor McConvey
Fight Parkinson's, Surrey Hills, VIC, Australia

Bradley McDaniels
Department of Rehabilitation and Health Services, University of North Texas, Denton, TX, United States

Janis Miyasaki
Parkinson and Movement Disorders Program and the Complex Neurologic Symptoms Clinic (Neuropalliative Care), University of Alberta, Edmonton, AB, Canada

Aleksandra M. Podlewska
Department of Neurosciences, Institute of Psychiatry, Psychology & Neuroscience; Department of Basic and Clinical Neuroscience, Division of Neuroscience, King's College London; Parkinson's Foundation Centre of Excellence, King's College Hospital NHS Foundation Trust, London, United Kingdom

Andrea Porzionato
Institute of Human Anatomy, Department of Neuroscience; Center for Neurodegenerative Disease Research (CESNE), University of Padova, Padova, Italy

Vanessa Raeder
Parkinson's Foundation Centre of Excellence, King's College Hospital, NHS Foundation Trust, London, United Kingdom; Department of Neurology, Technical University Dresden, Dresden, Germany; Department of Basic and Clinical Neuroscience, Institute of Psychiatry, Psychology & Neuroscience, King's College London, London, United Kingdom

Mayela Rodriguez-Violante
Instituto Nacional de Neurologia y Neurocirugia, Ciudad de México, Mexico

Anna Roszmann
Department of Neuro-Psychiatric Nursing, Medical University of Gdańsk, Gdańsk, Poland

Silvia Rota
Department of Basic and Clinical Neuroscience, Institute of Psychiatry, Psychology & Neuroscience, King's College London; Parkinson's Foundation Centre of Excellence, King's College Hospital NHS Foundation Trust, London, United Kingdom

Katarina Rukavina
Institute of Psychiatry, Psychology & Neuroscience at King's College London and King's College Hospital NHS Foundation Trust; Parkinson's Foundation Centre of Excellence, King's College Hospital NHS Foundation Trust, London, United Kingdom

Michele Sandre
Parkinson and Movement Disorders Unit, Department of Neuroscience, Centre for Rare Neurological Diseases (ERN-RND); Center for Neurodegenerative Disease Research (CESNE), University of Padova, Padova, Italy

Indu Subramanian
David Geffen School of Medicine, UCLA, Department of Neurology; PADRECC, West Los Angeles, Veterans Administration, Los Angeles, CA, United States

David Sulzer
Departments of Psychiatry, Neurology, Pharmacology, Columbia University Medical Center, New York State Psychiatric Institute, New York, United States

Eng-King Tan
Duke-NUS Medical School, National Neuroscience Institute, Singapore, Singapore

Daniel J. van Wamelen
Department of Neurosciences; Department of Basic and Clinical Neuroscience, Institute of Psychiatry, Psychology & Neuroscience; Department of Basic and Clinical Neuroscience, Division of Neuroscience, King's College London; Parkinson's Foundation Centre of Excellence, King's College Hospital NHS Foundation Trust, London, United Kingdom;

Department of Neurology, Centre of Expertise for Parkinson & Movement Disorders, Donders Institute for Brain, Cognition and Behaviour, Radboud University Medical Center, Nijmegen, the Netherlands

Yi-Min Wan
Department of Basic and Clinical Neuroscience, Institute of Psychiatry, Psychology & Neuroscience, King's College London; Parkinson's Foundation Centre of Excellence, King's College Hospital NHS Foundation Trust, London, United Kingdom; Department of Psychiatry, Ng Teng Fong General Hospital, Singapore, Singapore

CHAPTER ONE

Parkinsonism associated with viral infection

Irina Ivan[a,†], Laura Irincu[a,†], Ştefania Diaconu[a,b,*], and Cristian Falup-Pecurariu[a,b]

[a]County Clinic Hospital, Braşov, Romania
[b]Faculty of Medicine, Transilvania University, Braşov, Romania
*Corresponding author: e-mail address: stefi_diaconu@yahoo.com

Contents

Abstract

There are several known causes of secondary parkinsonism, the most common being head trauma, stroke, medications, or infections. A growing body of evidence suggests that viral agents may trigger parkinsonian symptoms, but the exact pathological mechanisms are still unknown. In some cases, lesions or inflammatory processes in the basal ganglia or substantia nigra have been found to cause reversible or permanent impairment of the dopaminergic pathway, leading to the occurrence of extrapyramidal symptoms.

This chapter reviews current data regarding the viral agents commonly associated with parkinsonism, such as Epstein Barr virus (EBV), hepatitis viruses, human immunodeficiency virus (HIV), herpes viruses, influenza virus, coxsackie virus, and Severe Acute Respiratory Syndrome Coronavirus-2 (SARS-CoV-2). We present possible risk factors, proposed pathophysiology mechanisms, published case reports, common associations, and prognosis in order to offer a concise overview of the viral spectrum involved in parkinsonism.

† These authors contributed equally to this work.

International Review of Neurobiology, Volume 165
ISSN 0074-7742
https://doi.org/10.1016/bs.irn.2022.07.005
Copyright © 2022 Elsevier Inc.
All rights reserved.

1. Introduction

Pathogenesis of Parkinson's disease (PD) has been assumed to involve both environmental factors and genetics, while aggregation of the neuronal protein alpha-synuclein is considered necessary for PD occurrence (Woulfe, Gray, Gray, Munoz, & Middeldorp, 2014). However, the type of environmental triggers which might lead to or unmask PD, has not been clarified yet.

The viral etiology of PD has been suggested by many researchers, although the exact mechanisms by which a viral infection can be associated with this disease remains unknown. More specifically, under certain circumstances, a direct viral insult of the nigral neurons or virally-induced autoimmune mechanisms have been thought to be implicated in PD pathogenesis (Woulfe et al., 2014). Meanwhile, parkinsonian features have also been observed in some cases of viral encephalitis, including coxsackie B virus, Japanese encephalitis virus, poliovirus, human immunodeficiency virus (HIV), measles virus and Central European tickborne encephalitis virus (Hsieh, Lue, & Lee, 2002).

We are presenting the main viruses involved in parkinsonism.

2. Epstein Barr virus

Epstein Barr virus (EBV) is an ubiquitous human gamma herpesvirus which infects and establishes latency in 90–95% of the human adult population worldwide. Acute infection, which can manifest as infectious mononucleosis, appears in children and adolescents.

EBV encephalitis is generally considered a self-limited condition, associated with few sequelae, and it is hardly ever that EBV causes injury to the substantia nigra directly, leading to acute Parkinsonism. The powerful and complex immune response is defined by the generation of cellular and humoral immune effectors directed against multiple viral antigens (Hislop, Taylor, Sauce, & Rickinson, 2007; Middeldorp & Herbrink, 1988).

EBV seropositivity in people with PD (PwP) has been found significantly higher compared to the general population (Woulfe et al., 2014). Moreover, there are several published case reports, describing patients who developed acute parkinsonism after an acute EBV infection. Two models mediating the EBV neurological manifestations have been suggested, those being via a direct brain invasion or by a post-infectious, immune-mediated encephalitis (Roselli et al., 2006).

In this context, it is worth mentioning an interesting case report of acute parkinsonian syndrome, which developed during EBV-associated acute disseminated encephalomyelitis (ADEM) (Dimova, Bojinova, Georgiev, & Milanov, 2006), and which promptly responded to steroid medication. An unusual involvement of both putamina was observed, which was attributed to a reversible inflammatory process, and was considered the functional and morphological substrate of the extrapyramidal syndrome. It is probable that the putaminal immune-mediated damage probably led neurons to become non-reactive to nigrostriatal dopamine input. This reversible disturbance of the dopaminergic pathway in the absence of any substantia nigra lesions determined the transitory nature of parkinsonism (Dimova et al., 2006; Hsieh et al., 2002; Kim, Choi, & Lee, 1995; Mellon, Appleton, Gardner-Medwin, & Aynsley-Green, 1991).

Roselli and colleagues have also reported another reversible parkinsonian syndrome following an EBV infection, with symptoms subsiding about 2 months after their onset (Roselli et al., 2006). Serologic findings met the diagnostic criteria for a recent EBV infection. According to the authors, the absence of any brain Magnetic Resonance Imaging (MRI) changes and the negative result from the Polymerase Chain Reaction (PCR) test performed in the CSF cannot exclude the possibility of an EBV-related complication.

Hsieh and colleagues described a case of EBV encephalitis, presenting as a reversible form of parkinsonism with a brain Tc-99m hexamethylpropyleneamine oxime (HMPAO) single-photon emission computerized tomography (SPECT) revealing decreased perfusion of the right caudate nucleus, without any abnormalities in the brain MRI (Hsieh et al., 2002).

Guan and colleagues also presented a case of post-encephalitic parkinsonism, which met the serological criteria of an EBV infection, while the brain MRI exhibited bilateral lesions of the substantia nigra (Guan, Lu, & Zhou, 2012). Both patient's symptoms and MRI findings fully resolved after a 2-month therapeutic scheme with levodopa.

Although there are known cases of viral infections—associated parkinsonism, it is rare for a virus to be able to predominantly insult the substantia nigra and cause reversible parkinsonism. However, the above reports highlight that EBV may have a direct acute neurotropic effect on this type of neurons.

3. Hepatitis viruses

Hepatitis C virus (HCV) has been considered neurotropic due to the ability to replicate in the central nervous system (CNS) (Forton et al., 2001; Laskus et al., 2005; Radkowski et al., 2002; Tsai et al., 2016). However, data on HCV neurotropism is conflicting. Fletcher and colleagues indicated that important HCV receptors were present on the brain microvascular endothelial cells (Fletcher et al., 2012), suggesting that a HCV infection might affect the blood–brain barrier (BBB) integrity (Tsai et al., 2016). Chronic HCV infection can sometimes be associated with cognitive dysfunction, depression and fatigue, independent of hepatic encephalopathy (Laskus et al., 2005).

Parkinsonian symptoms are rarely described in patients with HCV infection (Tsai et al., 2016). HCV can produce dopaminergic neuronal loss by neuroinflammation, suggesting a possible link between HCV infection and PD (Kim et al., 2016; Tsai et al., 2016). Although this relationship remains largely speculative, the prevalence of anti-HCV antibody positivity in the PD group was found significantly higher when compared to the control group in a hospital-based, case-control study (Kim et al., 2016).

In contrast to HCV, the association between hepatitis B surface antigen (HBsAg) and PD has not been confirmed in the above study (Kim et al., 2016), similarly to other recent studies (Lilach, Fogel-Grinvald, & Israel, 2019; Tsai et al., 2016; Wijarnpreecha, Chesdachai, Jaruvongvanich, & Ungprasert, 2018). Interestingly though, a large, retrospective study suggested a significantly increased risk of PD among patients with hepatitis B (Pakpoor et al., 2017).

On the other hand, Golabi and colleagues, using data from subjects from the Medicare database, did not observe any correlation between HCV infections and PD (Golabi et al., 2017), which contradicts the results of previous studies. Some potential explanations might be the geographical differences of the enrolled subjects, the variability of the extra-hepatic symptoms of HCV infection and the variability of HCV genotype in each study.

Regarding antiviral treatment, it is of note that several studies have shown a slower rate of PD development in patients treated with Pegylated interferon and ribavirin (PEG-IFN/RBV) combination for hepatitis C (Lin et al., 2019; Ramirez-Zamora, Hess, & Nelson, 2019; Su et al., 2019).

4. Human immunodeficiency virus

Parkinsonism or parkinsonian features can occur in the context of HIV infection (Mirsattari, Power, & Nath, 1998). In these cases, the extrapyramidal syndrome is often atypical in presentation, with frequent lack of rest tremor, symmetrical signs of rigidity and bradykinesia. Early-onset gait difficulties and postural instability are also anticipated (Tse et al., 2004).

Parkinsonian symptoms can be observed in the absence of any detectable cause other than HIV infection. The symptoms often progress rapidly, causing severe parkinsonism. In addition, patients with acquired immunodeficiency syndrome (AIDS) are often sensitive to extrapyramidal side effects of some medications commonly used in AIDS, such as anti-emetics (e.g., metoclopramide) or neuroleptics (Hriso, Kuhn, Masdeu, & Grundman, 1991; Mirsattari et al., 1998; Tse et al., 2004). This vulnerability might be subsided by the virus potential to directly insult the basal ganglia or it could be the consequence of pre-existent subclinical nigral degeneration (Dehner, Spitz, & Pereira, 2016; Tse et al., 2004).

Animal models and pathology and neuroimaging studies have demonstrated that the substantia nigra and the basal ganglia are vulnerable to HIV infection (Mirsattari et al., 1998). Interestingly, some studies have supported a pathogenetic link between long-term anti-retroviral therapies and chronic HIV infection with the development of PD. Schneider and colleagues reported three cases of male HIV-infected patients treated with highly active antiretroviral therapy (HAART), who presented a parkinsonian syndrome at a relatively young age, which was clinically very similar to idiopathic PD (Schneider et al., 2010). Mirsattari and colleagues reported 115 patients with HIV infection, 5% of whom developed parkinsonism (Mirsattari et al., 1998). Treatment included administration of anti-retroviral agents, discontinuation of any neuroleptics and anticholinergics, but the clinical response varied considerably among subjects. The presence of extrapyramidal symptoms represents a poor prognostic sign in AIDS patients (Tse et al., 2004).

5. Herpes virus

An increasing number of studies show that PD pathophysiology may involve several immunological aspects (Yan et al., 2014). Injury of dopaminergic neurons can be inflicted through immunological cross-reactivity or molecular mimicry. Several researchers have suggested that the DNA of herpes simplex virus type 1 (HSV1) is present in abundance in latent form in the human brain. One possible explanation could be that the immune system may be less efficient as individuals age, allowing the virus to invade the brain more easily (Wozniak, Shipley, Combrinck, Wilcock, & Itzhaki, 2005). Molecular mimicry might constitute the underlying mechanism of autoimmunity, involving the membranes of dopaminergic neurons of pars compacta in the substantia nigra (Caggiu et al., 2016).

There has been a number of cohort studies, investigating a potential association between Herpes simplex infection and PD. More specifically, Bu and colleagues found that serum antibody levels against HSV-1 were significantly higher among PwP compared to the control group (Bu et al., 2015). In a case control study exploring potential associations of infectious diseases to PD, it was found that herpes simplex infections were significantly related to PD (Vlajinac et al., 2013). These results were further supported by Lai and colleagues, who also suggested that those infected with the varicella zoster virus (VZV) had a higher risk for PD development in the future (Lai et al., 2017). This risk was also highlighted in a nationwide cohort in Taiwan (Cheng et al., 2020). More specifically, it was found that among PD patients, those with a past VZV infection had an earlier PD onset compared to those without such a history. The authors suggested that both local and systemic inflammation (accompanied by neurological symptoms) may increase the risk of PD. However, a recent meta-analysis showed no significant associations between Herpes virus and PD (Wang et al., 2020).

Divergent results have been reported, showing an inverse association between herpes viruses and PD, as well with anti-herpes drugs (Camacho-Soto et al., 2021). A possible mechanism could be a neuroprotective effect of the alpha herpesviruses, as they continue to stay inactive in cells of the peripheral nervous system. In addition to the direct antiviral effect obtained by inhibiting viral DNA synthesis, antiviral drugs could reduce aberrant protein aggregation in PD. However, this observed inverse association between alpha herpesviruses and PD may be a pure coincidence (Camacho-Soto et al., 2021).

Lastly, a case of acute, but remissive, parkinsonism developing after an acute herpes simplex infection with secondary meningoencephalitis has also been reported (Solbrig & Nashef, 1993).

6. Influenza/H1N1

Influenza is the most common virus associated with PD, although the virus ability to spread in the CNS varies depending on the specific strain. Most influenza viruses inflicting humans are non-neurotropic (Smeyne, Noyce, Byrne, Savica, & Marras, 2021), including the Spanish flu (H1N1, 1918) (Mccall, Henry, Reid, & Taubenberger, 2001), the Asian flu (H2N2, 1957–1958) (Smeyne et al., 2021), the Hong Kong flu (H3N2, 1968) (Hosseini et al., 2018) and the Mexican or Swine flu (H1N1, 2009) (Sadasivan, Zanin, O'Brien, Schultz-Cherry, & Smeyne, 2015).

A large case control analysis performed using the UK-based General Practice Research Database found that, although past influenza infections were not associated with a higher risk of developing PD, they might predispose to parkinsonian symptoms, including tremor, especially in the period following such an infection (Toovey, Jick, & Meier, 2011). A recent, large case control study, exploring data from the Danish National Patient Registry, showed that influenza virus infections were associated with a diagnosis of PD more than 10 years after the infection. This association was even stronger when analysis was restricted to months of higher influenza activity. Authors concluded that, although the study methodology cannot suggest causality, further investigation is needed to explore such associations toward the viral hypothesis of PD (Cocoros et al., 2021).

Over time, some researchers have attempted to detect the influenza virus in the brain of patients with post-encephalitic parkinsonism (PEP) (Davis, Koster, & Cawthon, 2014). Brain autopsy studies performed on acute encephalitis lethargica (EL) cases, a controversial entity which has been characterized by acute and chronic parkinsonism, and on PEP cases have not confirmed the presence of influenza genetic material, thus, refuting the initial hypothesis that influenza virus might have mediated the EL outbreak or that influenza virus is neurotropic (Mccall et al., 2001).

Even if there is no direct penetration of the CNS by influenza viruses, cases of encephalitis with extrapyramidal features have been reported during each flu pandemic outbreaks (Smeyne et al., 2021). A higher risk of developing PD was also observed in individuals born during the Spanish flu pandemic (Henry, Smeyne, Jang, Miller, & Okun, 2010; Yamada, 1996).

Influenza-induced inflammation might also be responsible for the neuronal damage, offering a link between influenza and the reported neurological symptoms (Toovey et al., 2011). Influenza virus infections trigger a systemic inflammation, with high levels of cytokines and chemokines (Henry et al., 2010), leading to a "cytokine storm" (Ferrara, Abhyankar, & Gilliland, 1993). It is well known that these cytokines can pass through the blood–brain barrier leading to CNS inflammation and increasing the susceptibility of individuals to subsequent stressors (Sulzer, 2007). This is a hypothesis which may partially apply to the presumed link of the 1918 flu pandemic and the development of PEP.

6.1 Encephalitis lethargica

Constantin Freiherr von Economo described EL, as a complication induced by influenza infection after the 1918 Spanish flu pandemic. Clinical symptoms included somnolence, ophthalmoplegia, delirium and even coma. The malady had high mortality. People who survived manifested parkinsonian features, like rigidity, tremor, bradykinesia or hypomimia, and this constellation of symptoms was named PEP (Falup-Pecurariu, Diaconu, Falup-Pecurariu, Ciopleiaş, & Sîrbu, 2020; Hayase & Tobita, 1997).

CNS neurodegeneration, especially in the region of the substantia nigra pars compacta, could be a result of the microglial cells activation by the virus. There have been reports of acute, reversible parkinsonism after H1N1 vaccination. Parkinsonian symptoms improved with antiparkinsonian drugs administration (Falup-Pecurariu et al., 2020).

The onset of this syndrome has been speculated to be quite specific to the H1N1 flu strain from 1918, as they are epidemiologically connected (Hoffman & Vilensky, 2017). However, it seems there were lack of a significant association between influenza infection and Parkinson's Disease. Information was gathered from four small case-control studies. This implies that not all influenza viruses present a risk of affecting the CNS. There are some similarities between post-encephalitis patients and PD, but the age of onset was much younger in the first group (Wang et al., 2020).

7. Coxsackie virus

The RNA Coxsackie virus is a Picornavirus, highly contagious mostly among children which typically presents with digestive, skin and flu-like symptoms (Sin, Mangale, Thienphrapa, Gottlieb, & Feuer, 2015). Certain serotypes can cause severe neurological manifestations, such as meningitis (Rhoades, Tabor-Godwin, Tsueng, & Feuer, 2011), rhombencephalitis (Molimard, Baudou, Mengelle, Sevely, & Cheuret, 2019), ventriculitis (Erickson, Hoyle, Abramson, Hester, & Shetty, 2003) and myelitis (Banno, Shibata, Hasegawa, Matsuoka, & Okumura, 2021), most of them reported in children. Although a recent meta-analysis found no association between the Coxsackie virus infection and the risk of PD (Wang et al., 2020), there are some case reports of acute reversible parkinsonism (Poser, Huntley, & Poland, 1969), of meningoencephalitis complicated with unremitting parkinsonism (Walters, 2010) and of lethal encephalitis lethargica, which developed after the infection (Cree, Bernardini, Hays, & Lowe, 2003). In this latter case, the patient presented parkinsonian features and impaired level of consciousness, and the brain MRI findings showed bilateral lesions in the substantia nigra (Cree et al., 2003).

8. Flaviviruses

Japanese encephalitis virus (JEV), Saint Louis encephalitis (SLE) virus and West Nile virus (WNV) can be transmitted through the bite of mosquitoes (Griffiths, Turtle, & Solomon, 2014). JEV is endemic in some Asian regions, and even if some cases are asymptomatic, others can manifest with encephalitis, acute flaccid paralysis and other severe neurological symptoms (Misra & Kalita, 2010; Tiroumourougane, Raghava, & Srinivasan, 2002). The most affected regions by the JEV are the basal ganglia and the cerebral cortex (Diagana, Preux, & Dumas, 2007; Griffiths et al., 2014). Several extrapyramidal findings were observed in 20–60% of patients with a JEV infection (Misra & Kalita, 2010). The spectrum of movement disorders included parkinsonism, dystonia (Misra & Kalita, 2010), choreoathetosis, and opsoclonus-myoclonus (Griffiths et al., 2014). In one study including 50 patients infected with JEV, parkinsonism was observed in 16 cases, while parkinsonism and dystonia were reported in another 19 cases, with a worse outcome of the disease (Misra & Kalita, 2002). Parkinsonian features were reversible and included tremor and hypophonia, the latter being a common sequela in these case series (Misra & Kalita, 2002). Persistent movement disorders after JEV infection were also reported; 2 out of 15 patients presented parkinsonian symptoms at the 3–5 years follow-up with lesions in the substantia nigra (Murgod, Muthane, Ravi, Radhesh, & Desai, 2001). A certain link with PD pathogenesis was also observed in rat models, in which the JEV infection induced marked gliosis in the substantia nigra pars compacta and manifested clinically with reversible bradykinesia (Ogata, Tashiro, Nukuzuma, Nagashima, & Hall, 1997).

In most cases, SLE virus presents with flu-like symptoms (Oyer, Beckham, & Tyler, 2014), however, it has also been associated with meningitis and encephalitis, especially in children (Kaplan, Longhurst, & Randall, 1978), but also in elderly patients (Oyer et al., 2014). Manifestations of parkinsonism in the context of SLE virus infection are uncommonly observed. More specifically, in one case report, a 14-year-old patient developed seizures, psychiatric symptoms, moderate parkinsonism, prominent dysphagia and dystonia (Pranzatelli, Mott, Pavlakis, Conry, & Tate, 1994). The parkinsonian signs were non-progressive and reversible after some months. Brain MRI studies revealed bilateral lesions of the basal ganglia. In adults, one case was documented in a 21-year-old patient with aseptic meningitis, which was further complicated by high fever, ataxia and severe tremor. Bilateral T2-weighted hyperintensities involving the substantia nigra were found on brain MRI (Cerna, Mehrad, Luby, Burns, & Fleckenstein, 1999).

The WNV is known to disseminate into the CNS, probably via cerebral microvasculature or through axonal transport (Oyer et al., 2014). Parkinsonian symptoms constitute rare complications of a WNV infection (<10% cases) (Oyer et al., 2014). In contrast, postural and kinetic tremor have been observed in up to 90% of the infected patients (DeBiasi & Tyler, 2006). Parkinsonian features can manifest at the acute stage of a WNV infection, but have the tendency to be mild and remit over time (DeBiasi & Tyler, 2006). In some cases of persistent parkinsonism, symptoms were mild and not bothersome for patients (Sejvar & Marfin, 2006). Imaging studies have demonstrated bilateral lesions in the striatum and thalamus in some of these cases (Sejvar & Marfin, 2006; Solomon et al., 2003). Evidence from post-mortem studies have also identified lesions of the substantia nigra in patients infected with the WNV, although without a clear correlation to clinically manifested parkinsonism in all cases (Lenka, Kamat, & Mittal, 2019).

9. COVID-19

Since the first reported cases of respiratory infection with SARS-CoV-2 in 2019, several neurological manifestations have been described during the current pandemic (Harapan & Yoo, 2021). Due to hyposmia being one of the main symptoms of the Coronavirus Disease 2019 (COVID-19), which is also a well-known prodromal symptom of PD, the hypothesis of a possible link between these two entities has been raised (Boika, 2020). Up to now, 20 cases of parkinsonism following COVID-19 have been described (Boura & Chaudhuri, 2022). In more than half of these cases, parkinsonism developed in the context of encephalopathy, while four patients presented with post-infectious parkinsonism without encephalopathy and another four were diagnosed with idiopathic PD. Parkinsonism was transient in four cases and persistent in the remaining patients, while one patient succumbed to complications of being bedridden and incontinent. Eight patients demonstrated a significant response to dopaminergic therapy, while four patients responded to immunomodulatory/immunosuppressive therapy. Functional neuroimaging was performed on seven occasions (either dopamine transporter single-photon emission computerized tomography (SPECT) imaging with ioflupane I-123 injection (DaTscan) or 6-[18F]-L-3,4-dihydroxypheylalanine (F-FDOPA PET)), revealing a decreased dopaminergic uptake in one or both putamina in all cases. Several mechanisms have been suggested to mediate the potential connection between COVID-19 and PD, including the neuroinvasive potential of the virus, the generated neuroinflammatory response, which further leads to neurodegeneration (Chaná-Cuevas, Salles-Gándara, Rojas-Fernandez, Salinas-Rebolledo, & Milán-Solé, 2020; Onaolapo & Onaolapo, 2021) or the possibility that the viral infection might unmask an underlying prodromal PD (Merello, Bhatia, & Obeso, 2021). Close observation of the patients infected by SARS-CoV-2 and further prospective studies are necessary to explore a potential link between the novel coronavirus infection and parkinsonism.

10. Conclusion

In conclusion, certain infectious agents can be considered as possible risk factors for the appearance of parkinsonism or PD. Further research is needed to examine the magnitude of the risk for each pathogen and the role of inflammatory cytokines in the pathogenesis of PD. Other risk factors, such as genetics and exposure to environmental insults, also need to be taken into consideration in subsequent investigations. Understanding the pathogenesis of PD might support researchers to develop a clearer classification of PD variants, to discover new methods in order to early diagnose PD at the prodromal stage, to predict response to treatment, and adjust treatment towards the direction of a personalized approach for an improved quality of life for PwP (Titova & Chaudhuri, 2017).

References

Banno, F., Shibata, S., Hasegawa, M., Matsuoka, S., & Okumura, A. (2021). Acute flaccid myelitis presumably caused by coxsackie virus A10. *Pediatrics International*, *63*(1), 104–105. https://doi.org/10.1111/PED.14361.

Boika, A. V. (2020). A post-COVID-19 parkinsonism in the future? *Movement Disorders*, *35*(7), 1094. https://doi.org/10.1002/MDS.28117.

Boura, I., & Chaudhuri, K. R. (2022). Coronavirus disease 2019 and related parkinsonism: The clinical evidence thus far. *Movement Disorders Clinical Practice*. https://doi.org/10.1002/MDC3.13461.

Bu, X. L., Wang, X., Xiang, Y., Shen, L. L., Wang, Q. H., Liu, Y. H., et al. (2015). The association between infectious burden and Parkinson's disease: A case-control study. *Parkinsonism & Related Disorders*, *21*(8), 877–881. https://doi.org/10.1016/j.parkreldis.2015.05.015.

Caggiu, E., Paulus, K., Arru, G., Piredda, R., Sechi, G. P., & Sechi, L. A. (2016). Humoral cross reactivity between α-synuclein and herpes simplex-1 epitope in Parkinson's disease, a triggering role in the disease? *Journal of Neuroimmunology*, *291*, 110–114. https://doi.org/10.1016/j.jneuroim.2016.01.007.

Camacho-Soto, A., Faust, I., Racette, B. A., Clifford, D. B., Checkoway, H., & Nielsen, S. S. (2021). Herpesvirus infections and risk of Parkinson's disease. *Neurodegenerative Diseases*, *20*(2–3), 97–103. https://doi.org/10.1159/000512874.

Cerna, F., Mehrad, B., Luby, J. P., Burns, D., & Fleckenstein, J. L. (1999). St. Louis encephalitis and the substantia nigra: MR imaging evaluation. *AJNR. American Journal of Neuroradiology*, *20*(7), 1281.

Chaná-Cuevas, P., Salles-Gándara, P., Rojas-Fernandez, A., Salinas-Rebolledo, C., & Milán-Solé, A. (2020). The potential role of SARS-COV-2 in the pathogenesis of Parkinson's disease. *Frontiers in Neurology*, *11*, 1044. https://doi.org/10.3389/FNEUR.2020.01044/BIBTEX.

Cheng, C. M., Bai, Y. M., Tsai, C. F., Tsai, S. J., Wu, Y. H., Pan, T. L., et al. (2020). Risk of Parkinson's disease among patients with herpes zoster: A nationwide longitudinal study. *CNS Spectrums*, *25*(6), 797–802. https://doi.org/10.1017/S1092852919001664.

Cocoros, N. M., Svensson, E., Szépligeti, S. K., Vestergaard, S. V., Szentkuti, P., Thomsen, R. W., et al. (2021). Long-term risk of Parkinson disease following influenza and other infections. *JAMA Neurology*, *78*(12), 1461–1470. https://doi.org/10.1001/JAMANEUROL.2021.3895.

Cree, B. C., Bernardini, G. L., Hays, A. P., & Lowe, G. (2003). A fatal case of coxsackievirus B4 meningoencephalitis. *Archives of Neurology*, *60*(1), 107–112. https://doi.org/10.1001/ARCHNEUR.60.1.107.

Davis, L. E., Koster, F., & Cawthon, A. (2014). Neurologic aspects of influenza viruses. In *Vol. 123. Handbook of clinical neurology* (pp. 619–645). Elsevier B.V. https://doi.org/10.1016/B978-0-444-53488-0.00030-4.

DeBiasi, R. L., & Tyler, K. L. (2006). West Nile virus meningoencephalitis. *Nature Clinical Practice Neurology*, *2*(5), 264–275. https://doi.org/10.1038/ncpneuro0176.

Dehner, L. F., Spitz, M., & Pereira, J. S. (2016). Parkinsonism in HIV infected patients during antiretroviral therapy – Data from a Brazilian tertiary hospital. *Brazilian Journal of Infectious Diseases*, *20*(5), 499–501. https://doi.org/10.1016/j.bjid.2016.05.008.

Diagana, M., Preux, P. M., & Dumas, M. (2007). Japanese encephalitis revisited. *Journal of the Neurological Sciences*, *262*(1), 165–170. https://doi.org/10.1016/J.JNS.2007.06.041.

Dimova, P. S., Bojinova, V., Georgiev, D., & Milanov, I. (2006). Acute reversible parkinsonism in Epstein-Barr virus-related encephalitis lethargica-like illness. *Movement Disorders*, *21*(4), 564–566. https://doi.org/10.1002/mds.20742.

Erickson, L. S., Hoyle, G., Abramson, J., Hester, L. C., & Shetty, A. K. (2003). Coxsackie B1 infection associated with ventriculitis. *Pediatric Infectious Disease Journal*, *22*(8), 750–751. https://doi.org/10.1097/01.INF.0000078162.99789.CA.

Falup-Pecurariu, C., Diaconu, Ş., Falup-Pecurariu, O., Ciopleiaş, B., & Sîrbu, C. A. (2020). Acute reversible parkinsonism post-influenza infection. *Acta Neurologica Belgica*, *120*(3), 723–724. Springer https://doi.org/10.1007/s13760-019-01215-2.

Ferrara, J. L., Abhyankar, S., & Gilliland, D. G. (1993). Cytokine storm of graft-versus-host disease: A critical effector role for interleukin-1. *Transplantation Proceedings*, *25*(1 Pt 2), 1216–1217.

Fletcher, N. F., Wilson, G. K., Murray, J., Hu, K., Lewis, A., Reynolds, G. M., et al. (2012). Hepatitis C virus infects the endothelial cells of the blood-brain barrier. *Gastroenterology*, *142*(3), 634–643. https://doi.org/10.1053/J.GASTRO.2011.11.028.

Forton, D. M., Allsop, J. M., Main, J., Foster, G. R., Thomas, H. C., & Taylor-Robinson, S. D. (2001). Evidence for a cerebral effect of the hepatitis C virus. *Lancet (London, England)*, *358*(9275), 38–39. https://doi.org/10.1016/S0140-6736(00)05270-3.

Golabi, P., Otgonsuren, M., Sayiner, M., Arsalla, A., Gogoll, T., & Younossi, Z. M. (2017). The prevalence of parkinson disease among patients with hepatitis C infection. *Annals of Hepatology*, *16*(3), 342–348. https://doi.org/10.5604/16652681.1235476.

Griffiths, M. J., Turtle, L., & Solomon, T. (2014). Japanese encephalitis virus infection. *Handbook of Clinical Neurology, 123*, 561–576. https://doi.org/10.1016/B978-0-444-53488-0.00026-2.

Guan, J., Lu, Z., & Zhou, Q. (2012). Reversible parkinsonism due to involvement of substantia nigra in Epstein-Barr virus encephalitis. *Movement Disorders, 27*(1), 156–157. https://doi.org/10.1002/mds.23935.

Harapan, B. N., & Yoo, H. J. (2021). Neurological symptoms, manifestations, and complications associated with severe acute respiratory syndrome coronavirus 2 (SARS-CoV-2) and coronavirus disease 19 (COVID-19). *Journal of Neurology, 268*(9), 3059–3071. https://doi.org/10.1007/S00415-021-10406-Y.

Hayase, Y., & Tobita, K. (1997). Influenza virus and neurological diseases. *Psychiatry and Clinical Neurosciences, 51*(4).

Henry, J., Smeyne, R. J., Jang, H., Miller, B., & Okun, M. S. (2010). Parkinsonism and neurological manifestations of influenza throughout the 20th and 21st centuries. *Parkinsonism & Related Disorders, 16*(9), 566–571. https://doi.org/10.1016/j.parkreldis.2010.06.012.

Hislop, A. D., Taylor, G. S., Sauce, D., & Rickinson, A. B. (2007). Cellular responses to viral infection in humans: Lessons from Epstein-Barr virus. *Annual Review of Immunology, 25*, 587–617. https://doi.org/10.1146/ANNUREV.IMMUNOL.25.022106.141553.

Hoffman, L. A., & Vilensky, J. A. (2017). Encephalitis lethargica: 100 years after the epidemic. *Brain: A Journal of Neurology, 140*(8), 2246–2251. https://doi.org/10.1093/BRAIN/AWX177.

Hosseini, S., Wilk, E., Michaelsen-Preusse, K., Gerhauser, I., Baumgartner, W., Geffers, R., et al. (2018). Long-term neuroinflammation induced by influenza a virus infection and the impact on hippocampal neuron morphology and function. *The Journal of Neuroscience: The Official Journal of the Society for Neuroscience, 38*(12), 3060–3080. https://doi.org/10.1523/JNEUROSCI.1740-17.2018.

Hriso, E., Kuhn, T., Masdeu, J. C., & Grundman, M. (1991). Extrapyramidal symptoms due to dopamine-blocking agents in patients with AIDS encephalopathy. *The American Journal of Psychiatry, 148*(11), 1558–1561. https://doi.org/10.1176/AJP.148.11.1558.

Hsieh, J. C., Lue, K. H., & Lee, Y. L. (2002). Parkinson-like syndrome as the major presenting symptom of Epstein-Barr virus encephalitis [3]. *Archives of Disease in Childhood, 87*(4), 358–359. https://doi.org/10.1136/adc.87.4.358.

Kaplan, A. M., Longhurst, W. L., & Randall, D. L. (1978). St. Louis encephalitis in children. *The Western Journal of Medicine, 128*(4), 279–281.

Kim, J. S., Choi, I. S., & Lee, M. C. (1995). Reversible parkinsonism and dystonia following probable mycoplasma pneumoniae infection. *Movement Disorders: Official Journal of the Movement Disorder Society, 10*(4), 510–512. https://doi.org/10.1002/MDS.870100419.

Kim, J. M., Jang, E. S., Ok, K., Oh, E. S., Kim, K. J., Jeon, B., et al. (2016). Association between hepatitis C virus infection and Parkinson's disease. *Movement Disorders, 31*(10), 1584–1585. John Wiley and Sons Inc. https://doi.org/10.1002/mds.26755.

Lai, S. W., Lin, C. H., Lin, H. F., Lin, C. L., Lin, C. C., & Liao, K. F. (2017). Herpes zoster correlates with increased risk of Parkinson's disease in older people a population-based cohort study in Taiwan. *Medicine (United States), 96*(7). https://doi.org/10.1097/MD.0000000000006075.

Laskus, T., Radkowski, M., Adair, D. M., Wilkinson, J., Scheck, A. C., & Rakela, J. (2005). Emerging evidence of hepatitis C virus neuroinvasion. *AIDS (London, England), 19*(SUPPL. 3). https://doi.org/10.1097/01.AIDS.0000192083.41561.00.

Lenka, A., Kamat, A., & Mittal, S. O. (2019). Spectrum of movement disorders in patients with neuroinvasive West Nile virus infection. *Movement Disorders Clinical Practice, 6*(6), 426–433. https://doi.org/10.1002/MDC3.12806.

Lilach, G., Fogel-Grinvald, H., & Israel, S. (2019). Hepatitis B and C virus infection as a risk factor for Parkinson's disease in Israel-a nationwide cohort study. *Journal of the Neurological Sciences*, *398*, 138–141. https://doi.org/10.1016/j.jns.2019.01.012.

Lin, W. Y., Lin, M. S., Weng, Y. H., Yeh, T. H., Lin, Y. S., Fong, P. Y., et al. (2019). Association of Antiviral Therapy with risk of Parkinson disease in patients with chronic hepatitis C virus infection. *JAMA Neurology*, *76*(9), 1019–1027. https://doi.org/10.1001/jamaneurol.2019.1368.

Mccall, S., Henry, J. M., Reid, A. H., & Taubenberger, J. K. (2001). Influenza RNA not detected in archival brain tissues from acute encephalitis lethargica cases or in postencephalitic Parkinson cases. *Journal of Neuropathology and Experimental Neurology*, *60*(7), 696–704. https://doi.org/10.1093/JNEN/60.7.696.

Mellon, A. F., Appleton, R. E., Gardner-Medwin, D., & Aynsley-Green, A. (1991). Encephalitis lethargica-like illness in a five-year-old. *Developmental Medicine and Child Neurology*, *33*(2), 158–161. https://doi.org/10.1111/J.1469-8749.1991.TB05095.X.

Merello, M., Bhatia, K. P., & Obeso, J. A. (2021). SARS-CoV-2 and the risk of Parkinson's disease: Facts and fantasy. *The Lancet Neurology*, *20*(2), 94–95. https://doi.org/10.1016/S1474-4422(20)30442-7.

Middeldorp, J. M., & Herbrink, P. (1988). Epstein-Barr virus specific marker molecules for early diagnosis of infectious mononucleosis. *Journal of Virological Methods*, *21*(1–4), 133–146. https://doi.org/10.1016/0166-0934(88)90060-2.

Mirsattari, S. M., Power, C., & Nath, A. (1998). Parkinsonism with HIV infection. *Movement Disorders*, *13*(4).

Misra, U. K., & Kalita, J. (2002). Prognosis of Japanese encephalitis patients with dystonia compared to those with parkinsonian features only. *Postgraduate Medical Journal*, *78*, 238–241.

Misra, U. K., & Kalita, J. (2010). Overview: Japanese encephalitis. *Progress in Neurobiology*, *91*(2), 108–120. https://doi.org/10.1016/J.PNEUROBIO.2010.01.008.

Molimard, J., Baudou, E., Mengelle, C., Sevely, A., & Cheuret, E. (2019). Coxsackie B3-induced rhombencephalitis. *Archives de Pédiatrie*, *26*(4), 247–248. https://doi.org/10.1016/J.ARCPED.2019.02.013.

Murgod, U. A., Muthane, U. B., Ravi, V., Radhesh, S., & Desai, A. (2001). Persistent movement disorders following Japanese encephalitis. *Neurology*, *57*(12), 2313–2315. https://doi.org/10.1212/WNL.57.12.2313.

Ogata, A., Tashiro, K., Nukuzuma, S., Nagashima, K., & Hall, W. W. (1997). A rat model of Parkinson's disease induced by Japanese encephalitis virus. *Journal of Neurovirology*, *3*(2), 141–147. https://doi.org/10.3109/13550289709015803.

Onaolapo, A., & Onaolapo, O. (2021). COVID-19, the brain, and the future: Is infection by the novel coronavirus a harbinger of neurodegeneration? *CNS & Neurological Disorders - Drug Targets*, *21*(9), 818–829. https://doi.org/10.2174/1871527321666211222162811.

Oyer, R. J., Beckham, J. D., & Tyler, K. L. (2014). West Nile and St. Louis encephalitis viruses. *Handbook of Clinical Neurology*, *123*, 433–447. https://doi.org/10.1016/B978-0-444-53488-0.00020-1.

Pakpoor, J., Noyce, A., Goldacre, R., Selkihova, M., Mullin, S., Schrag, A., et al. (2017). Viral hepatitis and Parkinson disease: A national record-linkage study. *Neurology*, *88*(17), 1630–1633. https://doi.org/10.1212/WNL.0000000000003848.

Poser, C. M., Huntley, C. J., & Poland, J. D. (1969). Para-encephalitic parkinsonism. Report of an acute case due to coxsackie virus type B 2 and re-examination of the etiologic concepts of postencephalitic parkinsonism. *Acta Neurologica Scandinavica*, *45*(2), 199–215.

Pranzatelli, M. R., Mott, S. H., Pavlakis, S. G., Conry, J. A., & Tate, E. D. (1994). Clinical spectrum of secondary parkinsonism in childhood: A reversible disorder. *Pediatric Neurology*, *10*(2), 131–140. https://doi.org/10.1016/0887-8994(94)90045-0.

Radkowski, M., Wilkinson, J., Nowicki, M., Adair, D., Vargas, H., Ingui, C., et al. (2002). Search for hepatitis C virus negative-strand RNA sequences and analysis of viral sequences in the central nervous system: Evidence of replication. *Journal of Virology*, *76*(2), 600–608. https://doi.org/10.1128/JVI.76.2.600-608.2002.

Ramirez-Zamora, A., Hess, C. W., & Nelson, D. R. (2019). Is interferon therapy for hepatitis C infection a treatable risk factor for Parkinson disease? *JAMA Neurology*, *76*(9), 1006–1007. American Medical Association. https://doi.org/10.1001/jamaneurol.2019.1377.

Rhoades, R. E., Tabor-Godwin, J. M., Tsueng, G., & Feuer, R. (2011). Enterovirus infections of the central nervous system. *Virology*, *411*(2), 288–305. https://doi.org/10.1016/J.VIROL.2010.12.014.

Roselli, F., Russo, I., Fraddosio, A., Aniello, M. S., De Mari, M., Lamberti, P., et al. (2006). Reversible parkinsonian syndrome associated with anti-neuronal antibodies in acute EBV encephalitis: A case report. *Parkinsonism & Related Disorders*, *12*(4), 257–260. https://doi.org/10.1016/j.parkreldis.2005.11.004.

Sadasivan, S., Zanin, M., O'Brien, K., Schultz-Cherry, S., & Smeyne, R. J. (2015). Induction of microglia activation after infection with the non-neurotropic a/CA/04/2009 H1N1 influenza virus. *PLoS One*, *10*(4). https://doi.org/10.1371/JOURNAL.PONE.0124047.

Schneider, S. A., Paisan-Ruiz, C., Quinn, N. P., Lees, A. J., Houlden, H., Hardy, J., et al. (2010). ATP13A2 mutations (PARK9) cause neurodegeneration with brain iron accumulation. *Movement Disorders*, *25*(8), 979–984. https://doi.org/10.1002/mds.22947.

Sejvar, J. J., & Marfin, A. A. (2006). Manifestations of West Nile neuroinvasive disease. *Reviews in Medical Virology*, *16*(4), 209–224. https://doi.org/10.1002/RMV.501.

Sin, J., Mangale, V., Thienphrapa, W., Gottlieb, R. A., & Feuer, R. (2015). Recent progress in understanding coxsackievirus replication, dissemination, and pathogenesis. *Virology*, *484*, 288–304. https://doi.org/10.1016/J.VIROL.2015.06.006.

Smeyne, R. J., Noyce, A. J., Byrne, M., Savica, R., & Marras, C. (2021). Infection and risk of Parkinson's disease. *Journal of Parkinson's Disease*, *11*(1), 31–43. https://doi.org/10.3233/JPD-202279.

Solbrig, M. V., & Nashef, L. (1993). Acute parkinsonism in suspected herpes simplex encephalitis. *Movement Disorders*, *8*(2), 233–234. https://doi.org/10.1002/MDS.870080226.

Solomon, T., Fisher, A. F., Beasley, D. W. C., Mandava, P., Granwehr, B. P., Langsjoen, H., et al. (2003). Natural and nosocomial infection in a patient with West Nile encephalitis and extrapyramidal movement disorders. *Clinical Infectious Diseases*, *36*(11), e140–e145. https://doi.org/10.1086/374936.

Su, T. H., Yang, H. C., Tseng, T. C., Chou, S. W., Lin, C. H., Liu, C. H., et al. (2019). Antiviral therapy in patients with chronic hepatitis C is associated with a reduced risk of parkinsonism. *Movement Disorders*, *34*(12), 1882–1890. https://doi.org/10.1002/mds.27848.

Sulzer, D. (2007). Multiple hit hypotheses for dopamine neuron loss in Parkinson's disease. *Trends in Neurosciences*, *30*(5), 244–250. https://doi.org/10.1016/J.TINS.2007.03.009.

Tiroumourougane, S. V., Raghava, P., & Srinivasan, S. (2002). Japanese viral encephalitis. *Postgraduate Medical Journal*, *78*(918), 205–215. https://doi.org/10.1136/PMJ.78.918.205.

Titova, N., & Chaudhuri, K. R. (2017). Personalized medicine and nonmotor symptoms in Parkinson's disease. *International Review of Neurobiology*, *134*, 1257–1281. https://doi.org/10.1016/BS.IRN.2017.05.015.

Toovey, S., Jick, S. S., & Meier, C. R. (2011). Parkinson's disease or Parkinson symptoms following seasonal influenza. *Influenza and Other Respiratory Viruses*, *5*(5), 328–333. https://doi.org/10.1111/J.1750-2659.2011.00232.X.

Tsai, H. H., Liou, H. H., Muo, C. H., Lee, C. Z., Yen, R. F., & Kao, C. H. (2016). Hepatitis C virus infection as a risk factor for Parkinson disease. *Neurology, 86*(9), 840–846. https://doi.org/10.1212/WNL.0000000000002307.

Tse, W., Cersosimo, M. G., Gracies, J. M., Morgello, S., Olanow, C. W., & Koller, W. (2004). Movement disorders and AIDS: A review. *Parkinsonism & Related Disorders, 10*(6), 323–334. https://doi.org/10.1016/j.parkreldis.2004.03.001.

Vlajinac, H., Dzoljic, E., Maksimovic, J., Marinkovic, J., Sipetic, S., & Kostic, V. (2013). Infections as a risk factor for Parkinson's disease: A case-control study. *International Journal of Neuroscience, 123*(5), 329–332. https://doi.org/10.3109/00207454.2012.760560.

Walters, J. H. (2010). Postencephalitic Parkinson syndrome after meningoencephalitis due to coxsackie virus group B, type 2. *The New England Journal of Medicine, 263*(15), 744–747. https://doi.org/10.1056/NEJM196010132631507.

Wang, H., Liu, X., Tan, C., Zhou, W., Jiang, J., Peng, W., et al. (2020). Bacterial, viral, and fungal infection-related risk of Parkinson's disease: Meta-analysis of cohort and case–control studies. *Brain and Behavior: A Cognitive Neuroscience Perspective, 10*(3). https://doi.org/10.1002/brb3.1549.

Wijarnpreecha, K., Chesdachai, S., Jaruvongvanich, V., & Ungprasert, P. (2018). Hepatitis C virus infection and risk of Parkinson's disease: A systematic review and meta-analysis. *European Journal of Gastroenterology and Hepatology, 30*(1), 9–13. https://doi.org/10.1097/MEG.0000000000000991.

Woulfe, J. M., Gray, M. T., Gray, D. A., Munoz, D. G., & Middeldorp, J. M. (2014). Hypothesis: A role for EBV-induced molecular mimicry in Parkinson's disease. *Parkinsonism & Related Disorders, 20*(7), 685–694. Elsevier Ltd. https://doi.org/10.1016/j.parkreldis.2014.02.031.

Wozniak, M. A., Shipley, S. J., Combrinck, M., Wilcock, G. K., & Itzhaki, R. F. (2005). Productive herpes simplex virus in brain of elderly normal subjects and Alzheimer's disease patients. *Journal of Medical Virology, 75*(2), 300–306. https://doi.org/10.1002/JMV.20271.

Yamada, T. (1996). Review viral etiology of Parkinson's disease: Focus on influenza a virus. *Parkinsonism & Related Disorders, 2*(3), 113–121.

Yan, J., Fu, Q., Cheng, L., Zhai, M., Wu, W., Huang, L., et al. (2014). Inflammatory response in Parkinson's disease (review). *Molecular Medicine Reports, 10*(5), 2223–2233. https://doi.org/10.3892/MMR.2014.2563/HTML.

CHAPTER TWO

Covid-19, nervous system pathology, and Parkinson's disease: Bench to bedside

Aron Emmi[a], Iro Boura[b], Vanessa Raeder[c,d,h], Donna Mathew[e], David Sulzer[f], James E. Goldman[g], and Valentina Leta[c,h,]*

[a]Institute of Human Anatomy, Department of Neuroscience, University of Padova, Padova, Italy
[b]Department of Neurology, University Hospital of Heraklion, Crete, Greece
[c]Parkinson's Foundation Centre of Excellence, King's College Hospital NHS Foundation Trust, London, United Kingdom
[d]Department of Neurology, Technical University Dresden, Dresden, Germany
[e]Neuroscience and Mental Health Research Institute, School of Medicine, Cardiff University, Cardiff, United Kingdom
[f]Departments of Psychiatry, Neurology, Pharmacology, Columbia University Medical Center, New York State Psychiatric Institute, New York, United States
[g]Department of Pathology and Cell Biology, and the Taub Institute for Research on Alzheimer's Disease and the Aging Brain, Vagelos College of Physicians and Surgeons, Columbia University and the New York Presbyterian Hospital, New York, NY, United States
[h]Department of Basic and Clinical Neuroscience, Institute of Psychiatry, Psychology & Neuroscience, King's College London, London, United Kingdom
*Corresponding author: e-mail address: valentina.leta@kcl.ac.uk

Contents

Abstract

Coronavirus disease 2019 (Covid-19) caused by Severe Acute Respiratory Syndrome Coronavirus 2 (SARS-CoV-2) infection is primarily regarded as a respiratory disease; however, multisystemic involvement accompanied by a variety of clinical manifestations, including neurological symptoms, are commonly observed. There is, however, little evidence supporting SARS-CoV-2 infection of central nervous system cells, and

International Review of Neurobiology, Volume 165
ISSN 0074-7742
https://doi.org/10.1016/bs.irn.2022.06.006
Copyright © 2022 Elsevier Inc.
All rights reserved.

neurological symptoms for the most part appear to be due to damage mediated by hypoxic/ischemic and/or inflammatory insults. In this chapter, we report evidence on candidate neuropathological mechanisms underlying neurological manifestations in Covid-19, suggesting that while there is mostly evidence against SARS-CoV-2 entry into brain parenchymal cells as a mechanism that may trigger Parkinson's disease and parkinsonism, that there are multiple means by which the virus may cause neurological symptoms.

1. Introduction

Severe Acute Respiratory Syndrome Coronavirus 2 (SARS-CoV-2) is a novel, highly contagious, and pathogenic strain of the Coronaviruses (CoV) family, which has caused the pandemic of coronavirus disease 2019 (Covid-19) (Hu, Guo, Zhou, & Shi, 2021; Lai, Shih, Ko, Tang, & Hsueh, 2020). Emerging in late 2019, SARS-CoV-2 has taken a toll on global morbidity and mortality as a major public health issue. Although manifesting primarily as a respiratory disease, there are numerous reports of neurological manifestations of Covid-19, either presenting as new symptoms or disorders (e.g., stroke, Guillain-Barré syndrome) or as the exacerbation of pre-existing symptoms of known chronic neurological conditions (Antonini, Leta, Teo, & Chaudhuri, 2020; Kubota & Kuroda, 2021; Leta et al., 2021; Ousseiran, Fares, & Chamoun, 2021). These observations, along with the detection of the virus in post-mortem tissue and cerebrospinal fluid (CSF) (Lewis et al., 2021; Mukerji & Solomon, 2021; Tandon et al., 2021) raise the question of whether SARS-CoV-2 enters the human brain or causes neurological symptoms because of the systemic illness, notably hypoxia and inflammation.

2. SARS-CoV-2 receptors

SARS-CoV-2 shares 80% of its genome with SARS-CoV and 50% with MERS-CoV, with homologies extending to the protein spike (S), which after cleavage by the transmembrane protease serine 2 (TMPRSS2), can bind host membrane proteins acting as viral receptors (Hoffmann et al., 2020; Xu et al., 2020).

The main viral receptor for SARS-CoV-2 and other CoV is angiotensin-converting enzyme 2 (ACE2), an enzyme that catalyzes the cleavage of angiotensin I into angiotensin 1–9 and angiotensin II into the vasodilator angiotensin 1–7 (Wang et al., 2020). ACE2 is a transmembrane protein that is widely expressed in human tissues, and while its expression in brain neurons and astrocytes has not been clearly demonstrated, it is expressed in brain vessels (Hamming et al., 2004) and hypoxic insult appears to upregulate its expression (Zhang et al., 2009).

Similarly to MERS-CoV, virus cellular endocytosis may be alternatively mediated by sialic acid residues, which are located on plasma membrane proteins of several type of cells, including neurons (Fantini, Di Scala, Chahinian, & Yahi, 2020) or by the lectin CD209L, which is a receptor for SARS-CoV (Jeffers et al., 2004), although such alternative receptors have not been clearly demonstrated for SARS-CoV-2.

3. Neuroinflammation in post-mortem Covid-19 brain

Neuroinflammation is prominent in Covid-19 and characterized by lympho-monocytic infiltrations, microglial activation and microglial nodules. Lympho-monocytic perivascular cuffing and infiltration have been reported by most authors, with a predominance of perivascular CD68+ monocytes/macrophages and CD8+ T-cells (Lee et al., 2021; Matschke et al., 2020; Meinhardt et al., 2021; Schwabenland et al., 2021; Thakur et al., 2021), although lymphocyte accumulation appears mild compared to that in viral encephalitis, such as from Herpes viruses.

In all studies examining microglial cells, authors identified prominent microglial activation with upregulation of MHC-II proteins (HLA-DR) and increased lysosomal activity (CD68+), while maintaining a homeostatic microglial marker TMEM119 (Deigendesch et al., 2020; Matschke et al., 2020; Meinhardt et al., 2021; Schwabenland et al., 2021; Thakur et al., 2021). Similarly, evidence of neuronophagia was documented by most authors with prominent involvement of the lower brainstem at the level of the medullary tegmentum, the midline raphe, inferior olivary nucleus, and the dorsal pons (Al-Dalahmah et al., 2020; Deigendesch et al., 2020; Matschke et al., 2020; Meinhardt et al., 2021; Schwabenland et al., 2021; Thakur et al., 2021). Microglial nodules with associated CD8+/CD3+ T-cell clusters were also documented in most, but not all, Covid-19 patients, in contrast to non-Covid-19 controls and ExtraCorporeal Membrane Oxygenation (ECMO) patients, who displayed no instances of microglial nodules and neuronophagia (Schwabenland et al., 2021), although we have observed activated microglia and nodules in the brainstems of patients with severe respiratory distress (JE Goldman, P Canoll, unpublished observations). Deep spatial profiling of the local immune response in Covid-19 brains by imaging mass spectrometry revealed significant immune activation in the medulla and olfactory bulb with a prominent role mediated by CD8+ T-cell—microglia crosstalk in the parenchyma (Schwabenland et al., 2021). Conversely, Deigendesch and colleagues found significant differences in HLA-DR+ activated microglia when comparing Covid-19 subjects to non-septic controls, but no differences were found with patients who had died under septic conditions; according to the authors, this may represent a histopathological correlate of critical illness-related encephalopathy and hypoxia, rather than a Covid-19-specific finding (Deigendesch et al., 2020).

Neurological symptoms of Covid-19 occur regularly without detection of SARS-CoV-2 in CSF samples, but there is evidence for autoantibody formation. In a cohort of 102 Covid-19 patients, where almost 60% presented some form of neurological symptoms, CSF anti-neuronal autoantibodies were detected in 35% of those tested (Fleischer et al., 2021). In a study of critically ill Covid-19 patients ($n=11$) with unexplained neurological sequelae, anti-neuronal autoantibodies were found in the CSF of all patients, as well as in their serum, suggesting that multiple autoantigens and a potential molecular mimicry to SARS-CoV-2 might mediate these symptoms, especially those related to hyperexcitability (myoclonus, seizures) (Franke et al., 2021). The above findings might guide clinicians to consider administering immunotherapy in selected patients.

Thus, while most studies are consistent on the overall neuroinflammatory conditions occurring in the context of Covid-19, with particular focus on microgliosis and microglia—T-cell crosstalk, disagreements on the source of inflammation are unresolved. In particular, it is not clear how much might be due to an ongoing systemic inflammation/cytokine storm, and whether and how frequent comorbidities, such as hypertension, diabetes, cardiovascular disease or neurodegenerative conditions, and hypoxic/ischemic damage may influence microgliosis and brain inflammation.

4. Little evidence of neuro-invasion in post-mortem Covid-19 cases

Neurotropism has been established for many species of the CoV family, including Middle East Respiratory Syndrome (MERS-CoV) and severe acute respiratory syndrome (SARS-CoV-1) (Zubair et al., 2020), while there have been reports for persistent chronic infections of particular CoV strains in human neuronal cell lines or in the brain of patients with neurodegenerative disorders (Arbour et al., 1999; Murray, Brown, Brian, & Cabirac, 1992). SARS-CoV-2 has been speculated to possess a neuro-invasive potential (Koyuncu, Hogue, & Enquist, 2013; Verstrepen, Baisier, & De Cauwer, 2020; Yachou, El Idrissi, Belapasov, & Ait Benali, 2020), and has been claimed in experimental models (Bullen et al., 2020). However, while there is significant evidence for neuroinflammation in the brain induced by SARS-CoV-2, as detailed here, neuropathological findings at this juncture do not support significant infection of brain cells.

To date, the presence of viral peptides in brain, determined by immunostaining, has been reported only in a small set of patients and in a few cells, and has not been constantly reproduced throughout available studies (Lee et al., 2021; Matschke et al., 2020; Meinhardt et al., 2021; Schwabenland et al., 2021; Solomon et al., 2020; Thakur et al., 2021; Yang et al., 2021). While most studies agree on the detection of low or undetectable viral RNA levels using RT-PCR analyses, detection of viral RNA in blood and blood vessels of the investigated samples cannot be excluded. In contrast, immunoperoxidase and immunofluorescent staining, imaging mass spectrometry and in situ hybridization that provide for the spatial localization of viral proteins/RNA have failed to consistently detect viral antigens or RNA. Matschke and colleagues found sparse immunoreactive cells throughout the brainstem, without specific topographic localization, but also detected distinct immunoreactivity of the glossopharyngeal and vagus nerve bundles (CN IX-X) in a subset of patients, and suggested SARS-CoV-2 retrograde spread through these cranial nerves toward the medulla (Matschke et al., 2020). This is also supported by the detection of SARS-CoV-2 antigen and genomic sequences in the carotid body of Covid-19 subjects (Porzionato et al., 2021). However, there was no apparent association between the presence of viral antigens and neuropathological changes.

Several studies have detected viral proteins and RNA in the olfactory mucosa through immunohistochemistry, immunofluorescence and in-situ hybridization, as well as electron microscopy (Khan et al., 2021; Meinhardt et al., 2021; Zazhytska et al., 2022). Meinhardt et al. observed SARS-CoV-2 spike protein in primary olfactory neurons of the olfactory mucosa, suggesting an olfactory-transmucosal spread of the virus. This was supported by the detection of viral RNA at the level of the olfactory bulb and medulla, even though viral proteins were not detected in non-vascular cells. Furthermore, SARS-CoV-2 immunoreactive endothelial cells associated with microvascular injury and micro-thromboses were found in a subset of Covid-19 patients at the level of the medulla. Conversely, two recent studies found virus in sustentacular cells of the olfactory epithelium, but not in olfactory neurons (Khan et al., 2021; Zazhytska et al., 2022). The latter detected reorganization of nuclear architecture and downregulation of olfactory receptors, as well as their signaling pathways, in the olfactory neurons, suggesting a non-cell autonomous cause of anosmia (Zazhytska et al., 2022).

Schwabenland and colleagues detected viral antigen in ACE2-positive cells enriched in the vascular compartment (Schwabenland et al., 2021). According to the authors, this finding was linked to vascular proximity and ACE2 expression, and was correlated to the perivascular immune activation patterns of CD8+ and CD4+ T-cells and myeloid- and microglial-cell subsets, indicating a fundamental role of the vascular and perivascular compartment, as well as a blood–brain barrier impairment in mediating Covid-19-specific neuroinflammation.

Two studies of single nucleus RNA sequencing of brains of COVID-19 patients detected broad perturbations, with upregulation of genes involved in innate antiviral response and inflammation, microglia activation and neurodegeneration, but found no direct evidence of viral RNA (Fullard et al., 2021; Yang et al., 2021). Other authors have not detected viral proteins/RNA through immunohistochemistry or in situ hybridization, even though viral genomic sequences were found with RT-PCR assays (Lee et al., 2021; Solomon et al., 2020; Thakur et al., 2021).

Hence, SARS-CoV-2 invasion of the CNS remains to be definitively shown, and much evidence points against the presence of detectable virus, although there are reports of rare instances of the presence of viral antigens in brain cells.

5. SARS-CoV-2 and parkinsonism

Understandably, the idea that neuronal infection by SARS-CoV-2 might predispose to parkinsonism has been a source of great concern (Beauchamp, Finkelstein, Bush, Evans, & Barnham, 2020; Bouali-Benazzouz & Benazzouz, 2021; Brundin, Nath, & Beckham, 2020), although, fortunately, there is little evidence supporting this response. Although there are no reports of CoV-associated parkinsonism, antibodies against common CoV have been detected in the CSF of people with Parkinson's Disease (PwP, PD) in significantly higher titers compared to controls, while there was evidence of post-encephalitic parkinsonism in mice infected with a CoV strain (MHV-A59) (Fazzini, Fleming, & Fahn, 1992; Fishman et al., 1985). In a recent literature review, 20 cases of new-onset parkinsonism developing during or shortly after a SARS-CoV-2 infection have been described, presenting a variety of different subjacent mechanisms (Boura & Chaudhuri, 2022). Although these cases do not suffice to support an etiological association between the two entities, the concept of vigilance in the long-term is highlighted.

The concerns on Covid-19 influence on Parkinson's disease stem both from the 1917 Spanish flu/von Economo's encephalitis pandemics as well an association of numerous viruses with the development of transient or permanent parkinsonism (Jang, Boltz, Webster, & Smeyne, 2009), including reports that anti-viral treatment and vaccination are associated with a decreased risk of parkinsonism in humans and animal models (Lin et al., 2019; Sadasivan, Sharp, Schultz-Cherry, & Smeyne, 2017). It has been suggested that pathogens contribute to PD pathogenesis, particularly after the age of 50 in individuals with or without a susceptible genetic substrate (Beauchamp et al., 2020; Tanner et al., 1999). Braak and others have suggested that neurotropic pathogens may infect the CNS via the nasal or gastric pathway, both of which are sites of early pathology in PD (Hawkes, Del Tredici, & Braak, 2007; Klingelhoefer & Reichmann, 2015) and viral-related inflammation might render the CNS susceptible to preceding or subsequent stressors (Sulzer, 2007). Indeed, a meta-analysis demonstrated that a past history of an infection was associated with a 20% higher risk of presenting PD in the future, although this was significant only for bacterial and not viral infections (Meng, Shen, & Ji, 2019). An association between past CNS infections, particularly if multiple hospitalizations preceded, and a subsequent development of PD was described (Fang et al., 2012). Finally, epidemiological data from a large cohort of PD patients and controls indicates an increased PD frequency among occupations with a high risk for respiratory infections, such as teachers and health-care workers, (the "clustering of PD" theory) (Tsui, Calne, Wang, Schulzer, & Marion, 1999).

While there is little evidence at this time for a role for Covid-19 in increasing PD, there could be effects particularly on peripheral catecholamine systems. L-Dopa decarboxylase (DDC), an essential enzyme in the biosynthesis of dopamine and serotonin, is the most significantly coexpressed and coregulated gene with ACE2 in non-neuronal cell types, significantly affecting dopamine blood levels (Nataf, 2020). SARS-CoV-2 infection of monkey cell lines was found to induce downregulation of

DDC, an effect also noted with dengue and hepatitis C infections (Mpekoulis et al., 2021), pathogens associated with parkinsonism (Bopeththa & Ralapanawa, 2017; Tsai et al., 2016). DDC levels rose in nasopharyngeal tissues of asymptomatic or mild severity Covid-19 patients, while an inverse relationship was noted between SARS-CoV-2 RNA levels and DDC expression (Mpekoulis et al., 2021). Moreover, a dopamine D1 receptor agonist was found to suppress endotoxin-induced pulmonary inflammation in mice, suggesting that a potential protective role of dopamine in inflammation needs to be further explored (Bone, Liu, Pittet, & Zmijewski, 2017).

Vascular damage constitutes a recognized complication of Covid-19 (Siddiqi, Libby, & Ridker, 2021). A case of bilateral basal ganglia insult in the context of a thromboembolic encephalopathy without parkinsonism in a Covid-19 patient has been described (Haddadi, Ghasemian, & Shafizad, 2020).

6. Peripheral pathways that may contribute to neuroinflammation

6.1 Olfactory pathway

SARS-CoV-1 enters the brain of transgenic mice via the olfactory bulb, causing neuronal death without signs of encephalitis (Netland, Meyerholz, Moore, Cassell, & Perlman, 2008), while intranasal inoculation of MERS-CoV in transgenic mice led to high levels of the virus in the CNS, particularly in the thalamus and the brainstem (Li et al., 2016).

The characteristic clinical symptoms of hyposmia or anosmia in Covid-19, as well as hypogeusia or ageusia, which are quite prominent among affected patients (Guerrero et al., 2021) support the hypothesis of an involvement of the olfactory system (Lechien et al., 2020). An increased SARS-CoV-2 viral load in the nasal epithelium has been found by RT-PCR in situ hybridization and immunohistochemical staining methods (Meinhardt et al., 2021). Moreover, a prospective autopsy study in the Netherlands involving 21 patients who had succumbed to Covid-19 complications revealed extensive inflammation in the brain among other tissues, particularly focused in the olfactory bulbs and the medulla oblongata, although this study did not reveal presence of the virus in the brain (Schurink et al., 2020). Politi and colleagues described a case report of a Covid-19-positive woman with anosmia, who presented with subtle hyperintensities in the olfactory bulbs and the right gyrus rectus in brain magnetic resonance imaging (MRI), although the authors highlighted that this was not a consistent finding among Covid-19 patients with smell dysfunction (Politi, Salsano, & Grimaldi, 2020). A recently published paper, exploring the findings of brain scans before and after a SARS-CoV-2 infection in a large sample of UK Biobank participants, showed significant reductions in gray matter thickness and tissue-contrast in the orbitofrontal cortex and parahippocampal gyrus, along with prominent alterations in brain areas functionally connected to the primary olfactory cortex, suggestive of tissue damage (Douaud et al., 2022), although it is difficult to correlate the scan findings with cellular pathology. Furthermore, post-mortem brain MRI studies in 19 decedents of Covid-19 demonstrated an asymmetry in the olfactory bulbs in about 20% of the patients (Coolen et al., 2020). Post-mortem studies of Covid-19 patients detected the virus in the sustentacular cells of the olfactory epithelium, but not in the olfactory receptor neurons or the olfactory bulbs (Khan et al., 2021), as noted above.

6.2 Enteric pathways

With both respiratory and gastrointestinal symptoms being quite common in the context of Covid-19 (Cares-Marambio et al., 2021; Groff et al., 2021) and ACE2 being highly expressed in the alveolar epithelial type II cells and the intestinal endothelial cells (Williams et al., 2021), some researchers have suggested the potential of SARS-CoV-2 entry through the intestine (Lehmann et al., 2021; Mönkemüller, Fry, & Rickes, 2020; Zhang et al., 2020). RNA of the SARS-CoV-2 has been identified in stool samples and rectal swabs, even in cases with negative nasopharyngeal swabs (Tang, Schmitz, Persing, & Stratton, 2020). An autopsy study ($n=21$) revealed SARS-CoV-2-infected cells in the respiratory and gastrointestinal tract, among other tissues, along with extensive inflammation in the medulla oblongata, implying a potential effect on the respiratory control center (Schurink et al., 2020). Many studies report inflammation in the medulla (see above), but the link between peripheral organs and the brain is not clear. One possible way in which systemic inflammation can influence medullary nuclei is the activation of vagal medullary afferents by inflammatory molecules. Thus, vagal endings contain receptors to IL-1β and are activated by this cytokine, subsequently activating neurons of the solitary nucleus, which then activate sympathetic output from the vagal nucleus and the nucleus ambiguous through the vagus nerve (Pavlov & Tracey, 2012).

6.3 Vascular pathways

Non-neuronal pathways of SARS-CoV-2 that might lead to neuroinflammation must be considered, including the hematogenous route. SARS-CoV-2 can enter the blood stream and either access and damage the endothelium of brain vasculature to cross the BBB, or trigger an inflammatory response, leading to breakdown of the BBB (Meinhardt et al., 2021; Wan et al., 2021).

Cerebrovascular disease, including ischemic and hemorrhagic strokes, constitutes a severe and common complication in Covid-19 (Tsivgoulis et al., 2020). The brains of patients who did not survive Covid-19 show acute cerebrovascular disease, including thrombotic microangiopathy and endothelial injury, without any evidence of vasculitis (Hernández-Fernández et al., 2020; Wan et al., 2021). Accumulating data suggests that Covid-19 is connected to a hypercoagulable state, which is closely related to inflammation and predisposes to macro- and microvascular thrombosis, resulting in arterial and venous infarcts (Abou-Ismail, Diamond, Kapoor, Arafah, & Nayak, 2020).

ACE2 is expressed in cells of the vasculature in the human brain, although the specific cell types have not been defined (Hamming et al., 2004). TMPRSS2 and NRP1 are also present in the endothelial cells of the vasculature (Wan et al., 2021). SARS-CoV-2 endothelitis has also been confirmed in autopsy studies concerning other human tissues (lungs, kidney, heart, liver, small intestine) (Varga et al., 2020). Moreover, Pellegrini and colleagues showed a leakage across the BBB, using a model of the human choroid plexus epithelial cells infected by SARS-CoV-2 (Pellegrini et al., 2020), although one has to consider the results of such in vitro studies carefully, since they may not reflect what occurs in the living brain.

On the other hand, a SARS-CoV-2 infection can cause an excessive systemic inflammatory response after triggering a cytokine storm in the periphery, resulting in a BBB disruption (Sulzer et al., 2020). Researchers have suggested that SARS-CoV-2 may infect immune cells, using them as a "trojan horse" to invade the CNS via the impaired BBB or activate different populations of immune cells, which may infiltrate the CNS, causing a secondary cytokine storm and thus neurologic manifestations (Pezzini & Padovani, 2020; Wan et al., 2021; Williams et al., 2021).

6.4 Hypoxia and ischemia

Finally, hypoxic/ischemic changes of the brain due to respiratory abnormalities and hypoperfusion in the context of Covid-19 appear likely to play a crucial role in the development of secondary neurological manifestations (Sullivan & Fischer, 2021). Cases of hypoxic brain injury following Acute Respiratory Distress Syndrome (ARDS) have been described in the literature, including cases of parkinsonism, with the authors highlighting the possibility of silent hypoxia (Ayele et al., 2021; Fearon, Mikulis, & Lang, 2021; Radnis et al., 2020). However, the fact that neurological complications have been reported in the absence of any respiratory symptoms, suggest that alternative mechanisms mediate the virus-induced CNS insults (Ellul et al., 2020).

Hypoxic/ischemic pathology is particularly common in Covid-19, being reported in most studies (Matschke et al., 2020; Solomon et al., 2020; Thakur et al., 2021). Alterations were mostly widespread, and both acute and subacute findings could be appreciated at the level of the cortex, the basal ganglia, the hippocampus and, most notably, the brainstem. Hypoxic/ischemic damage can be associated with vascular pathology, with numerous reported cases presenting both ischemic and hemorrhagic infarcts (Lee et al., 2021; Matschke et al., 2020; Remmelink et al., 2020; Thakur et al., 2021). Reactive astrogliosis in the context of hypoxic/ischemic injuries was also commonly encountered, prominently involving the basal ganglia and the brainstem (Matschke et al., 2020; Meinhardt et al., 2021; Thakur et al., 2021). Microvascular pathology, fibrinogen leakage and small vessel thromboses were associated with ischemic and vascular pathology, representing a characteristic finding of the Covid-19 cohorts (Lee et al., 2021; Matschke et al., 2020; Meinhardt et al., 2021; Porzionato et al., 2021; Porzionato, Emmi, et al., 2021; Schwabenland et al., 2021). Small vessel platelet-enriched microthrombi were predominantly found in the basal ganglia, brainstem and cerebellum, often associated to microvascular injury and fibrinogen leakage. Several instances of SARS-CoV-2 immunoreactive endothelial cells in the context of vascular injury were also detected (Meinhardt et al., 2021), indicating a prominent role of SARS-CoV-2 infection in determining Covid-19 associated small vessel pathology.

7. Conclusion

Although Covid-19 caused by SARS-CoV-2 infection is mainly regarded as a respiratory disease, a variety of clinical manifestations, including neurological symptoms, often occur. The evidence available to date suggests that Covid-19 does not significantly contribute to the incidence of Parkinson's disease and that neurological clinical manifestations observed in the context of Covid-19 are mainly secondary to an indirect damage by SARS-CoV-2 in peripheral systems, in contrast to infection of CNS neurons and astrocytes.

References

Abou-Ismail, M. Y., Diamond, A., Kapoor, S., Arafah, Y., & Nayak, L. (2020). The hypercoagulable state in COVID-19: Incidence, pathophysiology, and management. *Thrombosis Research*, *194*, 101–115. https://doi.org/10.1016/j.thromres.2020.06.029.

Al-Dalahmah, O., Thakur, K. T., Nordvig, A. S., Prust, M. L., Roth, W., Lignelli, A., et al. (2020). Neuronophagia and microglial nodules in a SARS-CoV-2 patient with cerebellar hemorrhage. *Acta Neuropathologica Communications*, *8*(1), 147. https://doi.org/10.1186/s40478-020-01024-2.

Antonini, A., Leta, V., Teo, J., & Chaudhuri, K. R. (2020). Outcome of Parkinson's disease patients affected by COVID-19. *Movement Disorders*, *35*(6), 905–908. https://doi.org/10.1002/mds.28104.

Arbour, N., Côté, G., Lachance, C., Tardieu, M., Cashman, N. R., & Talbot, P. J. (1999). Acute and persistent infection of human neural cell lines by human coronavirus OC43. *Journal of Virology*, *73*(4), 3338–3350. https://doi.org/10.1128/jvi.73.4.3338-3350.1999.

Ayele, B. A., Demissie, H., Awraris, M., Amogne, W., Shalash, A., Ali, K., et al. (2021). SARS-COV-2 induced parkinsonism: The first case from the sub-Saharan Africa. *Clinical Parkinsonism & Related Disorders*, *5*, 100116. https://doi.org/10.1016/j.prdoa.2021.100116.

Beauchamp, L. C., Finkelstein, D. I., Bush, A. I., Evans, A. H., & Barnham, K. J. (2020). Parkinsonism as a third wave of the COVID-19 pandemic? *Journal of Parkinson's Disease*, *10*(4), 1343–1353. https://doi.org/10.3233/JPD-202211.

Bone, N. B., Liu, Z., Pittet, J.-F., & Zmijewski, J. W. (2017). Frontline science: D1 dopaminergic receptor signaling activates the AMPK-bioenergetic pathway in macrophages and alveolar epithelial cells and reduces endotoxin-induced ALI. *Journal of Leukocyte Biology*, *101*(2), 357–365. https://doi.org/10.1189/jlb.3HI0216-068RR.

Bopeththa, B., & Ralapanawa, U. (2017). Post encephalitic parkinsonism following dengue viral infection. *BMC Research Notes*, *10*(1), 655. https://doi.org/10.1186/s13104-017-2954-5.

Bouali-Benazzouz, R., & Benazzouz, A. (2021). Covid-19 infection and parkinsonism: Is there a link? *Movement Disorders*, *36*(8), 1737–1743. https://doi.org/10.1002/mds.28680.

Boura, I., & Chaudhuri, K. R. (2022). Coronavirus disease 2019 and related parkinsonism: The clinical evidence thus far. *Movement Disorders Clinical Practice*. https://doi.org/10.1002/mdc3.13461.

Brundin, P., Nath, A., & Beckham, J. D. (2020). Is COVID-19 a perfect storm for Parkinson's disease? *Trends in Neurosciences*, *43*(12), 931–933. https://doi.org/10.1016/j.tins.2020.10.009.

Bullen, C. K., Hogberg, H. T., Bahadirli-Talbott, A., Bishai, W. R., Hartung, T., Keuthan, C., et al. (2020). Infectability of human BrainSphere neurons suggests neurotropism of SARS-CoV-2. *ALTEX*, *37*(4), 665–671. https://doi.org/10.14573/altex.2006111.

Cares-Marambio, K., Montenegro-Jiménez, Y., Torres-Castro, R., Vera-Uribe, R., Torralba, Y., Alsina-Restoy, X., et al. (2021). Prevalence of potential respiratory symptoms in survivors of hospital admission after coronavirus disease 2019 (COVID-19): A systematic review and meta-analysis. *Chronic Respiratory Disease*, *18*. https://doi.org/10.1177/14799731211002240.

Coolen, T., Lolli, V., Sadeghi, N., Rovai, A., Trotta, N., Taccone, F. S., et al. (2020). Early postmortem brain MRI findings in COVID-19 non-survivors. *Neurology*, *95*(14), e2016–e2027. https://doi.org/10.1212/wnl.0000000000010116.

Deigendesch, N., Sironi, L., Kutza, M., Wischnewski, S., Fuchs, V., Hench, J., et al. (2020). Correlates of critical illness-related encephalopathy predominate postmortem COVID-19 neuropathology. *Acta Neuropathologica*, *140*(4), 583–586. https://doi.org/10.1007/s00401-020-02213-y.

Douaud, G., Lee, S., Alfaro-Almagro, F., Arthofer, C., Wang, C., McCarthy, P., et al. (2022). SARS-CoV-2 is associated with changes in brain structure in UK biobank. *Nature*. https://doi.org/10.1038/s41586-022-04569-5.

Ellul, M. A., Benjamin, L., Singh, B., Lant, S., Michael, B. D., Easton, A., et al. (2020). Neurological associations of COVID-19. *The Lancet Neurology*, *19*(9), 767–783. https://doi.org/10.1016/S1474-4422(20)30221-0.

Fang, F., Wirdefeldt, K., Jacks, A., Kamel, F., Ye, W., & Chen, H. (2012). CNS infections, sepsis and risk of Parkinson's disease. *International Journal of Epidemiology*, *41*(4), 1042–1049. https://doi.org/10.1093/ije/dys052.

Fantini, J., Di Scala, C., Chahinian, H., & Yahi, N. (2020). Structural and molecular modelling studies reveal a new mechanism of action of chloroquine and hydroxychloroquine against SARS-CoV-2 infection. *International Journal of Antimicrobial Agents*, *55*(5), 105960. https://doi.org/10.1016/j.ijantimicag.2020.105960.

Fazzini, E., Fleming, J., & Fahn, S. (1992). Cerebrospinal fluid antibodies to coronavirus in patients with Parkinson's disease. *Movement Disorders*, 7(2), 153–158. https://doi.org/10.1002/mds.870070210.

Fearon, C., Mikulis, D. J., & Lang, A. E. (2021). Parkinsonism as a sequela of SARS-CoV-2 infection: Pure hypoxic injury or additional COVID-19-related response? *Movement Disorders*, *36*(7), 1483–1484. https://doi.org/10.1002/mds.28656.

Fishman, P. S., Gass, J. S., Swoveland, P. T., Lavi, E., Highkin, M. K., & Weiss, S. R. (1985). Infection of the basal ganglia by a murine coronavirus. *Science*, *229*(4716), 877–879. https://doi.org/10.1126/science.2992088.

Fleischer, M., Köhrmann, M., Dolff, S., Szepanowski, F., Schmidt, K., Herbstreit, F., et al. (2021). Observational cohort study of neurological involvement among patients with SARS-CoV-2 infection. *Therapeutic Advances in Neurological Disorders*, *14*. https://doi.org/10.1177/17562864211993701. 17562864211993701.

Franke, C., Ferse, C., Kreye, J., Reincke, S. M., Sanchez-Sendin, E., Rocco, A., et al. (2021). High frequency of cerebrospinal fluid autoantibodies in COVID-19 patients with neurological symptoms. *Brain, Behavior, and Immunity*, *93*, 415–419. https://doi.org/10.1016/j.bbi.2020.12.022.

Fullard, J. F., Lee, H. C., Voloudakis, G., Suo, S., Javidfar, B., Shao, Z., et al. (2021). Single-nucleus transcriptome analysis of human brain immune response in patients with severe COVID-19. *Genome Medicine*, *13*(1), 118. https://doi.org/10.1186/s13073-021-00933-8.

Groff, A., Kavanaugh, M., Ramgobin, D., McClafferty, B., Aggarwal, C. S., Golamari, R., et al. (2021). Gastrointestinal manifestations of COVID-19: A review of what we know. *The Ochsner Journal*, *21*(2), 177–180. https://doi.org/10.31486/toj.20.0086.

Guerrero, J. I., Barragán, L. A., Martínez, J. D., Montoya, J. P., Peña, A., Sobrino, F. E., et al. (2021). Central and peripheral nervous system involvement by COVID-19: A systematic review of the pathophysiology, clinical manifestations, neuropathology, neuroimaging, electrophysiology, and cerebrospinal fluid findings. *BMC Infectious Diseases*, *21*(1), 515. https://doi.org/10.1186/s12879-021-06185-6.

Haddadi, K., Ghasemian, R., & Shafizad, M. (2020). Basal ganglia involvement and altered mental status: A unique neurological manifestation of coronavirus disease 2019. *Cureus*, *12*(4), e7869. https://doi.org/10.7759/cureus.7869.

Hamming, I., Timens, W., Bulthuis, M. L., Lely, A. T., Navis, G., & van Goor, H. (2004). Tissue distribution of ACE2 protein, the functional receptor for SARS coronavirus. A first step in understanding SARS pathogenesis. *The Journal of Pathology*, *203*(2), 631–637. https://doi.org/10.1002/path.1570.

Hawkes, C. H., Del Tredici, K., & Braak, H. (2007). Parkinson's disease: A dual-hit hypothesis. *Neuropathology and Applied Neurobiology*, *33*(6), 599–614. https://doi.org/10.1111/j.1365-2990.2007.00874.x.

Hernández-Fernández, F., Sandoval Valencia, H., Barbella-Aponte, R. A., Collado-Jiménez, R., Ayo-Martín, Ó., Barrena, C., et al. (2020). Cerebrovascular disease in patients with COVID-19: Neuroimaging, histological and clinical description. *Brain*, *143*(10), 3089–3103. https://doi.org/10.1093/brain/awaa239.

Hoffmann, M., Kleine-Weber, H., Schroeder, S., Krüger, N., Herrler, T., Erichsen, S., et al. (2020). SARS-CoV-2 cell entry depends on ACE2 and TMPRSS2 and is blocked by a clinically proven protease inhibitor. *Cell*, *181*(2), 271–280. https://doi.org/10.1016/j.cell.2020.02.052.

Hu, B., Guo, H., Zhou, P., & Shi, Z. L. (2021). Characteristics of SARS-CoV-2 and COVID-19. *Nature Reviews. Microbiology*, *19*(3), 141–154. https://doi.org/10.1038/s41579-020-00459-7.

Jang, H., Boltz, D. A., Webster, R. G., & Smeyne, R. J. (2009). Viral parkinsonism. *Biochimica et Biophysica Acta*, *1792*(7), 714–721. https://doi.org/10.1016/j.bbadis.2008.08.001.

Jeffers, S. A., Tusell, S. M., Gillim-Ross, L., Hemmila, E. M., Achenbach, J. E., Babcock, G. J., et al. (2004). CD209L (L-SIGN) is a receptor for severe acute respiratory syndrome coronavirus. *Proceedings of the National Academy of Sciences of the United States of America*, *101*(44), 15748–15753. https://doi.org/10.1073/pnas.0403812101.

Khan, M., Yoo, S. J., Clijsters, M., Backaert, W., Vanstapel, A., Speleman, K., et al. (2021). Visualizing in deceased COVID-19 patients how SARS-CoV-2 attacks the respiratory and olfactory mucosae but spares the olfactory bulb. *Cell*, *184*(24), 5932–5949. https://doi.org/10.1016/j.cell.2021.10.027.

Klingelhoefer, L., & Reichmann, H. (2015). Pathogenesis of Parkinson disease--the gut-brain axis and environmental factors. *Nature Reviews. Neurology*, *11*(11), 625–636. https://doi.org/10.1038/nrneurol.2015.197.

Koyuncu, O. O., Hogue, I. B., & Enquist, L. W. (2013). Virus infections in the nervous system. *Cell Host & Microbe*, *13*(4), 379–393. https://doi.org/10.1016/j.chom.2013.03.010.

Kubota, T., & Kuroda, N. (2021). Exacerbation of neurological symptoms and COVID-19 severity in patients with preexisting neurological disorders and COVID-19: A systematic review. *Clinical Neurology and Neurosurgery*, *200*, 106349. https://doi.org/10.1016/j.clineuro.2020.106349.

Lai, C. C., Shih, T. P., Ko, W. C., Tang, H. J., & Hsueh, P. R. (2020). Severe acute respiratory syndrome coronavirus 2 (SARS-CoV-2) and coronavirus disease-2019 (COVID-19): The epidemic and the challenges. *International Journal of Antimicrobial Agents*, *55*(3), 105924. https://doi.org/10.1016/j.ijantimicag.2020.105924.

Lechien, J. R., Chiesa-Estomba, C. M., De Siati, D. R., Horoi, M., Le Bon, S. D., Rodriguez, A., et al. (2020). Olfactory and gustatory dysfunctions as a clinical presentation of mild-to-moderate forms of the coronavirus disease (COVID-19): A multicenter European study. *European Archives of Oto-Rhino-Laryngology*, *277*(8), 2251–2261. https://doi.org/10.1007/s00405-020-05965-1.

Lee, M. H., Perl, D. P., Nair, G., Li, W., Maric, D., Murray, H., et al. (2021). Microvascular injury in the brains of patients with Covid-19. *The New England Journal of Medicine*, *384*(5), 481–483. https://doi.org/10.1056/NEJMc2033369.

Lehmann, M., Allers, K., Heldt, C., Meinhardt, J., Schmidt, F., Rodriguez-Sillke, Y., et al. (2021). Human small intestinal infection by SARS-CoV-2 is characterized by a mucosal infiltration with activated CD8(+) T cells. *Mucosal Immunology*, *14*(6), 1381–1392. https://doi.org/10.1038/s41385-021-00437-z.

Leta, V., Rodríguez-Violante, M., Abundes, A., Rukavina, K., Teo, J. T., Falup-Pecurariu, C., et al. (2021). Parkinson's disease and post–COVID-19 syndrome: The Parkinson's long-COVID Spectrum. *Movement Disorders*, *36*(6), 1287.

Lewis, A., Frontera, J., Placantonakis, D. G., Lighter, J., Galetta, S., Balcer, L., et al. (2021). Cerebrospinal fluid in COVID-19: A systematic review of the literature. *Journal of the Neurological Sciences*, *421*, 117316. https://doi.org/10.1016/j.jns.2021.117316.

Li, K., Wohlford-Lenane, C., Perlman, S., Zhao, J., Jewell, A. K., Reznikov, L. R., et al. (2016). Middle East respiratory syndrome coronavirus causes multiple organ damage and lethal disease in mice transgenic for human dipeptidyl peptidase 4. *The Journal of Infectious Diseases*, *213*(5), 712–722. https://doi.org/10.1093/infdis/jiv499.

Lin, W. Y., Lin, M. S., Weng, Y. H., Yeh, T. H., Lin, Y. S., Fong, P. Y., et al. (2019). Association of Antiviral Therapy with Risk of Parkinson disease in patients with chronic hepatitis C virus infection. *JAMA Neurology*, *76*(9), 1019–1027. https://doi.org/10.1001/jamaneurol.2019.1368.

Matschke, J., Lütgehetmann, M., Hagel, C., Sperhake, J. P., Schröder, A. S., Edler, C., et al. (2020). Neuropathology of patients with COVID-19 in Germany: A post-mortem case series. *The Lancet. Neurology*, *19*(11), 919–929. https://doi.org/10.1016/S1474-4422(20)30308-2.

Meinhardt, J., Radke, J., Dittmayer, C., Franz, J., Thomas, C., Mothes, R., et al. (2021). Olfactory transmucosal SARS-CoV-2 invasion as a port of central nervous system entry in individuals with COVID-19. *Nature Neuroscience*, *24*(2), 168–175. https://doi.org/10.1038/s41593-020-00758-5.

Meng, L., Shen, L., & Ji, H. F. (2019). Impact of infection on risk of Parkinson's disease: A quantitative assessment of case-control and cohort studies. *Journal of Neurovirology*, *25*(2), 221–228. https://doi.org/10.1007/s13365-018-0707-4.

Mönkemüller, K., Fry, L., & Rickes, S. (2020). COVID-19, coronavirus, SARS-CoV-2 and the small bowel. *Revista Española de Enfermedades Digestivas*, *112*(5), 383–388. https://doi.org/10.17235/reed.2020.7137/2020.

Mpekoulis, G., Frakolaki, E., Taka, S., Ioannidis, A., Vassiliou, A. G., Kalliampakou, K. I., et al. (2021). Alteration of L-dopa decarboxylase expression in SARS-CoV-2 infection and its association with the interferon-inducible ACE2 isoform. *PLoS One*, *16*(6), e0253458. https://doi.org/10.1371/journal.pone.0253458.

Mukerji, S. S., & Solomon, I. H. (2021). What can we learn from brain autopsies in COVID-19? *Neuroscience Letters*, *742*, 135528. https://doi.org/10.1016/j.neulet.2020.135528.

Murray, R. S., Brown, B., Brian, D., & Cabirac, G. F. (1992). Detection of coronavirus RNA and antigen in multiple sclerosis brain. *Annals of Neurology*, *31*(5), 525–533. https://doi.org/10.1002/ana.410310511.

Nataf, S. (2020). An alteration of the dopamine synthetic pathway is possibly involved in the pathophysiology of COVID-19. *Journal of Medical Virology*, *92*(10), 1743–1744. https://doi.org/10.1002/jmv.25826.

Netland, J., Meyerholz, D. K., Moore, S., Cassell, M., & Perlman, S. (2008). Severe acute respiratory syndrome coronavirus infection causes neuronal death in the absence of encephalitis in mice transgenic for human ACE2. *Journal of Virology*, *82*(15), 7264–7275. https://doi.org/10.1128/jvi.00737-08.

Ousseiran, Z. H., Fares, Y., & Chamoun, W. T. (2021). Neurological manifestations of COVID-19: A systematic review and detailed comprehension. *The International Journal of Neuroscience*, *1-16*. https://doi.org/10.1080/00207454.2021.1973000.

Pavlov V.A. and Tracey K.J., 2012. The vagus nerve and the inflammatory reflex—Linking immunity and metabolism. *Nature Reviews Endocrinology*, *8*(12), 743–754. doi: 10.1038/nrendo.2012.189. PMID: 23169440; PMCID: PMC4082307.

Pellegrini, L., Albecka, A., Mallery, D. L., Kellner, M. J., Paul, D., Carter, A. P., et al. (2020). SARS-CoV-2 infects the brain choroid plexus and disrupts the blood-CSF barrier in human brain organoids. *Cell Stem Cell*, *27*(6), 951–961.e955. https://doi.org/10.1016/j.stem.2020.10.001.

Pezzini, A., & Padovani, A. (2020). Lifting the mask on neurological manifestations of COVID-19. *Nature Reviews Neurology*, *16*(11), 636–644. https://doi.org/10.1038/s41582-020-0398-3.

Politi, L. S., Salsano, E., & Grimaldi, M. (2020). Magnetic resonance imaging alteration of the brain in a patient with coronavirus disease 2019 (COVID-19) and anosmia. *JAMA Neurology*, *77*(8), 1028–1029. https://doi.org/10.1001/jamaneurol.2020.2125.

Porzionato, A., Stocco, E., Emmi, A., Contran, M., Macchi, V., Riccetti, S., et al. (2021). Hypopharyngeal ulcers in COVID-19: Histopathological and virological analyses - a case report. *Frontiers in Immunology*, *12*, 676828. https://doi.org/10.3389/fimmu.2021.676828.

Porzionato, A., Emmi, A., Contran, M., Stocco, E., Riccetti, S., Sinigaglia, A., et al. (2021). Case report: The carotid body in COVID-19: Histopathological and virological analyses of an autopsy case series. *Frontiers in Immunology*, *12*, 736529. https://doi.org/10.3389/fimmu.2021.736529.

Radnis, C., Qiu, S., Jhaveri, M., Da Silva, I., Szewka, A., & Koffman, L. (2020). Radiographic and clinical neurologic manifestations of COVID-19 related hypoxemia. *Journal of the Neurological Sciences*, *418*, 117119. https://doi.org/10.1016/j.jns.2020.117119.

Remmelink, M., De Mendonça, R., D'Haene, N., De Clercq, S., Verocq, C., Lebrun, L., et al. (2020). Unspecific post-mortem findings despite multiorgan viral spread in COVID-19 patients. *Critical Care*, *24*(1), 495. https://doi.org/10.1186/s13054-020-03218-5.

Sadasivan, S., Sharp, B., Schultz-Cherry, S., & Smeyne, R. J. (2017). Synergistic effects of influenza and 1-methyl-4-phenyl-1,2,3,6-tetrahydropyridine (MPTP) can be eliminated by the use of influenza therapeutics: Experimental evidence for the multi-hit hypothesis. *NPJ Parkinsons Disease*, *3*, 18. https://doi.org/10.1038/s41531-017-0019-z.

Schurink, B., Roos, E., Radonic, T., Barbe, E., Bouman, C. S. C., de Boer, H. H., et al. (2020). Viral presence and immunopathology in patients with lethal COVID-19: A prospective autopsy cohort study. *The Lancet Microbe*, *1*(7), e290–e299. https://doi.org/10.1016/S2666-5247(20)30144-0.

Schwabenland, M., Salié, H., Tanevski, J., Killmer, S., Lago, M. S., Schlaak, A. E., et al. (2021). Deep spatial profiling of human COVID-19 brains reveals neuroinflammation with distinct microanatomical microglia-T-cell interactions. *Immunity*, *54*(7), 1594–1610. https://doi.org/10.1016/j.immuni.2021.06.002.

Siddiqi, H. K., Libby, P., & Ridker, P. M. (2021). COVID-19 - A vascular disease. *Trends in Cardiovascular Medicine*, *31*(1), 1–5. https://doi.org/10.1016/j.tcm.2020.10.005.

Solomon, I. H., Normandin, E., Bhattacharyya, S., Mukerji, S. S., Keller, K., Ali, A. S., et al. (2020). Neuropathological features of Covid-19. *The New England Journal of Medicine*, *383*(10), 989–992. https://doi.org/10.1056/NEJMc2019373.

Sullivan, B. N., & Fischer, T. (2021). Age-associated neurological complications of COVID-19: A systematic review and meta-analysis. *Frontiers in Aging Neuroscience*, *13*, 653694. https://doi.org/10.3389/fnagi.2021.653694.

Sulzer, D. (2007). Multiple hit hypotheses for dopamine neuron loss in Parkinson's disease. *Trends in Neurosciences*, *30*(5), 244–250.

Sulzer, D., Antonini, A., Leta, V., Nordvig, A., Smeyne, R. J., Goldman, J. E., et al. (2020). COVID-19 and possible links with Parkinson's disease and parkinsonism: From bench to bedside. *NPJ Parkinson's disease*, *6*, 18. https://doi.org/10.1038/s41531-020-00123-0.

Tandon, M., Kataria, S., Patel, J., Mehta, T. R., Daimee, M., Patel, V., et al. (2021). A comprehensive systematic review of CSF analysis that defines neurological manifestations of COVID-19. *International Journal of Infectious Diseases*, *104*, 390–397. https://doi.org/10.1016/j.ijid.2021.01.002.

Tang, Y.-W., Schmitz, J. E., Persing, D. H., & Stratton, C. W. (2020). Laboratory diagnosis of COVID-19: Current issues and challenges. *Journal of Clinical Microbiology*, *58*(6), e00512–e00520. https://doi.org/10.1128/JCM.00512-20.

Tanner, C. M., Ottman, R., Goldman, S. M., Ellenberg, J., Chan, P., Mayeux, R., et al. (1999). Parkinson disease in twins: An etiologic study. *JAMA*, *281*(4), 341–346. https://doi.org/10.1001/jama.281.4.341.

Thakur, K. T., Miller, E. H., Glendinning, M. D., Al-Dalahmah, O., Banu, M. A., Boehme, A. K., et al. (2021). COVID-19 neuropathology at Columbia University Irving medical center/New York presbyterian hospital. *Brain*, *144*(9), 2696–2708. https://doi.org/10.1093/brain/awab148.

Tsai, H. H., Liou, H. H., Muo, C. H., Lee, C. Z., Yen, R. F., & Kao, C. H. (2016). Hepatitis C virus infection as a risk factor for Parkinson disease: A nationwide cohort study. *Neurology*, *86*(9), 840–846. https://doi.org/10.1212/wnl.0000000000002307.

Tsivgoulis, G., Palaiodimou, L., Zand, R., Lioutas, V. A., Krogias, C., Katsanos, A. H., et al. (2020). COVID-19 and cerebrovascular diseases: A comprehensive overview. *Therapeutic Advances in Neurological Disorders*, *13*. https://doi.org/10.1177/1756286420978004.

Tsui, J. K., Calne, D. B., Wang, Y., Schulzer, M., & Marion, S. A. (1999). Occupational risk factors in Parkinson's disease. *Canadian Journal of Public Health. Revue Canadienne de Sante Publique*, *90*(5), 334–337. https://doi.org/10.1007/BF03404523.

Varga, Z., Flammer, A. J., Steiger, P., Haberecker, M., Andermatt, R., Zinkernagel, A. S., et al. (2020). Endothelial cell infection and endotheliitis in COVID-19. *Lancet*, *395*(10234), 1417–1418. https://doi.org/10.1016/s0140-6736(20)30937-5.

Verstrepen, K., Baisier, L., & De Cauwer, H. (2020). Neurological manifestations of COVID-19, SARS and MERS. *Acta Neurologica Belgica*, *120*(5), 1051–1060. https://doi.org/10.1007/s13760-020-01412-4.

Wan, D., Du, T., Hong, W., Chen, L., Que, H., Lu, S., et al. (2021). Neurological complications and infection mechanism of SARS-COV-2. *Signal Transduction and Targeted Therapy*, *6*(1), 406. https://doi.org/10.1038/s41392-021-00818-7.

Wang, Q., Zhang, Y., Wu, L., Niu, S., Song, C., Zhang, Z., et al. (2020). Structural and functional basis of SARS-CoV-2 entry by using human ACE2. *Cell*, *181*(4), 894–904.e899. https://doi.org/10.1016/j.cell.2020.03.045.

Williams, A., Branscome, H., Khatkar, P., Mensah, G. A., Al Sharif, S., Pinto, D. O., et al. (2021). A comprehensive review of COVID-19 biology, diagnostics, therapeutics, and disease impacting the central nervous system. *Journal of Neurovirology*, *27*(5), 667–690. https://doi.org/10.1007/s13365-021-00998-6.

Xu, X., Chen, P., Wang, J., Feng, J., Zhou, H., Li, X., et al. (2020). Evolution of the novel coronavirus from the ongoing Wuhan outbreak and modeling of its spike protein for risk of human transmission. *Science China. Life Sciences*, *63*(3), 457–460. https://doi.org/10.1007/s11427-020-1637-5.

Yachou, Y., El Idrissi, A., Belapasov, V., & Ait Benali, S. (2020). Neuroinvasion, neurotropic, and neuroinflammatory events of SARS-CoV-2: Understanding the neurological manifestations in COVID-19 patients. *Neurological Sciences*, *41*(10), 2657–2669. https://doi.org/10.1007/s10072-020-04575-3.

Yang, A. C., Kern, F., Losada, P. M., Agam, M. R., Maat, C. A., Schmartz, G. P., et al. (2021). Dysregulation of brain and choroid plexus cell types in severe COVID-19. *Nature*, *595*(7868), 565–571. https://doi.org/10.1038/s41586-021-03710-0.

Zazhytska, M., Kodra, A., Hoagland, D. A., Frere, J., Fullard, J. F., Shayya, H., et al. (2022). Non-cell-autonomous disruption of nuclear architecture as a potential cause of COVID-19-induced anosmia. *Cell*, *185*(6), 1052–1064.e1012. https://doi.org/10.1016/j.cell.2022.01.024.

Zhang, R., Wu, Y., Zhao, M., Liu, C., Zhou, L., Shen, S., et al. (2009). Role of HIF-1alpha in the regulation ACE and ACE2 expression in hypoxic human pulmonary artery smooth muscle cells. *American Journal of Physiology. Lung Cellular and Molecular Physiology*, *297*(4), L631–L640. https://doi.org/10.1152/ajplung.90415.2008.

Zhang, H., Kang, Z., Gong, H., Xu, D., Wang, J., Li, Z., et al. (2020). Digestive system is a potential route of COVID-19: An analysis of single-cell coexpression pattern of key proteins in viral entry process. *Gut*, *69*(6), 1010. https://doi.org/10.1136/gutjnl-2020-320953.

Zubair, A. S., McAlpine, L. S., Gardin, T., Farhadian, S., Kuruvilla, D. E., & Spudich, S. (2020). Neuropathogenesis and neurologic manifestations of the coronaviruses in the age of coronavirus disease 2019: A review. *JAMA Neurology*, 77(8), 1018–1027. https://doi.org/10.1001/jamaneurol.2020.2065.

CHAPTER THREE

Prevalence and outcomes of Covid-19 in Parkinson's disease: Acute settings and hospital

Conor Fearon[a] and Alfonso Fasano[a,b,c,*]
[a]Edmond J. Safra Program in Parkinson's Disease, Morton and Gloria Shulman Movement Disorders Clinic, Toronto Western Hospital–UHN, Division of Neurology, University of Toronto, Toronto, ON, Canada
[b]Krembil Research Institute, Toronto, ON, Canada
[c]Center for Advancing Neurotechnological Innovation to Application (CRANIA), Toronto, ON, Canada
*Corresponding author: e-mail address: alfonso.fasano@uhn.ca

Contents

Abstract

The global explosion of COVID-19 necessitated the rapid dissemination of information regarding SARS-CoV-2. Hence, COVID-19 prevalence and outcome data in Parkinson's disease patients were disseminated at a time when we only had part of the picture. In this chapter we firstly discuss the current literature on the prevalence of COVID-19 in people with PD. We then discuss outcomes from COVID-19 in people with PD, specifically risk of hospitalization and mortality. Finally, we discuss specific contributing and confounding factors which may put PD patients at higher or lower risk from COVID-19.

International Review of Neurobiology, Volume 165
ISSN 0074-7742
https://doi.org/10.1016/bs.irn.2022.03.001
Copyright © 2022 Elsevier Inc.
All rights reserved.

1. Introduction

The coronavirus disease 2019 (Covid-19) pandemic, caused by the severe acute respiratory syndrome coronavirus 2 (SARS-CoV-2), has spread globally with a previously unseen rapidity. Its dominance around the world has meant that prevalence and outcome data from almost any population could be obtained within months of the pandemic. This data could be pulled from hospital admissions, hospital deaths, community prevalence and nursing homes. The global explosion of Covid-19 has necessitated rapid dissemination of information regarding SARS-CoV-2. In addition, the ease with which reasonably sized data set could be drawn together has meant that Covid-19 prevalence and outcome data in Parkinson's disease (PD) patients has been disseminated at a time when we only had part of the picture. Many people with PD (PwP) were concerned about the Covid-19 risk and it would always be challenging to assess this risk in the absence of good quality data. As each new publication came out, we have gained a better idea of the Covid-19 prevalence and outcomes in PD, although we are continuing to learn and understand this data.

The heterogeneity of PD, its ability to affect both young and old, and its relatively slow disease progression from highly active and independent functioning to highly dependent advanced disease means that prevalence and outcome data must be carefully considered within the population from which those data is drawn, taking confounding factors into account. For example, it should not be surprising that we might find differences in prevalence and outcomes between a cohort of PwP with advanced PD admitted acutely to hospital and a cohort of early-stage PwP who undertake a telephone survey in the community and who may well be isolating at home during the peak lockdown period of the first wave of the Covid-19 pandemic in case PD poses a significant risk to them in the setting of Covid-19.

The first goal of this chapter is to discuss the current literature on the prevalence of Covid-19 in PwP. We will then discuss outcomes from Covid-19 in PwP, specifically risk of hospitalization and mortality. Finally, we will discuss specific contributing and confounding factors which may put PwP at higher or lower risk from Covid-19.

Prior to this discussion, it is important to note that lockdowns during the pandemic had a profound effect on PwP. This aspect is beyond the scope of this chapter, but has been covered in detail elsewhere (Brown et al., 2020;

Fearon & Fasano, 2021; Prasad et al., 2020). Suffice to say that motor and non-motor symptoms of PD worsen during lockdowns (Fabbri et al., 2021), anxiety heightens (Shalash et al., 2020), impulsive-compulsive behaviors become more common (Yule, Pickering, McBride, & Poliakoff, 2021), sleep problems become more common (Kumar & Gupta, 2021), access to medications, physiotherapy and exercise become more problematic (Cheong et al., 2020; van der Kolk et al., 2019). Hence, the effect of the Covid-19 *pandemic* on PwP is a far more complex consideration than the impact of SARS-CoV-2 infection alone. Nevertheless, we will direct our focus toward the latter, merely acknowledging that it forms part of a bigger picture.

Firstly, however, we will discuss an important question. The interest in prevalence and outcomes in PwP had started even before we had any data on these parameters. Why, therefore, would we expect prevalence and outcomes of SARS-CoV-2 infection to be different in PwP compared with the general population?

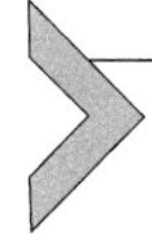

2. Why should PwP be more or less susceptible to Covid-19?

Particular interest has surrounded SARS-CoV-2 infection in PwP since the beginning of the Covid-19 pandemic. A major driver of this interest was the emergence of post-encephalitic parkinsonism ("encephalitis lethargica") in the wake of the last major global viral pandemic, the H1N1 "Spanish" flu of 1914–1918 (von Economo, 1918). In the decades that followed the pandemic, it is estimated that up to 1 million people worldwide may have developed encephalitis lethargica (Ravenholt & Foege, 1982). In spite of the fact that the causal role of H1N1 in the pathogenesis of encephalitis lethargica is debated (Hoffman & Vilensky, 2017), the knowledge that certain viruses can demonstrate specific neurotropism for the basal ganglia, in addition to the fear that a pandemic of similar degree could lead to another epidemic of parkinsonism has forever been etched on the minds of the scientific and medical community. Even in the few years prior to the Covid-19 pandemic, vigilance for the return of encephalitis lethargica persisted in historic discussions on the topic (Hoffman & Vilensky, 2017).

It is not surprising, therefore, that immediately after the first reports of Covid-19 spreading globally, it was suggested that SARS-CoV-2 may have the potential to cause neurodegeneration and that we might expect a significant increase in the incidence of post-infectious parkinsonism in the decades to come (Lippi, Domingues, Setz, Outeiro, & Krisko, 2020). Although clearly different viruses, SARS-CoV-2 and H1N1 viruses do share some common pathophysiological mechanisms, including mitochondrial dysfunction, lipid metabolism, proteostasis and stress responses, all of which have also been implicated in the pathophysiology of PD (Lippi et al., 2020). Nevertheless, this concept was debated vigorously in the scientific literature at the earlies stages of the pandemic (Gonzalez-Latapi, Fearon, Fasano, & Lang, 2021; Merello, Bhatia, & Obeso, 2021).

Even if SARS-CoV-2 did have the potential to lead to post-infectious parkinsonism, it does not necessarily follow that PwP would then be more susceptible to SARS-CoV-2 infection than the general population. However, further links between SARS-CoV-2 and PD were identified early on, particularly that dopamine may be involved in the pathophysiology of Covid-19. An alteration in dopamine synthetic pathways may be implicated in the pathophysiology of SARS-CoV-2, raising the question of whether alterations in dopamine metabolism could modulate the effect of SARS-CoV-2 infection. A co-expression link was also demonstrated between *DDC*, the gene which encodes for dopa decarboxylase (a major enzyme in dopamine synthetic pathways) and *ACE2*, the gene encoding one of the main receptors on the SARS-CoV-2 virus (Nataf, 2020). This is important because ACE2 is found in high quantities in dopaminergic neurons, which are decreased in PwP.

Further evidence that PwP were an important cohort to study in the early stages of the pandemic arose from the possibility that amantadine, an antiviral agent used relatively widely in the treatment of PD, could be protective against Covid-19. Amantadine is FDA-approved for the treatment of influenza A and its antiviral properties led its early consideration as a treatment for Covid-19 (Araújo, Aranda-Martínez, & Aranda-Abreu, 2020). A number of mechanisms of action of amantadine targeting SARS-CoV-2 have been proposed to limit the toxic effects of Covid-19, including disrupting lysosomal gene expression and viral replication (Aranda Abreu, Hernández Aguilar, Herrera Covarrubias, & Rojas Durán, 2020; Smieszek, Przychodzen, & Polymeropoulos, 2020).

Finally, it was then postulated that α-synuclein may itself be protective against Covid-19 (Ait Wahmane et al., 2020). Neuronal expression of α-synuclein inhibits viral central nervous system (CNS) invasion and restricts the ability of RNA viruses to replicate in the brain (Massey & Beckham, 2016). Physiological α-synuclein has myriad roles, which includes immune cell recruitment and protection against pro-inflammatory responses to infections (Labrie & Brundin, 2017). Depletion of α-synuclein in knockout mice leads to B- and T-cell deficiencies (Xiao, Shameli, Harding, Meyerson, & Maitta, 2014). Given this, it was proposed that peripheral overexpression of α-synuclein in PD may prevent neuroinvasion of SARS-CoV-2 (Ait Wahmane et al., 2020).

Taken together, this data has led to the expectation that PwP may be differently susceptible compared to the general population when it comes to risk of contracting Covid-19 and the severity of that illness. This idea had emerged long before we had relevant data on Covid-19 and PD risk. So, was this hypothesis borne out by the data which followed? We will explore this in the sections which follow.

3. Covid-19 prevalence in Parkinson's disease

It is important to state from the outset that whether there is an increased risk of Covid-19 in PwP remains a matter of debate. A number of small and larger studies have tried to answer this question, through case-control study design or by comparing prevalence in their selected group of PwP with prevalence in the general population in that region over a similar time period. Particularly in the early stages of the pandemic, study design and recruitment were undertaken at times in a rapid and ad hoc fashion out of necessity. This led to conflicting results scattered across a number of small studies initially, with critical mass for systematic reviews and meta-analyses only being reached more recently. The studies which report prevalence rates of SARS-CoV-2 among PwP are presented in Table 1. We will discuss these studies below and finally review the findings of the recent systematic reviews and meta-analyses.

The first reported neurological presentation of Covid-19 was in a PD patient (Filatov, Sharma, Hindi, & Espinosa, 2020). At that time, it was clear that Covid-19 affected elderly patients with chronic conditions to a greater extent, but it was not known whether PD itself was a specific risk factor for contracting SARS-CoV-2.

A large telephone-based case-control study from Lombardy in Italy (the region from which one-third of Italian cases of SARS-CoV-2 occurred during the first wave of the pandemic) demonstrated no difference in Covid-19 rates between 1486 PwP and 1207 familial controls (7.1% vs 7.6%) (Fasano, Cereda, et al., 2020). The PwP who contracted Covid-19 were younger, more likely to suffer from respiratory disease and obesity than unaffected patients. In addition, patients who avoided infection were more likely to be taking vitamin D supplementation. Although the PD group were

Table 1 Studies reporting prevalence figures relating to COVID-19 in PD.

Reference	Study Design	Total PD sample	PD	Controls	Risk factors
Fasano, Cereda, et al. (2020), Fasano, Elia, et al. (2020)	Phone survey	1486	7.1%	7.6%	Reduced risk from fewer weekly outings in PD cohort
Brown et al. (2020)	Online survey	5429	0.9%	1.8%	Smoking, heart disease, age, male sex
Del Prete et al. (2020)	Phone survey	740	0.9%	NA	Hypertension, diabetes,? age
Santos García et al. (2020)	Phone survey	568	2.6%	NA	Less advanced disease (possibly due to cocooning in more vulnerable advance), heart disease, amantadine (protective)
Salari, Etemadifar, Ashrafi, Ommi et al. (2021), Salari, Etemadifar, Zali, Aminzade, et al. (2021), Salari, Etemadifar, Zali, Medghalchi, et al. (2021)	Seroprevalence case-control study in asymptomatic individuals	90	25.6%	12.4%	None
Cilia et al. (2020)	Community-based case control study	141	8.5%	NA	No increased risk with advancing age or disease duration compared with COVID-19-negative
Artusi et al. (2020)	Phone survey	1407	0.57%	0.63%	Unclear if similar rates are due to lack of increased risk or due to increased self-isolation in at risk patients
Salari, Etemadifar, Ashrafi, Ommi, et al. (2021), Salari, Etemadifar, Zali, Aminzade, et al. (2021), Salari, Etemadifar, Zali, Medghalchi, et al. (2021)	Phone survey	647	11.28%	15.39%	None

older than their familial controls, they were less likely to be hospitalized. It was noted by the authors that the PD cohort appeared to be self-isolating to a greater degree than their family members who were surveyed (as measured by a lower number of weekly outings). A possible explanation for this finding may have been that families were cocooning the at-risk member of family (and as a result venturing out into the community relatively more frequently to run errands etc.). Irrespective of why this occurred, it may have underestimated true risk of being infected by SARS-CoV-2 in the PD cohort. At a similar time in the pandemic, a semi-structured telephone-based study in Germany reported that 73% of PwP had changed their behavior since the start of the Covid-19 pandemic in order to reduce the risk of infection (Zipprich, Teschner, Witte, Schönenberg, & Prell, 2020). Of these, 99% performed at least one specific preventive behavior, and 86.9% had reduced social contacts and stayed home. The proportion was higher in older patients, probably in light of higher perceived risk.

This theory might be further supported by a similar phone-based survey of 568 Spanish patients (Santos García et al., 2020). They found that 2.6% of patients reported a confirmed diagnosis of Covid-19. The Covid-19-negative group were more likely to experience motor fluctuations (61% vs 35.7%, $P=0.052$) and hallucinations (23.4% vs 0%, $P=0.025$), and there was a trend toward more prevalent dementia and behavioral disorders among the group who did not contract Covid-19. The authors similarly hypothesized that stricter prevention measures in this group with more complex PD may have led to these findings, however the Covid-19-positive cohort was small.

A study from Piedmont, the region in Italy with the second highest number of Covid-19 cases after Lombardy, aimed to estimate the prevalence of Covid-19 in a large population of PwP and compare the infection rate with the general population of the same region (Artusi et al., 2020). They performed a telephone survey of 1407 PwP, asking for a laboratory-confirmed diagnosis of Covid-19. They found eight PwP (0.57%) who were Covid-19-positive. The prevalence in Piedmont in the general population of Piedmont over the same period was 0.63%. The authors point out once again that it cannot be determined whether this is due to equivalent risk of contracting Covid-19 or due to higher self-isolation in the PD cohort masking a greater underlying risk.

Another telephone-based study of PwP in Tuscany, Italy demonstrated a Covid-19 prevalence of 0.9% among the 740 PwP who responded (Del Prete et al., 2020). There was no non-PD control group in the study, but estimated prevalence of Covid-19 in the region during that period was 0.18%–0.25%. All patients in the study who contracted Covid-19 were over the age of 60. However, the age profile of the Covid-19-negative group was similar. When comparing Covid-19-positive and Covid-19-negative patients, the former reported a significantly higher prevalence of hypertension ($P<0.001$) and diabetes ($P=0.049$), while no significant differences were found for the other comorbidities. There were no reported differences in anti-parkinsonian medications.

Hence increasing age does not appear to predispose to Covid-19 in PwP, as supported by another community-based case-control study from Lombardy, which found 12 of 141 PwP had tested positive for Covid-19 and when compared to the Covid-19-negative PD group, they were neither older nor of longer disease duration (Cilia et al., 2020). However, it is important to note that this study was likely underpowered to examine such a difference.

In an effort to reach an even larger cohort of PwP, Brown et al. surveyed users of Fox Insight, an online study that involves thousands of people with and without PD (Brown et al., 2020). 7209 users responded (5429 PwP and 1452 without PD). This represented 74% of all active Fox Insight users. 51 PwP (0.9%) and 26 (1.8%) respondents without PD reported a diagnosis of Covid-19. Covid-19-positive PwP were more likely to smoke (5.9% vs 1.6%, $P=0.048$) and have a history of heart disease (20% vs 8.2%, $P=0.008$) than those who were Covid-19-negative, and were more likely to be older, male, and less likely to have lung disease than Covid-19-positive respondents who did not have PD. Importantly, control subjects who contracted Covid-19 more often reported working in healthcare or as another essential worker which may have increased their exposure. The Covid-19-positive PD group were more likely to have a close contact with Covid-19 compared to the Covid-19-negative PD group (59% vs 7.7%, $P<0.001$). Hence general exposure risk may have greatly impacted the above findings.

Using patients' reports as a means of estimating prevalence has limitations. The higher prevalence reported in the Lombardy studies above could be influenced by the fact that patients without molecular test confirmation were included, leading to a possible overestimation of prevalence. On the other hand, patient reports may not account for patients who may have been exposed to SARS-CoV-2 and remained asymptomatic or pauci-symptomatic. To this end, Salari et al. used commercially available enzyme-linked immunosorbent assay (ELISA) kits to estimate the seroprevalence of SARS-CoV-2 among PwP who did not have the symptomatic infection during the third wave of Covid-19 in Iran (Salari, Etemadifar, Zali, Medghalchi, et al., 2021).

Ninety PwP and 97 healthy controls were tested and 25.6% of PwP and 12.4% of controls tested positive for SARS-CoV-2 IgG antibody. Risk factors for those who were seropositive were not explored, but the authors reported that the proportion of PwP with positive IgG test who had no direct contact with Covid-19 patients was significantly higher than that of the same individuals in the control group.

In an effort to clarify the incongruity of the reported studies, Chambergo-Michilot et al. provided an analysis of the factors associated with Covid-19 prevalence in PwP (Chambergo-Michilot, Barros-Sevillano, Rivera-Torrejón, De la Cruz-Ku, & Custodio, 2021). The authors included six studies (four case-controlled studies and two cross-sectional studies) in their systematic review and meta-analysis. Two of these case-control studies and one cross-sectional study were felt not to have controlled their analysis for important factors. The pooled prevalence of Covid-19 in PD was 2% They found that the following factors were associated with contracting Covid-19 in PwP: obesity (OR: 1.79, 95% CI: 1.07–2.99, I^2: 0%), any pulmonary disease (OR: 1.92, 95% CI: 1.17–3.15, I^2: 0%) and Covid-19 contact (OR: 41.77, 95% CI: 4.77–365.56, I^2: 0%) and that vitamin D supplementation was associated with a lower odds ratio of contracting Covid-19 (OR: 0.50, 95% CI: 0.30–0.83, I^2: 0%). They did not find any significant association between Covid-19 in PD and hypertension, diabetes, cardiopathy, cancer, any cognitive problem, dementia, chronic obstructive pulmonary disease, renal or hepatic disease, smoking, and tremor.

El-Qushayri et al. subsequently performed a systematic review and meta-analysis of Covid-19 in PwP (El-Qushayri et al., 2021). Four studies which reported prevalence of Covid-19 in a total of 6878 PwP were included in the analysis. The pooled prevalence rate of Covid-19 was estimated at 2.12% (95% CI: 0.75–5.98). There was, however, significant heterogeneity among the pooled studies ($I^2 = 95\%$; $P < 0.001$) with the prevalence estimates in the included studies ranging from 0.94% to 8.51%. The authors examined possible co-founders by comparing comorbidities among Covid-19-positive PwP and Covid-19-negative PwP. They found significantly higher rates of diabetes mellitus (OR: 2.12, 95% CI: 1.06–4.23; $P = 0.033$) and immune compromise (OR: 2.06, 95% CI: 1.08–3.94; $P = 0.029$) among the Covid-19-positive group. No significant differences were observed in hypertension, obesity, dementia, chronic pulmonary disease, malignancy, or cardiomyopathy between groups. The effect of age or other demographic factors was not examined.

A third systematic review and meta-analysis which included a greater number of studies found a pooled prevalence of Covid-19 in PD cases of

5% (95% CI: 4%–6%) ($I^2 = 98.1\%$, $P < 0.001$), higher than the pooled prevalence reported in the two previous systematic reviews (Khoshnood et al., 2021). Another systematic review very clearly demonstrated that prevalence varied according to geography, with a prevalence of 2.6% in Spain, 0.9% in the US, and variable values in different regions of Italy, ranging from 7.1% to 8.5% in Lombardy, 0.9% in Tuscany, and 0.6% in Piedmont and in the Bologna district (Artusi et al., 2021).

In summary, there is no consistent evidence that people with PD are at greater risk from contracting Covid-19 compared with the general population. The region from which the data were drawn appear to play a significant role, as does the presence of comorbidities. The main confounder in community studies may be that PwP potentially self-isolated to a greater degree than the general population who had no perceived added risk from Covid-19. This behavior may have masked an increased predisposition to SARS-CoV-2 infection, but we simply cannot say with certainty from the current data. Next, we will consider what happens to PwP if they do contract Covid-19.

4. Covid-19 outcomes in Parkinson's disease

Intuitively, older patients with advanced PD, impaired cough reflex and respiratory muscle involvement would be particularly vulnerable to a severe acute respiratory syndrome. Muscular weakness can appear in PD, and could contribute to respiratory failure, which leads to death (Baille et al., 2018). However, data from the pre-Covid-19 era of elderly patients with pneumonia showed that patients with parkinsonism had significantly lower in-hospital mortality than those without (Jo et al., 2018). Hence, the above factors may not pose as great a risk for PwP with Covid-19 as one might expect. We will consider hospitalization and mortality outcomes separately.

4.1 Hospitalization

Similar to the prevalence data above, hospitalization data was reported in a number of small studies with varying rates reported. For example, the aforementioned Spanish telephone-based study suggested that approximately 33% of PwP who contracted Covid-19 were hospitalized, although the number of Covid-19-positive PD cases in this study was small (15 cases) (Santos García et al., 2020). However, a systematic review incorporating 13 studies reported a similar hospitalization rate of 28.6% among PwP with Covid-19 (Artusi et al., 2021). A more recent systematic review and meta-analysis which pooled

17 studies found a pooled prevalence of hospitalization in cases with Covid-19 infection of 49% (95%CI: 29%–52%) (I^2: 93.5%, $P<0.001$) (Khoshnood et al., 2021).

It is important to note that PwP have an increased risk of being admitted to hospital, not only from Covid-19, but also due to PD-related complications (Temlett & Thompson, 2006). Furthermore, PwP are more likely to have comorbidities which further increases their risk of being admitted to hospital. The meta-analysis performed by Chambergo-Michelot and colleagues found hospitalization to be independently associated with Covid-19 and PD (OR: 11.78, 95% CI: 6.27–22.12, I^2: 0%) (Chambergo-Michilot et al., 2021).

Salari et al. found that although PwP were hospitalized because of the severity of Covid-19 more frequently than the rates in the general population, the rate of hospitalization among these patients was not significantly different from a simultaneously collected age-matched control group suggesting that advancing age may play a role (Salari, Etemadifar, Zali, Aminzade, et al., 2021).

Similarly, advanced disease, as expected, contributes to risk of hospitalization from Covid-19. The Fox Insight-based study demonstrated that longer PD-duration was associated with a higher risk of pneumonia, supplemental oxygen requirement and hospitalization among PwP who contracted Covid-19 (44% among disease duration >9 years vs 14% among disease duration ≤ 9 years, OR 5.44, 95% CI 1.04–30.5, $P=0.043$) (Brown et al., 2020). Mortality rate was not addressed in this study.

Interestingly, another study from Bologna, Italy of 696 subjects with PD, 184 with parkinsonism and 8590 control subjects found a 3-month hospitalization rate for Covid-19 of 0.6% in PD, 3.3% in parkinsonism, and 0.7% in controls (Vignatelli et al., 2021). This suggests that PD per se is not a risk factor for hospitalization, but that atypical parkinsonism may be a significant risk factor. The parkinsonism group (albeit a presumably heterogeneous group) will have included a significantly higher proportion of atypical parkinsonism who one expects to have a higher prevalence of dysphagia, cognitive impairment and gait impairment, all of which may contribute to hospitalization (and mortality). No studies specifically examining this risk in atypical parkinsonism has been undertaken to date.

4.2 Mortality

Seventeen studies have examined mortality in PD in the setting of Covid-19. The number of Covid-19 cases in the studies ranged from two to 694 cases, while reported mortality rates ranged from 5.2% to 100% (Table 2).

Table 2 Studies reporting mortality figures relating to COVID-19 in PD.

Reference	Study design	Total PD sample	PD	Controls	Risk factors
Antonini, Leta, Teo, and Chaudhuri (2020), Antonini, Leta, Teo, and Ray Chaudhuri (2020)	Case series	10 (10 COVID+)	40%	NA	Age, disease duration, use of advanced therapies
Hainque and Grabli (2020)	Case report	2 (2 COVID+)	100%	NA	STN DBS?
Kobylecki et al. (2020)	Hospitalized patients in a single center	58 (3 COVID+)	5.2%	NA	NA
Fasano, Cereda, et al. (2020), Fasano, Elia, et al. (2020)	Phone survey	1486 (105 COVID+)	5.7%	7.6%	NA
Artusi et al. (2020)	Phone survey	1407 (8 COVID+)	75%	NA	NA
Del Prete et al. (2020)	Phone survey	740 (7 COVID+)	14%	NA	NA
Fasano, Cereda, et al. (2020), Fasano, Elia, et al. (2020)	Multi-center case series	117 (117 COVID+)	19.7%	NA	Dementia, hypertension, disease duration
Sainz-Amo et al. (2020)	Single-center case series	211 (33 COVID+)	21%	NA	Cancer, hospital admission (no DA use, dementia)[a]
Vignatelli et al. (2021)	Prospective cohort study	696 (4 hospitalized with COVID-19)	25%	39%	Control group was matched for age and comorbidities
Salari, Etemadifar, Ashrafi, Ommi, et al. (2021), Salari, Etemadifar, Zali, Aminzade, et al. (2021), Salari, Etemadifar, Zali, Medghalchi, et al. (2021)	Hospitalized pts. patients in two referral centers	87 (87 COVID+)	35.6%	16.6%	Dementia (Age)

Parihar et al. (2021)	Retrospective review of admitted pts. (1 hospital in NYC)	70 (53 COVID +)	35.8%	NA	Age > 70, advanced PD, meds reduction, Black race
Nwabuobi et al. (2021)	Retrospective review of admitted pts. (1 hospital in NYC)	25 (25 COVID +)	32%	26%	Encephalopathy during admission
Xu et al. (2021)	Cross-sectional online survey (1 center)	46 (46 COVID +)	13%	NA	None
Zhai et al. (2021)	Retrospective review of admitted pts. (1 hospital in Wuhan)	10 (10 severe COVID)	30%	40.6	None
Fathi et al. (2021)	Multicenter study of hospitalized patients	259 (259 COVID +)	35.1%	29.5%	None
Salari, Etemadifar, Ashrafi, Ommi, et al. (2021), Salari, Etemadifar, Zali, Aminzade, et al. (2021), Salari, Etemadifar, Zali, Medghalchi, et al. (2021)	Phone survey	647 (73 COVID +)	10.9%	NA	None
Zhang et al. (2020)	COVID-19 Medical record database analysis	694 (694 COVID +)	21.3%	5.5%	None

[a]Lack of DA use and dementia did not survive the multivariate analysis.
Abbreviations: DA: Dopamine agonist; STN DBS: subthalamic nucleus deep brain stimulation.

As mentioned above, the first reported case of neurological complications of Covid-19 was in a patient with PD (Filatov et al., 2020). This 74-year-old man presented with encephalopathy in the setting of Covid-19 pneumonia and was admitted to the intensive care unit critically ill. Importantly this patient also had atrial fibrillation, a prior cardioembolic stroke and chronic obstructive pulmonary disease. This single case highlights the multiple confounding factors which must be considered when assessing the contribution of PD itself to outcomes from Covid-19 in PD.

The earliest assessment of Covid-19 outcomes in PD was a small case series of 10 Covid-19-positive PwP from the UK and Italy (Antonini, Leta, Teo, & Chaudhuri, 2020). Four of these patients died leading to an early, but worrying implication that PwP had a 40% risk of dying if they contracted Covid-19. This caused significant concerns when it was published (Antonini, Leta, Teo, & Ray Chaudhuri, 2020; Raphael, 2020) and it is likely that this impacted behavior in PwP, leading to extra caution and self-isolation. However, all patients in that study had advanced disease and age (mean age of 78.3 years). Two of the four patients who died in this series had an advanced therapy in the form of intrajejunal levodopa. However, two of the six patients who recovered also had advanced therapies (one had undergone deep brain stimulation, while another had an intrajejunal levodopa pump), suggesting that advanced therapies alone may not be a good predictor of mortality. This series highlighted early two important confounders in Covid-19 outcomes of PwP, namely advanced age and advanced disease (albeit extract from a small sample size).

The early concern that PwP with advanced therapies might be at higher risk was further compounded by a case series of just two PwP treated with subthalamic nucleus deep brain stimulation (STN-DBS), whose SARS-CoV-2 infection presented atypically and had poor outcomes (Hainque & Grabli, 2020). Both patients died within days of acute respiratory distress syndrome. However, Salari et al. found no risk of worse outcomes in 44 of 647 PwP who had undergone DBS surgery in their study (Salari, Etemadifar, Zali, Aminzade, et al., 2021). The relatively small numbers of infected patients included in these studies clearly highlight the difficulties in calculating meaningful estimates for outcomes.

Subsequently, a number of studies reported a *lower* mortality rate from Covid-19 among PwP. Analysis of admissions to hospital in PwP during the early stages of the pandemic with a three-year period prior to the pandemic revealed 13 deaths (22.4% of hospitalizations) during the pandemic compared with 6.5% of hospitalizations previously (Kobylecki, Jones, Lim, Miller, & Thomson, 2020). However, only three deaths during the pandemic period related to Covid-19 (in-hospital mortality 5.2%). This highlights the important effect of delay in seeking medical attention for other illnesses (e.g., cardiac issues), which may be reflected better by total excess mortality during the pandemic in PwP rather than Covid-19-related deaths alone.

A telephone survey of 1486 PwP in Italy (105 Covid-19 positive) similarly reported no significant difference in mortality from Covid-19 among PwP when compared to familial controls (5.7% vs 7.6%, $P=0.20$) (Fasano, Cereda, et al., 2020). Interestingly, another Italian study found that mild-to-moderate Covid-19 may be contracted independently of age and PD duration and that PwP with mid-stage PD do not seem to have an overall worse outcome than non-PD population (Cilia et al., 2020). Two further Italian studies reported much higher mortality rates of 14% (1 of 7 Covid-19-positive PwP) and 75% (6 of 8 Covid-19-positive PwP) respectively (Artusi et al., 2020; Del Prete et al., 2020). The demographic and clinical features of the examined cohorts must be considered, in particular whether patients with advanced PD and those living in nursing homes or other long-term care facilities were included. From the current data it appears that those studies with lower reported mortality had included fewer institutionalized patients and patients with advanced disease.

Supporting this, a multi-center study of 117 community-dwelling PwP with Covid-19 from 21 tertiary referral centers in Italy, Iran, Spain and the UK examined predictors of outcomes (Fasano, Elia, et al., 2020). Overall mortality was 19.7% and predictors of poor outcome included coexistent dementia (26.1% vs 8.5%, $P=0.049$) and duration of PD (11.7 ± 8.8 vs 6.6 ± 5.4 years, $P=0.029$). In addition, there was a trend toward increased mortality with hypertension (63.6% vs 37.6%, $P=0.054$). Thus, once again, patients with advanced PD are most at risk, although the overall mortality was lower than in previous studies. Given that much of these data are derived from Italian patients, it is helpful to consider what the mortality from Covid-19 was in the general population in these regions. Case-fatality rates of individuals dying in relation to Covid-19 in Italy over similar periods were 9.5% for people over 50 years of age and 12.8% for people over 70 years of age (Onder, Rezza, & Brusaferro, 2020).

A few studies have compared outcomes in Covid-19-positive PwP with Covid-19-positive controls. An Iranian study of 12,909 hospitalized Covid-19-positive individuals (87 of whom had PD) from two university hospitals demonstrated a case fatality of 35.6% among PwP who contracted Covid-19 compared to 19.8% among those who did not have PD (Salari, Etemadifar, Ashrafi, Ommi, et al., 2021). Although there was no significant difference in age between the two cohorts, Alzheimer's disease as an underlying condition was more frequent in deceased PwP in comparison to survived PD patients ($P<0.01$). Examination of clinical characteristics and mortality in 25 PwP admitted to a large referral center in New York revealed that 72% of these patients had comorbid hypertension and 48% had mild cognitive impairment or dementia (Nwabuobi et al., 2021). The authors found that the mortality rate in PwP did not differ significantly from age-matched controls (32% vs 26%, $P=0.743$). Interestingly, 44% of them presented with altered mental status and individuals who died were more likely to have encephalopathy during their admission (88% vs 35%; $P<0.03$). A study of cases of severe Covid-19 admitted to a hospital in Wuhan found mortality among the 10 PwP was non-significantly lower than those without PD (30% vs 46%, $P>0.05$) (Zhai et al., 2021). The authors highlight that PwP with older age, longer PD duration, and late-stage PD may be highly susceptible to critically severe Covid-19 infection and poor outcome.

With regards to other factors which might predict poor outcomes from Covid-19 in PD, a comparison of 29 PwP with severe Covid-19 (hospitalized or death) with 182 PwP who contracted mild Covid-19 or were Covid-19-negative found a positive association between worse outcome and institutionalization (28% vs 5%, $P<0.0001$), dementia (38% vs 15%, $P=0.0026$) and concomitant cancer (10% vs 2%, $P=0.0353$) (Sainz-Amo et al., 2020). They also demonstrated a negative association with dopamine agonist use (17% vs 74%, $P=0.0155$), although one might argue that dopamine agonists are less frequently used in older patients or those with advanced disease. The overall mortality in this study was 21%.

Another study examining 70 PwP admitted to a single hospital in New York found that the 53 patients who were Covid-19-positive had a significantly higher mortality than the 17 Covid-19-negative patients who were admitted to hospital (35.8% vs 5.9%, $P=0.028$) (Parihar, Ferastraoaru, Galanopoulou, Geyer, & Kaufman, 2021). PwP older than 70 years of age, those with advanced disease, those with reductions in their medications, and non-Hispanics (largely comprised of Black/African-Americans) had a statistically significant higher mortality rate, if infected with SARS-CoV-2. The latter point suggests that socioeconomic factors may contribute to mortality in this setting. Importantly, PD did not increase mortality rates from SARS-CoV-2 infection when age was controlled. Similarly, Xu et al. surveyed 46 PD patients who reported Covid-19 positivity to their movement disorders specialist at a university institution (Xu et al., 2021). They found a mortality rate of 13%, but did not find sufficient evidence that PD is an independent risk factor for severe Covid-19 and death. A study which included PwP, patients with a diagnosis of parkinsonism and a control group matched to the PD cohort (and therefore of advanced age and with comparable comorbidities) found a high case fatality rate for those hospitalized with Covid-19 in all groups (PD: 25%; parkinsonism 50%; matched controls 39%) (Vignatelli et al., 2021).

One study specifically aimed at comparing the case fatality of Covid-19-positive PwP with Covid-19-positive PwP by analysis of the TriNetX COVID-19 research network, a health research database with deidentified medical records of over 50 million patients (Zhang, Schultz, Aldridge, Simmering, & Narayanan, 2020). Among 78,355 Covid-19 patients without PD, 5.5% died compared with 21.3% of the 694 PwP with Covid-19. The PD and non-PD groups had different age distribution (median age 79 vs 50), sex distribution (female 39.8% vs 55.3%), and racial composition (African American 9.7% vs 19.7%). Logistic regression demonstrated that the mortality risk in PwP is significantly higher than that of the general population (odds ratio 1.27), even when controlling for these age, sex, and race differences. To address residual confounders, the authors then matched five Covid-19 patients without PD to each PD patient with the exact age, sex, and race and performed a conditional logistic regression which demonstrated that PwP had a significantly higher risk of dying from Covid-19 compared to patients without PD (OR = 1.30, 95% CI: 1.13–1.49, $P < 0.001$). However, crucially the database lacked information on comorbid risk factors for Covid-19 which could influence mortality.

Hence, mortality estimates in PwP with Covid-19 are broadly spread with figures ranging from 5.2% to 100% (Table 2). More recently, four systematic reviews and meta-analyses have helped to clarify risk. El-Qushayri and colleagues performed a systematic review which included 13 studies, encompassing 8649 PD patients and 88,710 controls (El-Qushayri et al., 2021). Hospitalization was reported in eight studies (263 patients) with a pooled rate of 39.89% for PwP (95% CI: 27.09–58.73). Length of hospital stay was comparable in PwP with Covid-19 and Covid-19-positive patients without PD (cumulative mean difference: 2.69, 95% CI: −6.99–12.37; $P = 0.586$). The pooled rate of Intensive Care Unit (ICU) admission was 4.7% (95% CI: 1.56–14.16). The total mortality rate for PwP was found to be 25.1% (95% CI: 16.37–38.49) with no significant differences between Covid-19 patients with PD and without PD (OR: 1.42, 95% CI: 0.26; 7.70; $P = 0.687$). It is important to note that, there was significant heterogeneity observed among studies with respect to rates of hospitalization and mortality. Moreover, these analyses do not consider age, other demographic factors, or geographic factors including race/ethnicity or comorbidities.

Another systematic review calculated a pooled mortality rate of 18.9% in PwP with Covid-19 (Artusi et al., 2021). Importantly, from six of the studies included which reported cases in nursing homes, 46.8% of Covid-19-positive PwP were living in a nursing home. A third systematic review and meta-analysis included 12 studies with 103,874 Covid-19 patients to examine in-hospital outcomes of PwP with COVID-19 (Putri et al., 2021). They assessed not only mortality in these patients, but also severe Covid-19. The latter was defined as any of the following: (1) respiratory distress (≥30 breaths per min); (2) oxygen saturation at rest ≤93%; (3) ratio of the partial pressure of arterial oxygen (PaO2) to a fractional concentration of oxygen inspired air (fiO2) ≤300 mmHg; or (4) critical complication (respiratory failure, septic shock, and or multiple organ dysfunction/failure) or admission into ICU. They found that PD was associated both with severe Covid-19 (OR 2.61 (95% CI 1.98–3.43), $P<0.00001$), and mortality (RR 2.63 (95% CI 1.50–4.60), $P=0.0007$]). The authors subsequently performed a meta-regression which determined that the above findings were affected by age ($P=0.05$), but not by gender ($P=0.46$), hypertension ($P=0.44$), diabetes ($P=0.58$), or dementia ($P=0.23$). Finally, the meta-analysis which included the largest number of studies to date (21 studies) found pooled prevalence of mortality in PD Covid-19 cases was 12% (95% CI: 10%–14%) ($I^2=97.6\%$, $P<0.001$) (Khoshnood et al., 2021).

In summary, PwP are at significant risk of poor outcomes and death from SARS-CoV-2 infection. Based on the currently available studies and meta-analyses, the mortality rate appears to be most likely in the range of 18%–25%. However, whether this risk is specific to PD itself or is largely driven by advanced age and other comorbidities is unclear. If we have learned anything from the heterogeneity PD and the specifics of the studies described in this chapter, it is reasonable to assume that pooled mortality rates do not capture the nuances of which PwP are at risk and why. We will discuss some of these factors in more detail in the next section.

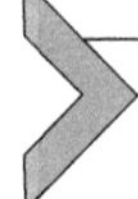

5. Reported modulators of Covid-19 risk in Parkinson's disease

5.1 Canonical Covid-19 risk factors

In large studies of the general population, hypertension, diabetes, pulmonary disease, obesity and immunosuppression have been consistently shown to be independent risk factors for Covid-19 (Alqahtani et al., 2020; de Siqueira et al., 2020; Zheng et al., 2020). These risk factors are so common in the general population that it is almost impossible to adjust for these confounders when studying any specific population with Covid-19. In many of the studies outlined in this chapter, authors have noted that cardiovascular comorbidities are considerable contributors to the Covid-19 risk in PD (Antonini, Leta, Teo, & Chaudhuri, 2020; Artusi et al., 2020; Bhidayasiri, Virameteekul, Kim, Pal, & Chung, 2020; Cilia et al., 2020; Del Prete et al., 2020; Fasano, Elia, et al., 2020). Diabetes is a frequent comorbidity in PwP (Bhidayasiri et al., 2020). Furthermore, Covid-19-positive PwP have been shown in one study to be more likely to suffer from chronic obstructive pulmonary disease than their Covid-19 negative counterparts (Fasano, Cereda, et al., 2020). In fact, an analysis of hospital data in Germany of in-patients with PD from the year 2018 found a strikingly high prevalence of what we now know to be Covid-19 risk comorbidities (Richter, Bartig, Krogias, & Tönges, 2020). Approximately half of PwP had hypertension, 15%–20% had diabetes mellitus, 10%–15% had cardiovascular disease and 6%–15% had chronic kidney disease. Hence these canonical risk factors for Covid-19 continually seep into these PD studies - the contribution, albeit poorly quantified, is undeniable.

5.2 Vitamin D

As mentioned above vitamin D has been highlighted to potentially reduce the risk of Covid-19 in one study (Fasano, Cereda, et al., 2020). This was, however, further consolidated by a meta-analysis which found vitamin D supplementation to be associated with a lower odds ratio of contracting Covid-19 among PwP (OR: 0.50, 95% CI: 0.30–0.83, I^2: 0%) (Chambergo-Michilot et al., 2021). This supports the independent suggestion that vitamin D could reduce the risk of Covid-19 infection in the general population via a number of mechanisms, including reducing the concentration of pro-inflammatory cytokines (Mitchell, 2020). Vitamin D deficiency may contribute to Covid-19 susceptibility (Ilie, Stefanescu, & Smith, 2020; Mitchell, 2020) and is a common comorbidity in PD (Ding et al., 2013). This has prompted some authors to propose vitamin D replacement as a therapy against both Covid-19 and PD (Azzam, Ghozy, & Azab, 2022; Behl et al., 2021; Hribar, Cobbold, & Church, 2020). It is clear that dedicated studies are needed to examine the effect of vitamin D on Covid-19 risk in PwP.

5.3 Age, disease severity and dementia

Early data from Wuhan showed clear evidence of the association between Covid-19 and older, frail patients with multiple pathologies (Meng et al., 2020). It is clear that increasing age is also a significant risk factor for

Covid-19 and PD prevalence increases with age, but other confounders may also be at play when examining Covid-19 prevalence in PD. Age greater than 60 is a significant determinant of mortality from Covid-19 in the general population (Bonanad et al., 2020). It is precisely in this age group that we see the highest prevalence of PD. Older people tend to present more atypically in the setting of SARS-CoV-2 infection (D'Adamo, Yoshikawa, & Ouslander, 2020). This phenomenon may be even more prominent in older people with PD where worsening of motor symptoms may be the only presenting symptom and, hence, mask the symptoms of Covid-19, leading to delayed treatment and poor outcomes (Fearon & Fasano, 2021). Furthermore, typical Covid-19 symptoms, such as fatigue, anosmia, subjective changes in temperature or limb pain, are also part of the spectrum of non-motor PD symptoms, which these patients experience every day. Anosmia, which was one of the earliest symptoms which people were told to look out for following the emergence of SARS-CoV-2 is present in over 96% of PwP (Tarakad & Jankovic, 2017).

However, the case fatality does seem to be higher in PwP compared to the general population of similar age (4.5% in people over 60 years of age (Verity et al., 2020)). Hence, other factors such as cognitive impairment, institutionalization and frailty may contribute to the observed increased risk. Indeed, dementia is an age-independent risk factor for severity and death in Covid-19 in hospitalized patients (Tahira, Verjovski-Almeida, & Ferreira, 2021). For comparison, taking all patients with dementia (which will include some PwP) the prevalence of Covid-19 is 13% (Bianchetti et al., 2020). A retrospective case-control study from Spain compared demographics of 39 Covid-19-positive PwP to 172 Covid-19-negative PD controls (Sainz-Amo et al., 2020). Covid-19-positive cases were more likely to be institutionalized and have co-existent dementia (36% vs 14%, $P=0.0013$). A comparison of mortality in a large number of hospitalized patients (259 with PD) in a multi-center study found no difference in the 28-day mortality risk between PwP and other patients (Fathi, Taghizadeh, Mojtahedi, Zargar Balaye Jame, & Markazi Moghaddam, 2021). They did, however, find a significant increase in 28-day mortality risk among the 363 patients with Alzheimer's disease also included in the analysis.

A significant number of studies have suggested that advanced PD may be a prominent risk factor for poor outcomes from Covid-19 (Antonini, Leta, Teo, & Chaudhuri, 2020; Artusi et al., 2020; Brown et al., 2020; Fasano, Elia, et al., 2020; Zhai et al., 2021). On the other hand, mild-to-moderate Covid-19 appears to be contracted independently of age and disease duration in PD and mid-stage PwP have a similar outcome from Covid-19 than the non-PD population (Cilia et al., 2020). Another study similarly found no difference in age or disease duration in symptomatic Covid-19 PwP compared to those who were Covid-19-negative. However, the finding that patients with parkinsonism have a much higher risk of hospitalization from Covid-19 compared to PwP or controls suggests that disease complexity and progression may play a significant role in determining outcomes (Vignatelli et al., 2021).

An important group of patients who were frequently omitted from the studies outlined in this chapter are those in residential care. A systematic review published in April 2021 estimated that from six studies in the literature which included nursing home residents 46.8% of Covid-positive PwP in the literature (n = 22/47) were living in a nursing home (Artusi et al., 2021). In a retrospective observational study of 18 patients (13 PD, 5 non-PD) hospitalized in a rehabilitation unit, 23% of the PwP had advanced therapies (one intra-jejunal levodopa, one STN-DBS, one MRI-guided focused ultrasound thalamotomy) (Sorbera et al., 2021). The authors found that 77% of PD and 60% of non-PD tested positive for Covid-19 and that there was no difference in Covid-19 disease course between the PD and non-PD groups. Another study of 12 residential care patients who were hospitalized due to a SARS-CoV-2 outbreak in the care home, reported that all patients had mild Covid-19 symptoms (Buccafusca et al., 2021). Most of the PwP had a long disease duration and multiple comorbidities, however none required ICU admission or died. However, a large prospective cohort study of 1538 residents in Dutch nursing homes, 14 of whom had PD, found a higher mortality among those with PD when adjusted for age, sex and comorbidities (HR 1.49, 95% CI 1.11–2.00, $P=0.007$) (Rutten et al., 2020).

5.4 Parkinson's disease medications

As mentioned above, the potentially protective role of amantadine, as an antiviral agent, naturally led to its study in PD cohorts. A telephone-based study found that none of the 568 patients receiving amantadine (16.5%) in the study developed Covid-19 (Santos García et al., 2020). Similarly a small number of Covid-19-positive PwP taking amantadine did not manifest symptoms of the disease in another study (Rejdak & Grieb, 2020) and other case reports have reported similar results (Aranda-Abreu, Aranda-Martínez, & Araújo, 2020; Cortés Borra, 2020). The potential role for amantadine in Covid-19 treatment is further supported by a drug screen gene expression study suggesting that amantadine might be effective in reducing replication and infectivity of SARS-CoV-2 (Smieszek et al., 2020). The protective role of amantadine has not been confirmed by other studies (Fasano, Elia, et al., 2020).

Two studies have demonstrated that patients with a worse outcome were less likely to be on dopamine agonists (Fasano, Elia, et al., 2020; Sainz-Amo et al., 2020). However, this most likely reflects the fact they are more rarely prescribed in patients with advanced disease and higher baseline frailty. Other studies have not demonstrated differences in concomitant medication use (Cilia et al., 2020; Fasano, Cereda, et al., 2020), although they have found that PwP and COVID-19 might require higher doses of levodopa.

6. Conclusion

There have been extensive attempts to clarify two important questions in PwP since the start of the Covid-19 pandemic: (1) Are PwP at higher risk for contracting Covid-19; and (2) if they do contract Covid-19, is PD a specific risk factor for poor outcomes from the infection? Unfortunately, the rapidity with which Covid-19 has swept across the globe has made constructing large well-designed studies to examine these questions in a rigorous fashion problematic.

The data collected so far does not indicate that PD itself is a specific risk factor for developing Covid-19. A broad range of prevalence rates have been reported to date and much of this may relate to the specific study design (e.g., seroprevalence vs self-reported phone-based questionnaire) or geographical factors. One major confounder has been that PwP may potentially self-isolated to a greater degree than the general population, hence, reducing their risk of exposure. This behavior may have masked a true increased predisposition to a SARS-CoV-2 infection.

PwP are at significant risk of poor outcomes and death from a SARS-CoV-2 infection. However, whether this risk is specific to PD itself or is largely driven by advanced age, advanced disease and frailty or by other comorbidities is unclear. Irrespective of the mechanisms of the risk of poor outcomes from Covid-19, PwP are a cohort for whom exposure to SARS-CoV-2 should be minimized to avoid these outcomes. The unfortunate downside of minimizing exposure risk in PwP is the negative effects of isolation. Social isolation is a known risk factor for poor outcomes in the general population (Pantell et al., 2013). The pandemic has led to significantly worse stress, depression, anxiety, physical activity and quality of life for PwP than healthy controls (Shalash et al., 2020). Lack of access to physiotherapy and aerobic exercise removes a crucial intervention which protects against progression of parkinsonian symptoms in these patients (van der Kolk et al., 2019). This co-existence of anxiety, stress, isolation and physical inactivity is a particularly detrimental combination for PwP (Helmich & Bloem, 2020) and worsening of motor and non-motor symptoms of PD have been demonstrated (Fabbri et al., 2021) in addition to worsening of standard cognitive test scores (Palermo et al., 2020). We will undoubtedly see the long-term negative outcomes of prolonged isolation and limited access to exercise and social interaction in people who may have, for example, young-onset PD and may be at the early stages of a progressive neurodegenerative illness. Like all considerations which have been undertaken since the start of the Covid-19 pandemic, one must weigh up these negative consequences of isolation in PwP with the risk of poor outcome outlined above. As a final point, it is important to note that the current data is almost entirely drawn from the pre-vaccination era of the pandemic and a re-evaluation of the Covid-19 risk in vaccinated PwP is desperately needed. An early population study from the UK indicates that PD still increases the risk of poor outcome among vaccinated patients (Hippisley-Cox et al., 2021).

References

Ait Wahmane, S., Achbani, A., Ouhaz, Z., Elatiqi, M., Belmouden, A., & Nejmeddine, M. (2020). The possible protective role of α-synuclein against severe acute respiratory syndrome coronavirus 2 infections in patients with Parkinson's disease. *Movement Disorders*. https://doi.org/10.1002/mds.28185.

Alqahtani, J. S., Oyelade, T., Aldhahir, A. M., Alghamdi, S. M., Almehmadi, M., Alqahtani, A. S., et al. (2020). Prevalence, severity and mortality associated with COPD and smoking in patients with COVID-19: A rapid systematic review and meta-analysis. *PLoS One*, *15*(5), e0233147. https://doi.org/10.1371/journal.pone.0233147.

Antonini, A., Leta, V., Teo, J., & Chaudhuri, K. R. (2020). Outcome of Parkinson's disease patients affected by COVID-19. *Movement Disorders*, *35*(6), 905–908. https://doi.org/10.1002/mds.28104.

Antonini, A., Leta, V., Teo, J., & Ray Chaudhuri, K. (2020). Reply to: "Concerns raised by publication of Antonini et al., 'Outcome of Parkinson disease patients affected by covid-19.'". *Movement Disorders*, *35*(8), 1298. https://doi.org/10.1002/mds.28183.

Aranda Abreu, G. E., Hernández Aguilar, M. E., Herrera Covarrubias, D., & Rojas Durán, F. (2020). Amantadine as a drug to mitigate the effects of COVID-19. *Medical Hypotheses*, *140*. https://doi.org/10.1016/j.mehy.2020.109755, 109755.

Aranda-Abreu, G. E., Aranda-Martínez, J. D., & Araújo, R. (2020). Use of amantadine in a patient with SARS-CoV-2. *Journal of Medical Virology*. https://doi.org/10.1002/jmv.26179.

Araújo, R., Aranda-Martínez, J. D., & Aranda-Abreu, G. E. (2020). Amantadine treatment for people with COVID-19. *Archives of Medical Research*. https://doi.org/10.1016/j.arcmed.2020.06.009.

Artusi, C. A., Romagnolo, A., Imbalzano, G., Marchet, A., Zibetti, M., Rizzone, M. G., et al. (2020). COVID-19 in Parkinson's disease: Report on prevalence and outcome. *Parkinsonism & Related Disorders*, *80*, 7–9. https://doi.org/10.1016/j.parkreldis.2020.09.008.

Artusi, C. A., Romagnolo, A., Ledda, C., Zibetti, M., Rizzone, M. G., Montanaro, E., et al. (2021). COVID-19 and Parkinson's disease: What do we know so far? *Journal of Parkinson's Disease*, *11*(2), 445–454. https://doi.org/10.3233/JPD-202463.

Azzam, A. Y., Ghozy, S., & Azab, M. A. (2022). Vitamin D and its' role in Parkinson's disease patients with SARS-CoV-2 infection. A review article. *Interdisciplinary Neurosurgery*, *27*, 101441. https://doi.org/10.1016/j.inat.2021.101441.

Baille, G., Perez, T., Devos, D., Deken, V., Defebvre, L., & Moreau, C. (2018). Early occurrence of inspiratory muscle weakness in Parkinson's disease. *PLoS One*, *13*(1), e0190400. https://doi.org/10.1371/journal.pone.0190400.

Behl, T., Kumar, S., Sehgal, A., Singh, S., Sharma, N., Chirgurupati, S., et al. (2021). Linking COVID-19 and Parkinson's disease: Targeting the role of vitamin-D. *Biochemical and Biophysical Research Communications*, *583*, 14–21. https://doi.org/10.1016/j.bbrc.2021.10.042.

Bhidayasiri, R., Virameteekul, S., Kim, J.-M., Pal, P. K., & Chung, S.-J. (2020). COVID-19: An early review of its global impact and considerations for Parkinson's disease patient care. *Journal of Movement Disorders*, *13*(2), 105–114. https://doi.org/10.14802/jmd.20042.

Bianchetti, A., Rozzini, R., Guerini, F., Boffelli, S., Ranieri, P., Minelli, G., et al. (2020). Clinical presentation of COVID19 in dementia patients. *The Journal of Nutrition, Health & Aging*, *24*(6), 560–562. https://doi.org/10.1007/s12603-020-1389-1.

Bonanad, C., García-Blas, S., Tarazona-Santabalbina, F., Sanchis, J., Bertomeu-González, V., Fácila, L., et al. (2020). The effect of age on mortality in patients with COVID-19: A meta-analysis with 611,583 subjects. *Journal of the American Medical Directors Association*, *21*(7), 915–918. https://doi.org/10.1016/j.jamda.2020.05.045.

Brown, E. G., Chahine, L. M., Goldman, S. M., Korell, M., Mann, E., Kinel, D. R., et al. (2020). The effect of the COVID-19 pandemic on people with Parkinson's disease. *Journal of Parkinson's Disease*, *10*(4), 1365–1377. https://doi.org/10.3233/JPD-202249.

Buccafusca, M., Micali, C., Autunno, M., Versace, A. G., Nunnari, G., & Musumeci, O. (2021). Favourable course in a cohort of Parkinson's disease patients infected by SARS-CoV-2: A single-centre experience. *Neurological Sciences*, *42*(3), 811–816. https://doi.org/10.1007/s10072-020-05001-4.

Chambergo-Michilot, D., Barros-Sevillano, S., Rivera-Torrejón, O., De la Cruz-Ku, G. A., & Custodio, N. (2021). Factors associated with COVID-19 in people with Parkinson's disease: A systematic review and meta-analysis. *European Journal of Neurology*, *28*(10), 3467–3477. https://doi.org/10.1111/ene.14912.

Cheong, J. L.-Y., Goh, Z. H. K., Marras, C., Tanner, C. M., Kasten, M., Noyce, A. J., et al. (2020). The impact of COVID-19 on access to Parkinson's disease medication. *Movement Disorders*, *35*(12), 2129–2133. https://doi.org/10.1002/mds.28293.

Cilia, R., Bonvegna, S., Straccia, G., Andreasi, N. G., Elia, A. E., Romito, L. M., et al. (2020). Effects of COVID-19 on Parkinson's disease clinical features: A community-based case-control study. *Movement Disorders*. https://doi.org/10.1002/mds.28170.

Cortés Borra, A. (2020). Does amantadine have a protective effect against COVID-19? *Neurologia i Neurochirurgia Polska*, *54*(3), 284–285. https://doi.org/10.5603/PJNNS.a2020.0041.

D'Adamo, H., Yoshikawa, T., & Ouslander, J. G. (2020). Coronavirus disease 2019 in geriatrics and long-term care: The ABCDs of COVID-19. *Journal of the American Geriatrics Society*, *68*(5), 912–917. https://doi.org/10.1111/jgs.16445.

de Siqueira, J. V. V., Almeida, L. G., Zica, B. O., Brum, I. B., Barceló, A., & de Siqueira Galil, A. G. (2020). Impact of obesity on hospitalizations and mortality, due to COVID-19: A systematic review. *Obesity Research & Clinical Practice*. https://doi.org/10.1016/j.orcp.2020.07.005.

Del Prete, E., Francesconi, A., Palermo, G., Mazzucchi, S., Frosini, D., Morganti, R., et al. (2020). Prevalence and impact of COVID-19 in Parkinson's disease: Evidence from a multi-center survey in Tuscany region. *Journal of Neurology*. https://doi.org/10.1007/s00415-020-10002-6.

Ding, H., Dhima, K., Lockhart, K. C., Locascio, J. J., Hoesing, A. N., Duong, K., et al. (2013). Unrecognized vitamin D3 deficiency is common in Parkinson disease: Harvard Biomarker Study. *Neurology*, *81*(17), 1531–1537. https://doi.org/10.1212/WNL.0b013e3182a95818.

El-Qushayri, A. E., Ghozy, S., Reda, A., Kamel, A. M. A., Abbas, A. S., & Dmytriw, A. A. (2021). The impact of Parkinson's disease on manifestations and outcomes of Covid-19 patients: A systematic review and meta-analysis. *Reviews in Medical Virology*, e2278. https://doi.org/10.1002/rmv.2278.

Fabbri, M., Leung, C., Baille, G., Béreau, M., Brefel Courbon, C., Castelnovo, G., et al. (2021). A French survey on the lockdown consequences of COVID-19 pandemic in Parkinson's disease. The ERCOPARK study. *Parkinsonism & Related Disorders*, *89*, 128–133. https://doi.org/10.1016/j.parkreldis.2021.07.013.

Fasano, A., Cereda, E., Barichella, M., Cassani, E., Ferri, V., Zecchinelli, A. L., et al. (2020). COVID-19 in Parkinson's disease patients living in Lombardy, Italy. *Movement Disorders*. https://doi.org/10.1002/mds.28176.

Fasano, A., Elia, A. E., Dallocchio, C., Canesi, M., Alimonti, D., Sorbera, C., et al. (2020). Predictors of COVID-19 outcome in Parkinson's disease. *Parkinsonism & Related Disorders*, *78*, 134–137. https://doi.org/10.1016/j.parkreldis.2020.08.012.

Fathi, M., Taghizadeh, F., Mojtahedi, H., Zargar Balaye Jame, S., & Markazi Moghaddam, N. (2021). The effects of Alzheimer's and Parkinson's disease on 28-day mortality of COVID-19. *Revue Neurologique*. https://doi.org/10.1016/j.neurol.2021.08.002. S0035-3787(21)00653-6.

Fearon, C., & Fasano, A. (2021). Parkinson's disease and the COVID-19 pandemic. *Journal of Parkinson's Disease*, *11*(2), 431–444. https://doi.org/10.3233/JPD-202320.

Filatov, A., Sharma, P., Hindi, F., & Espinosa, P. S. (2020). Neurological complications of coronavirus disease (COVID-19): Encephalopathy. *Cureus*, *12*(3), e7352. https://doi.org/10.7759/cureus.7352.

Gonzalez-Latapi, P., Fearon, C., Fasano, A., & Lang, A. E. (2021). Parkinson's disease and COVID-19: Do we need to be more patient? *Movement Disorders*, *36*(2), 277. https://doi.org/10.1002/mds.28469.

Hainque, E., & Grabli, D. (2020). Rapid worsening in Parkinson's disease may hide COVID-19 infection. *Parkinsonism & Related Disorders*. https://doi.org/10.1016/j.parkreldis.2020.05.008.

Helmich, R. C., & Bloem, B. R. (2020). The impact of the COVID-19 pandemic on Parkinson's disease: Hidden sorrows and emerging opportunities. *Journal of Parkinson's Disease*, *10*(2), 351–354. https://doi.org/10.3233/JPD-202038.

Hippisley-Cox, J., Coupland, C. A., Mehta, N., Keogh, R. H., Diaz-Ordaz, K., Khunti, K., et al. (2021). Risk prediction of covid-19 related death and hospital admission in adults after covid-19 vaccination: National prospective cohort study. *BMJ (Clinical Research Ed.)*, *374*, n2244. https://doi.org/10.1136/bmj.n2244.

Hoffman, L. A., & Vilensky, J. A. (2017). Encephalitis lethargica: 100 years after the epidemic. *Brain*, *140*(8), 2246–2251. https://doi.org/10.1093/brain/awx177.

Hribar, C. A., Cobbold, P. H., & Church, F. C. (2020). Potential role of vitamin D in the elderly to resist COVID-19 and to slow progression of Parkinson's disease. *Brain Sciences*, *10*(5). https://doi.org/10.3390/brainsci10050284.

Ilie, P. C., Stefanescu, S., & Smith, L. (2020). The role of vitamin D in the prevention of coronavirus disease 2019 infection and mortality. *Aging Clinical and Experimental Research*, *32*(7), 1195–1198. https://doi.org/10.1007/s40520-020-01570-8.

Jo, T., Yasunaga, H., Michihata, N., Sasabuchi, Y., Hasegawa, W., Takeshima, H., et al. (2018). Influence of Parkinsonism on outcomes of elderly pneumonia patients. *Parkinsonism & Related Disorders*, *54*, 25–29. https://doi.org/10.1016/j.parkreldis.2018.03.028.

Khoshnood, R. J., Zali, A., Tafreshinejad, A., Ghajarzadeh, M., Ebrahimi, N., Safari, S., et al. (2021). Parkinson's disease and COVID-19: A systematic review and meta-analysis. *Neurological Sciences*, 1–9. https://doi.org/10.1007/s10072-021-05756-4.

Kobylecki, C., Jones, T., Lim, C. K., Miller, C., & Thomson, A. M. (2020). Phenomenology and outcomes of in-patients with Parkinson's disease during the coronavirus disease 2019 pandemic. *Movement Disorders*. https://doi.org/10.1002/mds.28205.

Kumar, N., & Gupta, R. (2021). Impact of COVID-19 pandemic on Parkinson's disease: A tale of fears and sorrows! *Annals of Indian Academy of Neurology*, *24*(2), 121–123. https://doi.org/10.4103/aian.AIAN_97_21.

Labrie, V., & Brundin, P. (2017). Alpha-synuclein to the rescue: Immune cell recruitment by alpha-synuclein during gastrointestinal infection. *Journal of Innate Immunity*, *9*(5), 437–440. https://doi.org/10.1159/000479653.

Lippi, A., Domingues, R., Setz, C., Outeiro, T. F., & Krisko, A. (2020). SARS-CoV-2: At the crossroad between aging and neurodegeneration. *Movement Disorders*, *35*(5), 716–720. https://doi.org/10.1002/mds.28084.

Massey, A. R., & Beckham, J. D. (2016). Alpha-synuclein, a novel viral restriction factor hiding in plain sight. *DNA and Cell Biology*, *35*(11), 643–645. https://doi.org/10.1089/dna.2016.3488.

Meng, L., Qiu, H., Wan, L., Ai, Y., Xue, Z., Guo, Q., et al. (2020). Intubation and ventilation amid the COVID-19 outbreak: Wuhan's experience. *Anesthesiology*, *132*(6), 1317–1332. https://doi.org/10.1097/ALN.0000000000003296.

Merello, M., Bhatia, K. P., & Obeso, J. A. (2021). SARS-CoV-2 and the risk of Parkinson's disease: Facts and fantasy. *The Lancet. Neurology*, *20*(2), 94–95. https://doi.org/10.1016/S1474-4422(20)30442-7.

Mitchell, F. (2020). Vitamin-D and COVID-19: Do deficient risk a poorer outcome? *The Lancet. Diabetes & Endocrinology*, *8*(7), 570. https://doi.org/10.1016/S2213-8587(20)30183-2.

Nataf, S. (2020). An alteration of the dopamine synthetic pathway is possibly involved in the pathophysiology of COVID-19. *Journal of Medical Virology*. https://doi.org/10.1002/jmv.25826.

Nwabuobi, L., Zhang, C., Henchcliffe, C., Shah, H., Sarva, H., Lee, A., et al. (2021). Characteristics and outcomes of Parkinson's disease individuals hospitalized with COVID-19 in a New York City Hospital System. *Movement Disorders Clinical Practice*. https://doi.org/10.1002/mdc3.13309.

Onder, G., Rezza, G., & Brusaferro, S. (2020). Case-fatality rate and characteristics of patients dying in relation to COVID-19 in Italy. *JAMA*, *323*(18), 1775–1776. https://doi.org/10.1001/jama.2020.4683.

Palermo, G., Tommasini, L., Baldacci, F., Del Prete, E., Siciliano, G., & Ceravolo, R. (2020). Impact of COVID-19 pandemic on cognition in Parkinson's disease. *Movement Disorders*. https://doi.org/10.1002/mds.28254.

Pantell, M., Rehkopf, D., Jutte, D., Syme, S. L., Balmes, J., & Adler, N. (2013). Social isolation: A predictor of mortality comparable to traditional clinical risk factors. *American Journal of Public Health*, *103*(11), 2056–2062. https://doi.org/10.2105/AJPH.2013.301261.

Parihar, R., Ferastraoaru, V., Galanopoulou, A. S., Geyer, H. L., & Kaufman, D. M. (2021). Outcome of hospitalized Parkinson's disease patients with and without COVID-19. *Movement Disorders Clinical Practice*. https://doi.org/10.1002/mdc3.13231.

Prasad, S., Holla, V. V., Neeraja, K., Surisetti, B. K., Kamble, N., Yadav, R., et al. (2020). Impact of prolonged lockdown due to COVID-19 in patients with Parkinson's disease. *Neurology India*, *68*(4), 792–795. https://doi.org/10.4103/0028-3886.293472.

Putri, C., Hariyanto, T. I., Hananto, J. E., Christian, K., Situmeang, R. F. V., & Kurniawan, A. (2021). Parkinson's disease may worsen outcomes from coronavirus disease 2019 (COVID-19) pneumonia in hospitalized patients: A systematic review, meta-analysis, and meta-regression. *Parkinsonism & Related Disorders*, *87*, 155–161. https://doi.org/10.1016/j.parkreldis.2021.04.019.

Raphael, K. G. (2020). Concerns raised by publication of Antonini et al., "Outcome of Parkinson disease patients affected by covid-19.". *Movement Disorders*, *35*(8), 1297. https://doi.org/10.1002/mds.28180.

Ravenholt, R. T., & Foege, W. H. (1982). 1918 influenza, encephalitis lethargica, parkinsonism. *Lancet (London, England)*, *2*(8303), 860–864. https://doi.org/10.1016/s0140-6736(82)90820-0.

Rejdak, K., & Grieb, P. (2020). Adamantanes might be protective from COVID-19 in patients with neurological diseases: Multiple sclerosis, parkinsonism and cognitive impairment. *Multiple Sclerosis and Related Disorders*, *42*, 102163. https://doi.org/10.1016/j.msard.2020.102163.

Richter, D., Bartig, D., Krogias, C., & Tönges, L. (2020). Letter to the editor: Risk comorbidities of COVID-19 in Parkinson's disease patients in Germany. *Neurological Research and Practice*, *2*, 22. https://doi.org/10.1186/s42466-020-00069-x.

Rutten, J. J. S., van Loon, A. M., van Kooten, J., van Buul, L. W., Joling, K. J., Smalbrugge, M., et al. (2020). Clinical suspicion of COVID-19 in nursing home residents: Symptoms and mortality risk factors. *Journal of the American Medical Directors Association*, *21*(12), 1791–1797.e1. https://doi.org/10.1016/j.jamda.2020.10.034.

Sainz-Amo, R., Baena-Álvarez, B., Pareés, I., Sánchez-Díez, G., Pérez-Torre, P., López-Sendón, J. L., et al. (2020). COVID-19 in Parkinson's disease: What holds the key? *Journal of Neurology*. https://doi.org/10.1007/s00415-020-10272-0.

Salari, M., Etemadifar, M., Ashrafi, F., Ommi, D., Aminzade, Z., & Tehrani Fateh, S. (2021). Parkinson's disease patients may have higher rates of Covid-19 mortality in Iran. *Parkinsonism & Related Disorders*, *89*, 90–92. https://doi.org/10.1016/j.parkreldis.2021.07.002.

Salari, M., Etemadifar, M., Zali, A., Aminzade, Z., Navalpotro-Gomez, I., & Fateh, S. T. (2021). Covid-19 in Parkinson's disease treated by drugs or brain stimulation. *Neurologia (Barcelona, Spain)*. https://doi.org/10.1016/j.nrl.2021.07.002.

Salari, M., Etemadifar, M., Zali, A., Medghalchi, A., Tehrani Fateh, S., Aminzade, Z., et al. (2021). Seroprevalence of SARS-CoV-2 in Parkinson's disease patients: A case-control study. *Movement Disorders*, *36*(4), 794–795. https://doi.org/10.1002/mds.28580.

Santos García, D., Oreiro, M., Pérez, P., Fanjul, G., Paz González, J. M., Feal Painceiras, M., et al. (2020). Impact of COVID-19 pandemic on Parkinson's disease: A cross-sectional survey of 568 Spanish patients. *Movement Disorders*. https://doi.org/10.1002/mds.28261.

Shalash, A., Roushdy, T., Essam, M., Fathy, M., Dawood, N. L., Abushady, E. M., et al. (2020). Mental health, physical activity, and quality of life in Parkinson's disease during COVID-19 pandemic. *Movement Disorders*, *35*(7), 1097–1099. https://doi.org/10.1002/mds.28134.

Smieszek, S. P., Przychodzen, B. P., & Polymeropoulos, M. H. (2020). Amantadine disrupts lysosomal gene expression: A hypothesis for COVID19 treatment. *International Journal of Antimicrobial Agents*, *55*(6), 106004. https://doi.org/10.1016/j.ijantimicag.2020.106004.

Sorbera, C., Brigandì, A., Cimino, V., Bonanno, L., Ciurleo, R., Bramanti, P., et al. (2021). The impact of SARS-COV2 infection on people in residential care with Parkinson disease or parkinsonisms: Clinical case series study. *PLoS One*, *16*(5), e0251313. https://doi.org/10.1371/journal.pone.0251313.

Tahira, A. C., Verjovski-Almeida, S., & Ferreira, S. T. (2021). Dementia is an age-independent risk factor for severity and death in COVID-19 inpatients. *Alzheimer's & Dementia: The Journal of the Alzheimer's Association*, *17*(11), 1818–1831. https://doi.org/10.1002/alz.12352.

Tarakad, A., & Jankovic, J. (2017). Anosmia and ageusia in Parkinson's disease. *International Review of Neurobiology, 133*, 541–556. https://doi.org/10.1016/bs.irn.2017.05.028.

Temlett, J. A., & Thompson, P. D. (2006). Reasons for admission to hospital for Parkinson's disease. *Internal Medicine Journal, 36*(8), 524–526. https://doi.org/10.1111/j.1445-5994.2006.01123.x.

van der Kolk, N. M., de Vries, N. M., Kessels, R. P. C., Joosten, H., Zwinderman, A. H., Post, B., et al. (2019). Effectiveness of home-based and remotely supervised aerobic exercise in Parkinson's disease: A double-blind, randomised controlled trial. *The Lancet. Neurology, 18*(11), 998–1008. https://doi.org/10.1016/S1474-4422(19)30285-6.

Verity, R., Okell, L. C., Dorigatti, I., Winskill, P., Whittaker, C., Imai, N., et al. (2020). Estimates of the severity of coronavirus disease 2019: A model-based analysis. *The Lancet Infectious Diseases, 20*(6), 669–677. https://doi.org/10.1016/S1473-3099(20)30243-7.

Vignatelli, L., Zenesini, C., Belotti, L. M. B., Baldin, E., Bonavina, G., Calandra-Buonaura, G., et al. (2021). Risk of hospitalization and death for COVID-19 in people with Parkinson's disease or parkinsonism. *Movement Disorders, 36*(1), 1–10. https://doi.org/10.1002/mds.28408.

von Economo, K. (1918). Die Encephalitis lethargica. *Wiener Klinische Wochenschrift, 30*, 581–585.

Xiao, W., Shameli, A., Harding, C. V., Meyerson, H. J., & Maitta, R. W. (2014). Late stages of hematopoiesis and B cell lymphopoiesis are regulated by α-synuclein, a key player in Parkinson's disease. *Immunobiology, 219*(11), 836–844. https://doi.org/10.1016/j.imbio.2014.07.014.

Xu, Y., Surface, M., Chan, A. K., Halpern, J., Vanegas-Arroyave, N., Ford, B., et al. (2021). COVID-19 manifestations in people with Parkinson's disease: A USA cohort. *Journal of Neurology*. https://doi.org/10.1007/s00415-021-10784-3.

Yule, E., Pickering, J. S., McBride, J., & Poliakoff, E. (2021). People with Parkinson's report increased impulse control behaviours during the COVID-19 UK lockdown. *Parkinsonism & Related Disorders, 86*, 38–39. https://doi.org/10.1016/j.parkreldis.2021.03.024.

Zhai, H., Lv, Y., Xu, Y., Wu, Y., Zeng, W., Wang, T., et al. (2021). Characteristic of Parkinson's disease with severe COVID-19: A study of 10 cases from Wuhan. *Journal of Neural Transmission, 128*(1), 37–48. https://doi.org/10.1007/s00702-020-02283-y.

Zhang, Q., Schultz, J. L., Aldridge, G. M., Simmering, J. E., & Narayanan, N. S. (2020). Coronavirus disease 2019 case fatality and Parkinson's disease. *Movement Disorders, 35*(11), 1914–1915. https://doi.org/10.1002/mds.28325.

Zheng, Z., Peng, F., Xu, B., Zhao, J., Liu, H., Peng, J., et al. (2020). Risk factors of critical & mortal COVID-19 cases: A systematic literature review and meta-analysis. *The Journal of Infection, 81*(2), e16–e25. https://doi.org/10.1016/j.jinf.2020.04.021.

Zipprich, H. M., Teschner, U., Witte, O. W., Schönenberg, A., & Prell, T. (2020). Knowledge, attitudes, practices, and burden during the COVID-19 pandemic in people with Parkinson's disease in Germany. *Journal of Clinical Medicine, 9*(6). https://doi.org/10.3390/jcm9061643.

CHAPTER FOUR

Covid-19 and Parkinson's disease: Acute clinical implications, long-COVID and post-COVID-19 parkinsonism

Valentina Leta[a,b,†], Iro Boura[a,b,c,†], Daniel J. van Wamelen[a,b,d], Mayela Rodriguez-Violante[e], Angelo Antonini[f], and Kallol Ray Chaudhuri[a,b,*]

[a]Department of Neurosciences, Institute of Psychiatry, Psychology & Neuroscience, King's College London, London, United Kingdom
[b]Parkinson's Foundation Centre of Excellence, King's College Hospital NHS Foundation Trust, London, United Kingdom
[c]Medical School, University of Crete, Heraklion, Crete, Greece
[d]Department of Neurology, Centre of Expertise for Parkinson & Movement Disorders, Donders Institute for Brain, Cognition and Behaviour, Radboud University Medical Center, Nijmegen, the Netherlands
[e]Instituto Nacional de Neurologia y Neurocirugia, Ciudad de México, Mexico
[f]Parkinson and Movement Disorders Unit, Department of Neuroscience, Centre for Rare Neurological Diseases (ERN-RND), University of Padova, Padova, Italy
*Corresponding author: e-mail address: ray.chaudhuri@kcl.ac.uk

Contents

Abstract

The Coronavirus Disease 2019 (Covid-19), caused by the Severe Acute Respiratory Syndrome Coronavirus-2 (SARS-CoV-2), has led to unprecedented challenges for the delivery of healthcare and has had a clear impact on people with chronic neurological conditions such as Parkinson's disease (PD). Acute worsening of motor and non-motor symptoms and long-term sequalae have been described during and after SARS-CoV-2

† These authors contributed equally to this work.

International Review of Neurobiology, Volume 165
ISSN 0074-7742
https://doi.org/10.1016/bs.irn.2022.04.004
Copyright © 2022 Elsevier Inc.
All rights reserved.

infections in people with Parkinson's (PwP), which are likely to be multifactorial in their origin. On the one hand, it is likely that worsening of symptoms has been related to the viral infection itself, whereas social restrictions imposed over the course of the Covid-19 pandemic might also have had such an effect. Twenty cases of post-Covid-19 para-infectious or post-infectious parkinsonism have been described so far where a variety of pathophysiological mechanisms seem to be involved; however, a Covid-19-induced wave of post-viral parkinsonism seems rather unlikely at the moment. Here, we describe the interaction between SARS-CoV-2 and PD in the short- and long-term and summarize the clinical features of post-Covid-19 cases of parkinsonism observed so far.

1. Introduction

The Coronavirus Disease 2019 (Covid-19), caused by the Severe Acute Respiratory Syndrome Coronavirus-2 (SARS-CoV-2), has led to major and unprecedented challenges for the delivery of healthcare with a clear impact on patients, both directly through symptoms and morbidity caused by an active SARS-CoV-2 infection and indirectly by effects of pandemic-related restrictions. From an initial focus on the symptomatology, morbidity, and mortality caused by a SARS-CoV-2 infection, the focus of research has moved towards the interaction of this virus with pre-existing conditions, including movement disorders. The specific impact of Covid-19 on movement disorders, also in relation to Parkinson's disease (PD), has increased steadily over the course of the pandemic (Ellul et al., 2020; Wood, 2020).

Many hypotheses have been put forward to explain the worsening of symptoms observed in people with PD (PwP) after contracting a SARS-CoV-2 infection. At first sight, at least some of the symptomatology might be explained by the effects of stress, anxiety, and isolation (van Wamelen, Wan, Chaudhuri, & Jenner, 2020), with the inevitable overlap related to restrictions imposed by the Covid-19 pandemic in combination with the often more advanced age of PwP compared to the general population. Some have suggested a role for infection-induced altered dopaminergic neurotransmission (Ait Wahmane et al., 2020; Araújo, Aranda-Martínez, & Aranda-Abreu, 2020; Nataf, 2020) and others that the infection might even lead to secondary neurodegeneration (Lippi, Domingues, Setz, Outeiro, & Krisko, 2020). Nevertheless, the relation between Covid-19 and PD is likely to be more complex than this and many uncertainties remain.

Here, we aim to review the available evidence and implications of Covid-19 in PwP in the acute phase of the infection and possible long-term effects, as well as mechanisms that might underlie the potential worsening of symptoms in PwP. We also summarize the available evidence on post-Covid-19 cases of parkinsonism.

2. The acute effects of SARS-CoV-2 infection in people with Parkinson's disease

Several studies have reported a worsening of motor and non-motor symptoms in PwP during the acute phase of SARS-CoV-2 infection, which often required therapy adjustments. Not unimportantly, although many PwP present with typical Covid-19 symptoms, some have a more atypical presentation with isolated worsening of parkinsonian symptoms (Fearon & Fasano, 2021). However, most of these studies consisted of case series and observational studies with relatively small participant cohort numbers only, while prospectively collected data is largely lacking. In addition, it is sometimes difficult to separate the effects of the SARS-CoV-2 infection per se from the consequences of social restrictions imposed over the course of the Covid-19 pandemic (Table 1).

Worsening of PD symptomatology is often observed over the course of an acute infection and might be mediated by several putative mechanisms of action, including altered central dopamine metabolism, pharmacodynamic and pharmacokinetic changes to dopaminergic medication, as well as direct effects of pathogens endotoxins (Brugger et al., 2015) (Fig. 1). Exacerbation of PD symptomatology over the course of an acute infection is usually transient, but might persist even after resolution of the infection, including persisting motor deterioration (Umemura et al., 2014). Whether the worsening of PD symptomatology in the acute phase of a SARS-CoV-2 infection does have peculiarities compared to the commonly observed exacerbations of motor and non-motor symptoms triggered by other infections remains unclear. The proposed neurotropism of SARS-CoV-2 is currently debatable, and it is possible to argue that systemic inflammation in combination with pandemic-related stress (van Wamelen et al., 2020) might contribute to the exacerbation of the PD symptomatology. Moreover, the pandemic-induced social isolation in association with reduced access to healthcare services and rehabilitation can per se have a detrimental impact on PD symptoms, including motor disability, anxiety and depression as it has been discussed in Chapters 9 and 10.

Table 1 Overview of studies looking at the acute effect of SARS-CoV-2 infection in people with Parkinson's disease.

Study population and design	Main findings	Reference
Case series • 10 Hospitalized PwP with a SARS-CoV-2 infection	Worsening of motor and non-motor symptoms, including: • Anxiety • Fatigue • Orthostatic hypotension • Cognitive impairment • Psychosis Therapy adjustment in half of the cases.	Antonini, Leta, Teo, and Chaudhuri (2020)
Community-based, case-control study • 12 Non-hospitalized PwP with SARS-CoV-2 infection • 36 Patients PwP without SARS-CoV-2 infection	Worsening of motor symptoms: • Increased daily OFF-time Worsening of non-motor symptoms: • Urinary problems • Fatigue Therapy adjustment in 1/3 of the cases.	Cilia et al. (2020)
Telephone survey • 8 PwP with a SARS-COV-2 infection.	Worsening of motor symptoms. Therapy adjustment in 1/4 of cases.	Artusi et al. (2020)
Case report • 2 PwP (with DBS) and a SARS-CoV-2 infection	Worsening of motor symptoms. Worsening of non-motor symptoms. • Cognitive impairment	Hainque and Grabli (2020)
Telephone survey, cross-sectional study • 15 PwP with a SARS-CoV-2 infection	• Hallucinations (23% in PwP with SARS-CoV-2 vs 0% in PwP without) • Motor fluctuations (61% in PwP with SARS-CoV-2 vs 36% in PwP without)	Santos-García et al. (2020)

• 553 PwP without a SARS-CoV-19 infection	• Dementia (16% in PwP with SARS-CoV-2 vs 7% in PwP without) • Behavioral problems (34% in PwP with SARS-CoV-2 vs 15% in PwP without)	
Case-control study • 7 PwP with a SARS-CoV-2 infection • 733 PwP without a SARS-CoV-2 infection	PwP did not experience a subjective worsening of symptoms during lockdown period. No description of possible differences between PwP with and without a SARS-CoV-2 infection.	Del Prete et al. (2021)
Case-control study • 51 PwP with a SARS-CoV-2 infection • 26 Healthy participants with a SARS-CoV-2 infection • 7158 PwP without a SARS-CoV-2 infection • 9736 Healthy participants without a SARS-CoV-2 infection	PwP with a SARS-CoV-2 infection: • Worsening of motor (63%) and non-motor (75%) symptoms. PwP without SARS-CoV-2 infection: • Worsening of motor (43%) and non-motor (52%) symptoms.	Brown et al. (2020)

PwP: people with Parkinson's disease; DBS: deep brain stimulation; SARS-CoV-2: severe acute respiratory syndrome coronavirus 2; vs: versus.

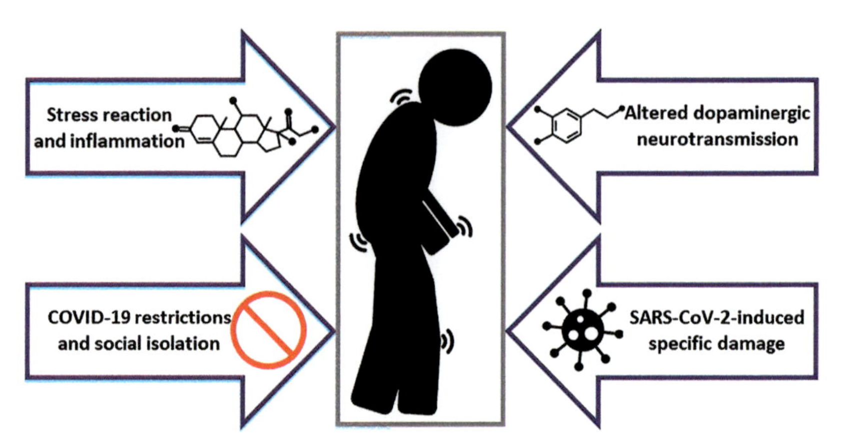

Fig. 1 Possible mechanisms for worsening of symptoms in the acute phase of COVID-19 in people with Parkinson's disease. Abbreviations: COVID-19: coronavirus disease 2019; SARS-CoV-2: severe acute respiratory syndrome coronavirus 2.

2.1 Dopaminergic signaling during infections

Given the clear role of dopamine in PD, it is not surprising to see that altered dopaminergic neurotransmission has been put forward as a possible culprit for symptoms worsening during infections, and SARS-CoV-2 infection in particular. In part, this motor deterioration might be caused by dosing or administration errors of dopaminergic medication drugs (Gerlach, Broen, & Weber, 2013), as well as changes in cognition and psychosis interfering with medication intake which could be particularly bothersome in unsupervised situations at home (Daley, Deane, Gray, Hill, & Myint, 2015). Nonetheless, it seems likely that other factors play a role in symptoms worsening during active SARS-CoV-2 infection, including disturbances in dopamine metabolism and dopaminergic signaling. For example, a retrospective analysis of 675 PwP showed that in 17 out of the 26 PwP who developed acute akinesia, infection was identified as the precipitating factor. With levodopa serum levels detected within normal range, it seems unlikely that malabsorption of the medication could explain this observation (Onofrj et al., 2009). More likely causes include alterations in dopaminergic drug transportation across the blood–brain barrier (BBB) where levodopa enters the brain through selective transporters (Okura, Ito, Ishiguro, Tamai, & Deguchi, 2007). Moreover, cytokines, which are released during infection, reduce the expression of type 2 vesicular monoamine transporters (VMAT2), involved in transferring cytosolic dopamine into vesicles (Kazumori et al., 2004). Also dopamine transporters, which are responsible for the reuptake of released dopamine, seem to be regulated by cytokines (Felger et al., 2013; Felger & Miller, 2012). Finally, at least in non-human primates, chronic exposure to interferon-α decreases presynaptic dopamine 2 receptor expression in the striatum (Felger et al., 2013; Felger & Miller, 2012). It could be reasoned that, similar to infections in general, the same mechanisms apply for SARS-CoV-2 infections.

2.2 The role of stress

The notion of stress, related changes in cortisol levels, and worsening of symptoms in PwP is not new. In fact, even Gowers in the late 19th century wrote 'Prolonged anxiety and severe emotional shock often precede the onset of PD' (Gowers, 1888). While motor performance often seems to improve under conditions of acute stress, chronic anxiety and stress on the other hand tend to characteristically aggravate motor symptoms in PwP, particularly resting tremor (Moore, Rose, & Grace, 2001). It is, therefore, of interest that a SARS-CoV-2 infection, in particular more severe infections, have consistently increased serum cortisol levels, as demonstrated in a recent meta-analysis and systematic review by Amiri-Dashatan and colleagues (Amiri-Dashatan, Koushki, Parsamanesh, & Chiti, 2022). Although the effects of increased cortisol levels on motor symptoms in PwP remain elusive (van Wamelen et al., 2020), higher serum cortisol levels appear to be more consistently associated with anxiety levels, risk-taking, sleep problems, and depressive symptoms in PwP (Breen et al., 2014; Djamshidian et al., 2011; Muller & Muhlack, 2007; Ruzicka et al., 2015). In the case of psychological stress, mainly induced by restrictions imposed during the Covid-19 pandemic, the overlap with the concept of exhaustion (a state of excessive fatigue and irritability often attributed to stress) is interesting (Clark, Ritz, Prescott, & Rod, 2013). Fatigue has been a commonly reported symptom in PwP during the pandemic, and the related vital exhaustion (a psychological response reflecting a breakdown of the adaption to stress) has been associated with a higher risk of hospitalization in PwP in an exposure-dependent manner (Clark et al., 2013), suggesting that at least some of the symptoms observed in PwP as related to Covid-19 might be stress-mediated. As the response to stress and the related regulation of cortisol levels are altered in PwP (van Wamelen et al., 2020), the pandemic and SARS-CoV-2 infection, as well as infections in general, might make them particularly prone to worsening of their symptoms in such situations.

3. Parkinson's disease and Long-COVID

While acute implications of Covid-19 in PwP have been widely reported (Fearon & Fasano, 2021), possible long-term sequelae of the viral infection in the general population, and specifically in patients with chronic neurogenerative diseases, such as PD, remain to be further elucidated. Given the potentially relevant implications in terms of clinical management and societal burden, on the 18th of December 2020, the National Institute for Health and Care Excellence (NICE) in conjunction with the Scottish Intercollegiate Guidelines Network (SIGN) and the Royal College of General Practitioners (RCGP) published the "Covid-19 rapid guideline" to identify, assess and manage possible long-term effects of Covid-19, often referred to as "Long-COVID" (NICE, 2020). The guidelines have been recently updated and provide recommendations to a variety of healthcare professionals about care for adults and children who have new or ongoing symptoms 4 weeks or more after the start of acute Covid-19 (NICE, 2020). Common long-COVID clinical manifestations are respiratory symptoms of breathlessness and cough, but also generalized, cardiovascular, psychological/psychiatric, neurological, gastrointestinal, dermatological, musculoskeletal, ear, nose and throat symptoms (Table 2) (NICE, 2020).

According to a recent metanalysis of 37 published articles, fatigue (16–64%), dyspnea (15–61%), cough (2–59%), arthralgia (8–55%), and thoracic pain (5–62%) have been reported as the most prevalent and persistent symptoms 4 weeks after a severe form of SARS-CoV-2 infection (i.e., requiring hospitalization), while for patients with milder forms (i.e. not requiring hospitalization) the persistence of the above mentioned symptoms was lower and 3% to 74% of the latter group of patients had prolonged smell and taste disorders (Nguyen et al., 2022). Risk factors for long-COVID symptoms were female sex, older age, presence of comorbidities and severity at the acute phase of the infection (Nguyen et al., 2022).

As far as long-COVID symptoms in PwP are concerned, only a few reports are available to the best of our knowledge. In addition to the case report of a 73-year-old male patient with a diagnosis of PD who developed severe and persistent oropharyngeal dysphagia after a severe form of Covid-19 (no dysphagia reported before Covid-19) (Boika, Sialitski, Chyzhyk, Ponomarev, & Fomina, 2021), a multicentre, international case series described the prevalence of persistent post-Covid-19 symptoms in 27 PwP (Leta et al., 2021). According to this series, 23 PwP (85%) developed

Table 2 Most common clinical manifestations of long-COVID (NICE, 2020).

Respiratory symptoms	Breathlessness
	Cough
Cardiovascular symptoms	Chest tightness
	Chest pain
	Palpitations
Generalized symptoms	Fatigue
	Fever
	Pain
Neurological symptoms	Cognitive impairment ('brain fog', loss of concentration or memory issues)
	Headache
	Sleep disturbance
	Peripheral neuropathy symptoms (pins and needles and numbness)
	Dizziness
	Delirium (in older populations)
	Mobility impairment
	Visual disturbance
Gastrointestinal symptoms	Abdominal pain
	Nausea and vomiting
	Diarrhea
	Weight loss and reduced appetite
Musculoskeletal symptoms	Joint pain
	Muscle pain
Ear, nose and throat symptoms	Tinnitus
	Earache
	Sore throat
	Dizziness
	Loss of taste and/or smell
	Nasal congestion

Continued

Table 2 Most common clinical manifestations of long-COVID (NICE, 2020).—cont'd

Respiratory symptoms	**Breathlessness**
	Cough
Dermatological symptoms	Skin rashes
	Hair loss
Psychological/psychiatric symptoms	Symptoms of depression
	Symptoms of anxiety
	Symptoms of post-traumatic stress disorder

prolonged post–Covid-19 symptoms and the most common long-term effects of Covid-19 were worsening of motor symptoms (52%), increased levodopa daily dose requirements (48%), fatigue (41%), cognitive disturbances (such as "brain fog", loss of concentration and memory deficits, 22%), and sleep disturbances (22%). Interestingly, a severe acute infection (as indicated by a history of hospitalization), was not a prerequisite for the development of persistent post–Covid-19 symptoms in PwP (Boika et al., 2021; Leta et al., 2021). Although recent evidence seems to suggest that chronic immunological changes might be involved in the long-COVID syndrome in PwP (Boika et al., 2021), whether viral illness-related worsening of pre-existing PD features, as well as lockdown-related stress combined with reduced access to health care services also contribute, needs to be further explored.

4. Post-covid-19 cases of Parkinsonism

The excessive dimensions of the Covid-19 pandemic have recently triggered discussions as to whether it could serve as a "perfect storm" for a post-Covid-19 emergence of new-onset parkinsonism in susceptible individuals (Beauchamp, Finkelstein, Bush, Evans, & Barnham, 2020; Brundin, Nath, & Beckham, 2020). Infectious agents, including viruses, have long been presumed to play a role in the pathogenesis of PD (Hawkes, Del Tredici, & Braak, 2007; Sulzer, 2007), while the association of numerous viruses with the development of transient or persistent parkinsonism has been well-documented in the literature (Jang, Boltz, Webster, & Smeyne, 2009; Xing, Marsili, & Truong, 2022). The idea of Covid-19 unmasking parkinsonism constitutes a source of concern to many and has been fuelled by historically documented parkinsonism cases appearing during the acute or chronic phase of encephalitis lethargica (EL), an entity which affected more than one million people in Europe, India and America from 1916 to 1930 (Hoffman & Vilensky, 2017). The causative substrate of EL constitutes one of the greatest medical mysteries with many researchers acknowledging an epidemiological link to the Spanish influenza, a pandemic with a death toll exceeding 40 million people globally in 1918–1919 (Hoffman & Vilensky, 2017). The Spanish influenza strains disappeared in 1933, coinciding with the end of the EL period, while people born between 1888 and 1924 were found to have an increased risk of developing PD compared to those born before or after this time period (Henry, Smeyne, Jang, Miller, & Okun, 2010).

Up to now, 20 cases have been reported in the literature with new-onset symptoms of parkinsonism (bradykinesia and/or rigidity), manifesting during or shortly after a diagnosis of Covid-19 (Table 3) (Akilli & Yosunkaya, 2021; Ayele et al., 2021; Cavallieri et al., 2021; Cohen et al., 2020; Faber et al., 2020; Fearon et al., 2021; Ghosh et al., 2021; Makhoul & Jankovic, 2021; Méndez-Guerrero et al., 2020; Morassi et al., 2021; Ong et al., 2022; Pilotto et al., 2021; Rao et al., 2022; Rass et al., 2021; Roy et al., 2021; Tiraboschi et al., 2021). Only one of these patients (patient 6) had mentioned prior symptoms of prodromal PD (constipation), while none of the above cases had a family history of PD. Hyposmia or anosmia in the context of Covid-19 was identified in nine patients, although the significance of this feature remains unclear given its high prevalence in SARS-CoV-2 infection. Genetic testing was performed on three occasions, with patients 9 and 10 testing positive for a heterozygous mutation in the genes of glucocerebrosidase (GBA) and leucine-rich repeat kinase 2 (LRRK2), respectively. A genetic substrate was not confirmed for patient 3.

In 11 of these cases, parkinsonism appeared in the context of encephalopathy (patients 2, 4, 5, 7–9, 11–14, 17), while four patients developed post-infectious parkinsonism without encephalopathy (patients 1, 18, 19, 20). The remaining four patients presented with parkinsonism and were clinically diagnosed with probable PD (patients 3, 6, 15, 16), as they bore significant similarities to typical PD. The severity of Covid-19 was mild in four cases, moderate in four cases and severe in 11 cases. Eight of these patients had to be admitted in the Intensive Care Unit (ICU) and, with one exception, parkinsonism was detected after their level of consciousness was improved and they were discharged from the ICU. Eleven patients underwent a lumbar puncture, which revealed signs of subjacent inflammation (increased protein, oligoclonal bands) in two cases (patients 14 and 17).

Table 3 Published papers of new-onset post-Covid-19 parkinsonism and patients' characteristics.

First author, date	Country	Patient's ID	Age, Gender	Past medical history	Covid severity	Onset	Encephalopathy
Faber et al. (2020)	Brazil	#1	35F	–	Mild	10 d	–
Méndez-Guerrero et al. (2020)	Spain	#2	58 M	Hypertension, hypercholesterolemia	Severe/ICU	38 d	+
Cohen et al. (2020)	Jerusalem	#3	45 M	Hypertension, asthma	Moderate	14–21 d	–
Pilotto et al. (2021)	Italy	#4	73 M			0	+
Akilli and Yosunkaya (2021)	Turkey	#5	72 M	Hypertension, DM, peripheral artery disease	Severe/ICU	2 d	+
Makhoul & Jankovic, 2021	USA	#6	64F		Mild	5 d	–
Roy, Song, Awad, and Zamudio (2021)	USA	#7	60 M	Hypertension, DM, hypercholesterolemia	Severe/ICU	8 d	+
Fearon, Mikulis, and Lang (2021)	Canada	#8	46 M		Severe/ICU	38 d	+
Tiraboschi et al. (2021)	Italy	#9	40F	Overweight	Severe/ICU	22 d	+
Rass et al. (2021)	Austria	#10			Severe/ICU	3 mo	
Ghosh et al. (2021)	India	#11	65F	Diabetes mellitus	Moderate	6 d	+
Ayele et al. (2021)	Ethiopia	#12	35F	–	Severe/ICU	7–14 d	+

Morassi et al. (2021)	Italy	#13	70F	Hypertension, anxiety-depressive disorder	Severe	31 d	+
		#14	73F	Hypertension, DM, anxiety-depressive disorder	Mild	0	+
Cavallieri et al. (2021)	Italy	#15	67 M		Moderate	4 mo	–
		#16	45 M	–	Mild	3 mo	–
Ong, Nor, Yusoff, and Sapuan (2022)	Malaysia	#17	31 M	–	Severe	6 d	+
Rao, Hidayathullah, Hegde, and Adhikari (2022)	India	#18	72 M	–	Severe	14 d	–
		#19	66 M	Hypertension, DM, seizures	Severe/ ICU	14 d	–
		#20	74 M		Moderate	21 d	–

Covid-19: Coronavirus Disease 2019; d: days; DM: diabetes mellitus; F: female; ICU: Intensive Care Unit; ID: identification number; M: male; mo: months; USA: United States of America.

Polymerase chain reaction (PCR) tests and cultures for various pathogens in the cerebrospinal fluid (CSF) were negative on all occasions. Interestingly, an acute SARS-CoV-2 infection of the central nervous system (CNS), which would be achieved by detecting the virus via PCR, was not confirmed on any occasion. Such findings in the CSF are consistent with the majority of Covid-19 patients, who have been investigated for various types of neurological manifestations (Neumann et al., 2020). Seven patients (patients 2, 3, 9, 11, 13, 14, 17) were screened for a range of serum and/or CSF antibodies related to autoimmune encephalitis with negative results on all occasions. Seven patients (patients 1, 2, 3, 6, 14, 15, 16) underwent dopaminergic uptake imaging (either 6-[18F]-L-fluoro-L-3,4-dihydroxyphenylalanine (F-FDOPA)-based positron emission tomography (PET) or dopamine transporter single-photon emission computerized tomography (SPECT) imaging with ioflupane I-123 injection (DaTscan)), with all of them having a decreased dopamine uptake either in one or both putamina, similarly to typical PD. Four patients (patients 1, 9, 13, 14) underwent a brain FDG (2-deoxy-2-[18F]fluoro-D-glucose)-based PET scan with only one of them exhibiting normal findings. A brain magnetic resonance imaging (MRI) was performed on all but three cases (patients 6, 10, 18), and was found abnormal on six occasions (Table 4).

Although duration of follow-up varied greatly, ranging from 1 month to 1 year and missing in almost half of the cases, most patients in this case series exhibited a favorable outcome (Table 5). Like already mentioned, 11 of them presented with encephalopathy, with the majority of them well-fitting the concept of general viral post-encephalitic parkinsonism, as has been already described in the past (Jang et al., 2009).

The frequency of encephalopathy in the context of Covid-19 seems to vary widely (7–69%) (Ellul et al., 2020), while in a large group of 129,008 SARS-CoV-2 positive patients of all ages, 138 cases of encephalitis have been confirmed, generating an incidence of 0.215% (Siow, Lee, Zhang, Saffari, & Ng, 2021). However, the definition of encephalitis might vary in different publications. Under these circumstances, corticosteroids, intravenous immune globulin (IVIG), convalescent plasma therapy, monoclonal antibodies administration or plasmapheresis have been included in the treating protocols across different clinical settings (usually with supportive imaging indications), recently including Covid-19-related encephalitis with concurrent neurological manifestations (Huo, Xu, & Wang, 2021; Sonneville, Klein, de Broucker, & Wolff, 2009). Five of the above cases were treated with immunomodulatory/ immunosuppressive therapy

Table 4 Imaging diagnostic means (if applicable) used in patients with new-onset post-Covid-19 parkinsonism.

Patient's ID	MRI	Dopaminergic uptake imaging	FDG PET scan
#1 (Faber et al., 2020)	Unremarkable findings.	DaTscan: ↓DA uptake of the L putamen*	Unremarkable findings.
#2 (Méndez-Guerrero et al., 2020)	Unremarkable findings.	DaTscan: ↓DA uptake of both putamina asymmetrically*	
#3 (Cohen et al., 2020)	Unremarkable findings.	F-FDOPA PET: ↓DA uptake of both putamina (L > R) & L caudate	
#4 (Pilotto et al., 2021)	↑T2 signal of the frontal lobes.		
#5 (Akilli & Yosunkaya, 2021)	Unremarkable findings.		
#6 (Makhoul & Jankovic, 2021)		DaTscan: ↓DA uptake of the R putamen*	
#7 (Roy et al., 2021)	Ischemic stroke in the basal ganglia and corona radiata.		
#8 (Fearon et al., 2021)	Oedema of the globus pallidus and microbleeds in cerebellar nuclei attributed to hypoxia. Atrophy of the above regions in subsequent imaging.		
#9 (Tiraboschi et al., 2021)	Unremarkable findings.		↑glu metabolism in the mesial temporal lobes & subthalamic nuclei (normalization of signal after IVIg).
#10 (Rass et al., 2021)			

Continued

Table 4 Imaging diagnostic means (if applicable) used in patients with new-onset post-Covid-19 parkinsonism.—cont'd

Patient's ID	MRI	Dopaminergic uptake imaging	FDG PET scan
#11 (Ghosh et al., 2021)	Symmetrical lesions of the caudate and putamen, sparing the globus pallidus, with ↑T2 signal and diffusion restriction, attributed to extra-pontine osmotic demyelination.		
#12 (Ayele et al., 2021)	Symmetrical, non-enhancing lesions with ↑T2 signal in the pallidum, possibly attributed to silent hypoxia.		
#13 (Morassi et al., 2021)	Unremarkable findings.	DaTscan: ↓DA uptake of both putamina asymmetrically*	Diffuse cortical hypo-metabolism, ↑glu metabolism in the mesial temporal lobes, basal ganglia, brainstem (indicative of encephalitis)
#14 (Morassi et al., 2021)	Unremarkable findings.		
#15 (Cavallieri et al., 2021)	Unremarkable findings.	DaTscan: ↓DA uptake of both putamina	
#16 (Cavallieri et al., 2021)	Unremarkable findings.	DaTscan: ↓DA uptake of both putamina	
#17 (Ong et al., 2022)	↑T2 signal of both thalami with hemosiderin deposition and patchy enhancement and ↑T2 signal of the pons attributed to ANEC.		
#18 (Rao et al., 2022)			
#19 (Rao et al., 2022)	Unremarkable findings.		
#20 (Rao et al., 2022)	Unremarkable findings.		

ANEC: Acute Necrotizing Encephalitis; Covid-19: Coronavirus Disease 2019; DA: dopamine; DaTscan: dopamine transporter single-photon emission computerized tomography (SPECT) imaging with ioflupane I-123 injection; FDG: 2-deoxy-2-[18F]fluoro-D-glucose; F-FDOPA: 6-[18F]-L-fluoro-L-3,4-dihydroxyphenylalanine; glu: glucose; ID: identification; IVIg: intravenous immunoglobulin therapy; L: left; MRI: magnetic resonance imaging; PET: positron emission tomography; R: right.

Table 5 Treatment, possible diagnosis and clinical course with follow-up assessment (if applicable) in patients with new-onset post-Covid-19 parkinsonism.

Patient's ID	Response to immunomodulatory/ immunosuppressive treatment	Response to anti-parkinsonian therapy	Possible diagnosis/ Clinical course—Follow-up
#1 (Faber et al., 2020)	–	200/50 mg of levodopa/ benserazide TD.	A probable diagnosis of Covid-related post-infectious parkinsonism without encephalopathy. Significant improvement after few days of therapy. No follow-up.
#2 (Méndez-Guerrero et al., 2020)	–	Adverse events & no clinical response to Apo test (3 mg initially, 2 mg after 5 day).	A probable diagnosis of Covid-related post-encephalitic parkinsonism. Significant, spontaneous improvement, although symptoms persisted (follow-up at 53 day).
#3 (Cohen et al., 2020)	High-dose of methylprednisolone without any consistent effect.	0.375 mg ER pramipexole OD, 2 mg biperiden resulted in parkinsonism improvement.	Authors suggested a diagnosis of probable PD. Improvement with dopaminergic therapy. No follow-up is specified.
#4 (Pilotto et al., 2021)	–	–	–
#5 (Akilli & Yosunkaya, 2021)	Convalescent plasma therapy (twice), patient improvement. Plasmapheresis due to ARDS.	–	A probable diagnosis of Covid-related post-encephalitic parkinsonism. No symptoms found at 2 month follow-up.
#6 (Makhoul & Jankovic, 2021)	–	–	Authors suggested that the patient's prodromal PD became symptomatic due to Covid-19 stress. No follow-up.
#7 (Roy et al., 2021)	–	Levodopa/ carbidopa & modafinil for 1mo with symptoms improvement.	Authors suggested a 'locked in'/parkinsonian state due to Covid-related encephalopathy and an acute ischemic stroke in the basal ganglia. Full recovery at 1mo follow-up.

Continued

Table 5 Treatment, possible diagnosis and clinical course with follow-up assessment (if applicable) in patients with new-onset post-Covid-19 parkinsonism.—cont'd

Patient's ID	Response to immunomodulatory/ immunosuppressive treatment	Response to anti-parkinsonian therapy	Possible diagnosis/ Clinical course—Follow-up
#8 (Fearon et al., 2021)	–	No response to 450 mg levodopa.	Probable diagnosis of hypoxic-ischemic injury. Parkinsonism persisted at 1y follow-up.
#9 (Tiraboschi et al., 2021)	2 IVIg cycles, significant improvement.	–	Authors suggested a diagnosis of immune-mediated Covid-related encephalopathy. Full resolution of clinical and imaging findings at 4mo follow-up.
#10 (Rass et al., 2021)	–	–	–
#11 (Ghosh et al., 2021)	Low doses of dexamethasone as per Covid-19 treatment protocol before parkinsonism manifestation.	Levodopa/carbidopa 100/25 mg BD, pramipexole 1.5 mg OD with parkinsonism improvement.	Probable diagnosis of osmotic demyelination due to hyperglycemic state, triggered by dexamethasone. Significant improvement at 4mo follow-up assessment, while on dopaminergic therapy.
#12 (Ayele et al., 2021)	–	Levodopa/ carbidopa 200/50 mg TD with parkinsonism improvement.	Probable diagnosis of silent hypoxia. Significant improvement on follow-up (not specified when) while on dopaminergic therapy.
#13 (Morassi et al., 2021)	1 IVIg cycle & corticosteroids followed by improvement.	Levodopa/ carbidopa 100/25 mg BD with moderate response of parkinsonism.	Possible immune-mediated substrate of encephalopathy. At 9mo follow-up, parkinsonism persisted, cognitive & ADL worsening.
#14 (Morassi et al., 2021)	2 cycles of corticosteroids and 1 cycle of IVIg.	Amantadine 100 mg BD, levodopa/ carbidopa 100/25 mg QD.	Probable post-encephalitic parkinsonism. No improvement with therapy, death 30d after discharge (aspiration pneumonia, bedsores).

#15 (Cavallieri et al., 2021)	–	–	Diagnosis of probable PD. No information on treatment or follow-up.
#16 (Cavallieri et al., 2021)	–	–	
#17 (Ong et al., 2022)	High doses of methylprednisolone.	Trihexyphenidyl 2mg TD.	Diagnosis of Covid-induced ANEC. Full resolution of parkinsonism and cognitive impairment.
#18 (Rao et al., 2022)	–	Levodopa 50mg TD, improvement.	Significant improvement at 4mo (pt 18), 1mo (pt 19), 6mo (pt 20) on dopaminergic therapy. Probable diagnosis of post-infectious parkinsonism (patients' characteristics did not account for PD).
#19 (Rao et al., 2022)	–	Levodopa/ carbidopa titration, improvement.	
#20 (Rao et al., 2022)	–	Levodopa/ carbidopa, improvement.	

ADL: activities of daily living; ANEC: acute necrotizing encephalopathy; Apo: apomorphine; ARDS: acute respiratory distress syndrome; BD: twice per day; Covid: Coronavirus disease 2019; d: day; ER: extended release; ID: identification; mg: milligrams; IVIg: intravenous immunoglobulin therapy; mo: month; OD: once daily; PD: Parkinson's disease; pt.: patient; QD: four times per day; TD: three times per day; y: year.

(Table 5) with three of them demonstrating full resolution of their symptoms (patients 5, 9, 17) and one of them (patient 13) exhibiting a good response to therapy, although parkinsonism and cognitive decline persisted in the follow-up assessment after 9 months. Interestingly, patient 17 was diagnosed with COVID-19-precipitated acute necrotizing encephalopathy (ANEC), a rare, but distinctive kind of typically virus-related acute encephalopathy, which has been classically managed with immunomodulatory/ immunosuppressive therapy, especially corticosteroids (Wu et al., 2015).

An immune-mediated substrate has been suggested in the above cases. SARS-CoV-2 has the potential of causing a cytokine release and triggering an excessive immune response in the periphery (Wan et al., 2021). Such processes might affect the BBB permeability, thus, allowing infected immune cells or the virus per se to invade the CNS and induce a secondary cytokine storm (Williams et al., 2021). Interestingly, increased proinflammatory cytokines and a high titre of anti-SARS-CoV-2 IgG antibodies were detected in the CSF of patient 9, which is a previously described, although rare phenomenon (Lewis et al., 2021). Whether these antibodies were locally produced or crossed the BBB due to the systemic inflammation remains unclear. Neuroinflammation has been suspected to promote neurodegeneration and significantly contribute to PD pathogenesis with midbrain dopamine neurons being particularly vulnerable to systemic inflammation due to high energy requirements (Pissadaki & Bolam, 2013; Tufekci, Meuwissen, Genc, & Genc, 2012).

Trials of dopaminergic therapy have also been classically attempted in cases of post-virus parkinsonism, whether encephalopathy was present or not (Bopeththa & Ralapanawa, 2017; Guan, Lu, & Zhou, 2012). Eleven patients in this case series were treated with dopaminergic therapy due to assumed post-Covid-19 parkinsonism, with (patients 2, 7, 8, 11–14) or without encephalopathy (patients 1, 18–20). One patient manifested full recovery, six of them exhibited a significant and one of them a moderate response of their parkinsonian features (Table 5), suggesting an underlying, occasionally reversible, impairment of the dopaminergic pathway. Secondary causes of parkinsonism should be investigated, although the underlying cause might not, ultimately, differentiate the therapeutic strategy followed. In two of the above cases (patients 8, 12) silent hypoxia was identified as the potential subjacent cause of parkinsonism due to the imaging findings; dopaminergic therapy led to significant improvement of one patient's symptoms, but had no clinical effect for the other. Parkinsonism in patient 11, who had a history of diabetes mellitus, was attributed to extra-pontine osmotic demyelination, triggered by uncontrolled hyperglycemia after dexamethasone administration in the context of Covid-19 treatment. Authors suggested that Covid-19-precipitated inflammation could have also contributed to the BBB disruption, although osmotic demyelination per se has been associated with de novo movement disorders due to impairment of the striato-thalamo-cortical network (de Souza, 2013). Symptoms in this occasion have also improved with dopaminergic therapy. Imaging studies in patient 7 revealed an ischemic stroke in the basal ganglia, which could also have accounted for the patient's acute hemiparkinsonism and hemiparesis. Finally, four patients had a history of neuroleptic drugs use prior to the emergence of parkinsonism (patients 9, 13, 14, 17), thus, the possibility of drug-induced parkinsonism cannot be overlooked.

The temporal proximity of new-onset parkinsonism with a Covid-19 diagnosis, along with the co-existent encephalopathy in some cases, led the authors to assume an etiological connection between them. In summary, different mechanisms underlying these cases of supposedly para- or post-infectious parkinsonism, including structural or functional impairment of the nigrostriatal pathway, inflammatory or vascular damage and unmasking of already active, though asymptomatic, prodromal PD, have been described in the literature (Merello, Bhatia, & Obeso, 2021). However, with more than 5300 confirmed cases of Covid-19 per 100,000 globally (as of February 2022) (Worldmeter.info, 2021) and an annual incidence of about 15 PD cases per 100,000 (Tysnes & Storstein, 2017), anticipating a parkinsonism wave based solely on the current 20 published cases may be premature, as the possibility of chance or exterior factors cannot be overlooked. Moreover, the popularity of the topic might have led to publication bias, favoring the publication of post-Covid parkinsonism cases, while others might argue that parkinsonism might be underdiagnosed, as full neurologic examination might not take place in the Covid-19 wards or Covid-19 patients are usually managed by non-neurological specialties. Finally, it is conceivable that the combination of inflammation, fever and respiratory symptoms may amplify clinical manifestations particularly in the premotor and prodromal phase of PD. Greater vigilance is recommended in order to timely recognize and address potential neurological manifestations of SARS-CoV-2, including parkinsonism, especially in the long-term.

5. Conclusions

SARS-CoV-2 and the restrictions imposed during the Covid-19 pandemic have had a clear impact on PwP. As expected, many have seen their symptoms worsen and it is likely that multiple factors have contributed to this. On the one hand, some of the symptomatology might be explained by the effects of reduced mobility, stress, anxiety, and isolation during the pandemic which may have negative consequences on PD symptoms, including gait dysfunction and risk of falls (Luis-Martínez et al., 2021); on the other hand, SARS-CoV-2 infection may also impact dopaminergic neurotransmission or could even be associated to secondary neurodegeneration.

The relation between Covid-19 and PD is likely to be more complex and many uncertainties remain, although some might argue that it provides an unprecedented opportunity to study the effects of viral infections on PwP and the short- and long-term impact of infection and social restrictions on symptoms of PD and quality of life.

References

Ait Wahmane, S., Achbani, A., Ouhaz, Z., Elatiqi, M., Belmouden, A., & Nejmeddine, M. (2020). The possible protective role of α-Synuclein against severe acute respiratory syndrome coronavirus 2 infections in patients with Parkinson's disease. *Movement Disorders*, *35*(8), 1293–1294. https://doi.org/10.1002/mds.28185.

Akilli, N. B., & Yosunkaya, A. (2021). Part of the Covid19 puzzle: Acute parkinsonism. *The American Journal of Emergency Medicine*, *47*, 333.e331–333.e333. https://doi.org/10.1016/j.ajem.2021.02.050.

Amiri-Dashatan, N., Koushki, M., Parsamanesh, N., & Chiti, H. (2022). Serum cortisol concentration and COVID-19 severity: A systematic review and meta-analysis. *Journal of Investigative Medicine*. https://doi.org/10.1136/jim-2021-001989.

Antonini, A., Leta, V., Teo, J., & Chaudhuri, K. R. (2020). Outcome of Parkinson's disease patients affected by COVID-19. *Movement Disorders*, *35*(6), 905–908. https://doi.org/10.1002/mds.28104.

Araújo, R., Aranda-Martínez, J. D., & Aranda-Abreu, G. E. (2020). Amantadine treatment for people with COVID-19. *Archives of Medical Research*, *51*(7), 739–740. https://doi.org/10.1016/j.arcmed.2020.06.009.

Artusi, C. A., Romagnolo, A., Imbalzano, G., Marchet, A., Zibetti, M., Rizzone, M. G., et al. (2020). COVID-19 in Parkinson's disease: Report on prevalence and outcome. *Parkinsonism & Related Disorders*, *80*, 7–9. https://doi.org/10.1016/j.parkreldis.2020.09.008.

Ayele, B. A., Demissie, H., Awraris, M., Amogne, W., Shalash, A., Ali, K., et al. (2021). SARS-COV-2 induced parkinsonism: The first case from the sub-Saharan Africa. *Clinical Parkinsonism & Related Disorders*, *5*, 100116. https://doi.org/10.1016/j.prdoa.2021.100116.

Beauchamp, L. C., Finkelstein, D. I., Bush, A. I., Evans, A. H., & Barnham, K. J. (2020). Parkinsonism as a third wave of the COVID-19 pandemic? *Journal of Parkinson's Disease*, *10*(4), 1343–1353. https://doi.org/10.3233/JPD-202211.

Boika, A. V., Sialitski, M. M., Chyzhyk, V. A., Ponomarev, V. V., & Fomina, E. G. (2021). Post-COVID worsening of a Parkinson's disease patient. *Clinical Case Reports*, *9*(7), e04409. https://doi.org/10.1002/ccr3.4409.

Bopeththa, B., & Ralapanawa, U. (2017). Post encephalitic parkinsonism following dengue viral infection. *BMC Research Notes*, *10*(1), 655. https://doi.org/10.1186/s13104-017-2954-5.

Breen, D. P., Vuono, R., Nawarathna, U., Fisher, K., Shneerson, J. M., Reddy, A. B., et al. (2014). Sleep and circadian rhythm regulation in early Parkinson disease. *JAMA Neurology*, *71*(5), 589–595. https://doi.org/10.1001/jamaneurol.2014.65.

Brown, E. G., Chahine, L. M., Goldman, S. M., Korell, M., Mann, E., Kinel, D. R., et al. (2020). The effect of the COVID-19 pandemic on people with Parkinson's disease. *Journal of Parkinson's Disease*, *10*(4), 1365–1377. https://doi.org/10.3233/jpd-202249.

Brugger, F., Erro, R., Balint, B., Kägi, G., Barone, P., & Bhatia, K. P. (2015). Why is there motor deterioration in Parkinson's disease during systemic infections-a hypothetical view. *NPJ Parkinson's Disease*, *1*, 15014. https://doi.org/10.1038/npjparkd.2015.14.

Brundin, P., Nath, A., & Beckham, J. D. (2020). Is COVID-19 a perfect storm for Parkinson's disease? *Trends in Neurosciences*, *43*(12), 931–933. https://doi.org/10.1016/j.tins.2020.10.009.

Cavallieri, F., Fioravanti, V., Toschi, G., Grisanti, S., Napoli, M., Moratti, C., et al. (2021). COVID-19 and Parkinson's disease: A casual association or a possible second hit in neurodegeneration? *Journal of Neurology*, *1–3*. https://doi.org/10.1007/s00415-021-10694-4.

Cilia, R., Bonvegna, S., Straccia, G., Andreasi, N. G., Elia, A. E., Romito, L. M., et al. (2020). Effects of COVID-19 on Parkinson's disease clinical features: A community-based case-control study. *Movement Disorders*, *35*(8), 1287–1292. https://doi.org/10.1002/mds.28170.

Clark, A. J., Ritz, B., Prescott, E., & Rod, N. H. (2013). Psychosocial risk factors, pre-motor symptoms and first-time hospitalization with Parkinson's disease: A prospective cohort study. *European Journal of Neurology*, *20*(8), 1113–1120. https://doi.org/10.1111/ene.12117.

Cohen, M. E., Eichel, R., Steiner-Birmanns, B., Janah, A., Ioshpa, M., Bar-Shalom, R., et al. (2020). A case of probable Parkinson's disease after SARS-CoV-2 infection. *Lancet Neurology*, *19*(10), 804–805. https://doi.org/10.1016/s1474-4422(20)30305-7.

Daley, D. J., Deane, K. H., Gray, R. J., Hill, R., & Myint, P. K. (2015). Qualitative evaluation of adherence therapy in Parkinson's disease: A multidirectional model. *Patient Preference and Adherence*, *9*, 989–998. https://doi.org/10.2147/ppa.S80158.

de Souza, A. (2013). Movement disorders and the osmotic demyelination syndrome. *Parkinsonism & Related Disorders*, *19*(8), 709–716. https://doi.org/10.1016/j.parkreldis.2013.04.005.

Del Prete, E., Francesconi, A., Palermo, G., Mazzucchi, S., Frosini, D., Morganti, R., et al. (2021). Prevalence and impact of COVID-19 in Parkinson's disease: Evidence from a multi-center survey in Tuscany region. *Journal of Neurology*, *268*(4), 1179–1187. https://doi.org/10.1007/s00415-020-10002-6.

Djamshidian, A., O'Sullivan, S. S., Papadopoulos, A., Bassett, P., Shaw, K., Averbeck, B. B., et al. (2011). Salivary cortisol levels in Parkinson's disease and its correlation to risk behaviour. *Journal of Neurology, Neurosurgery, and Psychiatry*, *82*(10), 1107–1111. https://doi.org/10.1136/jnnp.2011.245746.

Ellul, M. A., Benjamin, L., Singh, B., Lant, S., Michael, B. D., Easton, A., et al. (2020). Neurological associations of COVID-19. *The Lancet Neurology*, *19*(9), 767–783. https://doi.org/10.1016/S1474-4422(20)30221-0.

Faber, I., Brandão, P. R. P., Menegatti, F., de Carvalho Bispo, D. D., Maluf, F. B., & Cardoso, F. (2020). Coronavirus disease 2019 and parkinsonism: A non-post-encephalitic case. *Movement Disorders*, *35*(10), 1721–1722. https://doi.org/10.1002/mds.28277.

Fearon, C., & Fasano, A. (2021). Parkinson's disease and the COVID-19 pandemic. *Journal of Parkinson's Disease*, *11*(2), 431–444. https://doi.org/10.3233/jpd-202320.

Fearon, C., Mikulis, D. J., & Lang, A. E. (2021). Parkinsonism as a sequela of SARS-CoV-2 infection: Pure hypoxic injury or additional COVID-19-related response? *Movement Disorders*, *36*(7), 1483–1484. https://doi.org/10.1002/mds.28656.

Felger, J. C., & Miller, A. H. (2012). Cytokine effects on the basal ganglia and dopamine function: The subcortical source of inflammatory malaise. *Frontiers in Neuroendocrinology*, *33*(3), 315–327. https://doi.org/10.1016/j.yfrne.2012.09.003.

Felger, J. C., Mun, J., Kimmel, H. L., Nye, J. A., Drake, D. F., Hernandez, C. R., et al. (2013). Chronic interferon-α decreases dopamine 2 receptor binding and striatal dopamine release in association with anhedonia-like behavior in nonhuman primates. *Neuropsychopharmacology*, *38*(11), 2179–2187. https://doi.org/10.1038/npp.2013.115.

Gerlach, O. H., Broen, M. P., & Weber, W. E. (2013). Motor outcomes during hospitalization in Parkinson's disease patients: A prospective study. *Parkinsonism & Related Disorders*, *19*(8), 737–741. https://doi.org/10.1016/j.parkreldis.2013.04.017.

Ghosh, R., Ray, A., Roy, D., Das, S., Dubey, S., & Benito-León, J. (2021). Parkinsonism with akinetic mutism following osmotic demyelination syndrome in a SARS-CoV-2 infected elderly diabetic woman: A case report. *Neurología*. https://doi.org/10.1016/j.nrl.2021.09.007.

Gowers, W. R. (1888). A manual of diseases of the nervous system. *The American Journal of Psychology*, *1*(2), 346–347. https://doi.org/10.2307/1411377.

Guan, J., Lu, Z., & Zhou, Q. (2012). Reversible parkinsonism due to involvement of substantia nigra in Epstein-Barr virus encephalitis. *Movement Disorders*, *27*(1), 156–157. https://doi.org/10.1002/mds.23935.

Hainque, E., & Grabli, D. (2020). Rapid worsening in Parkinson's disease may hide COVID-19 infection. *Parkinsonism & Related Disorders*, *75*, 126–127. https://doi.org/10.1016/j.parkreldis.2020.05.008.

Hawkes, C. H., Del Tredici, K., & Braak, H. (2007). Parkinson's disease: A dual-hit hypothesis. *Neuropathology and Applied Neurobiology*, *33*(6), 599–614. https://doi.org/10.1111/j.1365-2990.2007.00874.x.

Henry, J., Smeyne, R. J., Jang, H., Miller, B., & Okun, M. S. (2010). Parkinsonism and neurological manifestations of influenza throughout the 20th and 21st centuries. *Parkinsonism & Related Disorders*, *16*(9), 566–571. https://doi.org/10.1016/j.parkreldis.2010.06.012.

Hoffman, L. A., & Vilensky, J. A. (2017). Encephalitis lethargica: 100 years after the epidemic. *Brain*, *140*(8), 2246–2251. https://doi.org/10.1093/brain/awx177.

Huo, L., Xu, K. L., & Wang, H. (2021). Clinical features of SARS-CoV-2-associated encephalitis and meningitis amid COVID-19 pandemic. *World Journal of Clinical Cases*, *9*(5), 1058–1078. https://doi.org/10.12998/wjcc.v9.i5.1058.

Jang, H., Boltz, D. A., Webster, R. G., & Smeyne, R. J. (2009). Viral parkinsonism. *Biochimica et Biophysica Acta*, *1792*(7), 714–721. https://doi.org/10.1016/j.bbadis.2008.08.001.

Kazumori, H., Ishihara, S., Rumi, M. A., Ortega-Cava, C. F., Kadowaki, Y., & Kinoshita, Y. (2004). Transforming growth factor-alpha directly augments histidine decarboxylase and vesicular monoamine transporter 2 production in rat enterochromaffin-like cells. *American Journal of Physiology. Gastrointestinal and Liver Physiology*, *286*(3), G508–G514. https://doi.org/10.1152/ajpgi.00269.2003.

Leta, V., Rodríguez-Violante, M., Abundes, A., Rukavina, K., Teo, J. T., Falup-Pecurariu, C., et al. (2021). Parkinson's disease and post-COVID-19 syndrome: The Parkinson's long-COVID Spectrum. *Movement Disorders, 36*(6), 1287–1289. https://doi.org/10.1002/mds.28622.

Lewis, A., Frontera, J., Placantonakis, D. G., Lighter, J., Galetta, S., Balcer, L., et al. (2021). Cerebrospinal fluid in COVID-19: A systematic review of the literature. *Journal of the Neurological Sciences, 421*, 117316. https://doi.org/10.1016/j.jns.2021.117316.

Lippi, A., Domingues, R., Setz, C., Outeiro, T. F., & Krisko, A. (2020). SARS-CoV-2: At the crossroad between aging and neurodegeneration. *Movement Disorders, 35*(5), 716–720. https://doi.org/10.1002/mds.28084.

Luis-Martínez, R., Di Marco, R., Weis, L., Cianci, V., Pistonesi, F., Baba, A., et al. (2021). Impact of social and mobility restrictions in Parkinson's disease during COVID-19 lockdown. *BMC Neurology, 21*(1), 332. https://doi.org/10.1186/s12883-021-02364-9.

Makhoul, K., & Jankovic, J. (2021). Parkinson's disease after COVID-19. *Journal of the Neurological Sciences, 422*, 117331. https://doi.org/10.1016/j.jns.2021.117331.

Méndez-Guerrero, A., Laespada-García, M. I., Gómez-Grande, A., Ruiz-Ortiz, M., Blanco-Palmero, V. A., Azcarate-Diaz, F. J., et al. (2020). Acute hypokinetic-rigid syndrome following SARS-CoV-2 infection. *Neurology, 95*(15), e2109–e2118. https://doi.org/10.1212/wnl.0000000000010282.

Merello, M., Bhatia, K. P., & Obeso, J. A. (2021). SARS-CoV-2 and the risk of Parkinson's disease: Facts and fantasy. *Lancet Neurology, 20*(2), 94–95. https://doi.org/10.1016/s1474-4422(20)30442-7.

Moore, H., Rose, H. J., & Grace, A. A. (2001). Chronic cold stress reduces the spontaneous activity of ventral tegmental dopamine neurons. *Neuropsychopharmacology, 24*(4), 410–419. https://doi.org/10.1016/s0893-133x(00)00188-3.

Morassi, M., Palmerini, F., Nici, S., Magni, E., Savelli, G., Guerra, U. P., et al. (2021). SARS-CoV-2-related encephalitis with prominent parkinsonism: Clinical and FDG-PET correlates in two patients. *Journal of Neurology, 1–8*. https://doi.org/10.1007/s00415-021-10560-3.

Muller, T., & Muhlack, S. (2007). Acute levodopa intake and associated cortisol decrease in patients with Parkinson disease. *Clinical Neuropharmacology, 30*(2), 101–106. https://doi.org/10.1097/01.wnf.0000240954.72186.91.

Nataf, S. (2020). An alteration of the dopamine synthetic pathway is possibly involved in the pathophysiology of COVID-19. *Journal of Medical Virology, 92*(10), 1743–1744. https://doi.org/10.1002/jmv.25826.

Neumann, B., Schmidbauer, M. L., Dimitriadis, K., Otto, S., Knier, B., Niesen, W. D., et al. (2020). Cerebrospinal fluid findings in COVID-19 patients with neurological symptoms. *Journal of the Neurological Sciences, 418*, 117090. https://doi.org/10.1016/j.jns.2020.117090.

Nguyen, N. N., Hoang, V. T., Dao, T. L., Dudouet, P., Eldin, C., & Gautret, P. (2022). Clinical patterns of somatic symptoms in patients suffering from post-acute long COVID: A systematic review. *European Journal of Clinical Microbiology & Infectious Diseases, 1-31*. https://doi.org/10.1007/s10096-022-04417-4.

NICE. (2020). *COVID-19 rapid guideline: Managing the long-term effects of COVID-19*. London: National Institute for Health and Care Excellence (NICE). https://pubmed.ncbi.nlm.nih.gov/33555768/.

Okura, T., Ito, R., Ishiguro, N., Tamai, I., & Deguchi, Y. (2007). Blood-brain barrier transport of pramipexole, a dopamine D2 agonist. *Life Sciences, 80*(17), 1564–1571. https://doi.org/10.1016/j.lfs.2007.01.035.

Ong, T. L., Nor, K. M., Yusoff, Y., & Sapuan, S. (2022). COVID-19 associated acute necrotizing encephalopathy presenting as parkinsonism and Myorhythmia. *Journal of Movement Disorders, 15*(1), 89–92. https://doi.org/10.14802/jmd.21063.

Onofrj, M., Bonanni, L., Cossu, G., Manca, D., Stocchi, F., & Thomas, A. (2009). Emergencies in parkinsonism: Akinetic crisis, life-threatening dyskinesias, and polyneuropathy during L-Dopa gel treatment. *Parkinsonism & Related Disorders*, *15*(Suppl 3), S233–S236. https://doi.org/10.1016/s1353-8020(09)70821-1.

Pilotto, A., Masciocchi, S., Volonghi, I., Crabbio, M., Magni, E., De Giuli, V., et al. (2021). Clinical presentation and outcomes of severe acute respiratory syndrome coronavirus 2-related encephalitis: The ENCOVID multicenter study. *The Journal of Infectious Diseases*, *223*(1), 28–37. https://doi.org/10.1093/infdis/jiaa609.

Pissadaki, E. K., & Bolam, J. P. (2013). The energy cost of action potential propagation in dopamine neurons: Clues to susceptibility in Parkinson's disease. *Frontiers in Computational Neuroscience*, 7, 13. https://doi.org/10.3389/fncom.2013.00013.

Rao, A. R., Hidayathullah, S. M., Hegde, K., & Adhikari, P. (2022). Parkinsonism: An emerging post COVID sequelae. *IDCases*, *27*, e01388. https://doi.org/10.1016/j.idcr.2022.e01388.

Rass, V., Beer, R., Schiefecker, A. J., Kofler, M., Lindner, A., Mahlknecht, P., et al. (2021). Neurological outcome and quality of life 3 months after COVID-19: A prospective observational cohort study. *European Journal of Neurology*, *28*(10), 3348–3359. https://doi.org/10.1111/ene.14803.

Roy, D., Song, J., Awad, N., & Zamudio, P. (2021). Treatment of unexplained coma and hypokinetic-rigid syndrome in a patient with COVID-19. *BML Case Reports*, *14*-(3). https://doi.org/10.1136/bcr-2020-239781.

Ruzicka, F., Jech, R., Novakova, L., Urgosik, D., Bezdicek, O., Vymazal, J., et al. (2015). Chronic stress-like syndrome as a consequence of medial site subthalamic stimulation in Parkinson's disease. *Psychoneuroendocrinology*, *52*, 302–310. https://doi.org/10.1016/j.psyneuen.2014.12.001.

Santos-García, D., Oreiro, M., Pérez, P., Fanjul, G., Paz González, J. M., Feal Painceiras, M. J., et al. (2020). Impact of coronavirus disease 2019 pandemic on Parkinson's disease: A cross-sectional survey of 568 Spanish patients. *Movement Disorders*, *35*(10), 1712–1716. https://doi.org/10.1002/mds.28261.

Siow, I., Lee, K. S., Zhang, J. J. Y., Saffari, S. E., & Ng, A. (2021). Encephalitis as a neurological complication of COVID-19: A systematic review and meta-analysis of incidence, outcomes, and predictors. *European Journal of Neurology*. https://doi.org/10.1111/ene.14913. doi:10.1111/ene.14913.

Sonneville, R., Klein, I., de Broucker, T., & Wolff, M. (2009). Post-infectious encephalitis in adults: Diagnosis and management. *The Journal of Infection*, *58*(5), 321–328. https://doi.org/10.1016/j.jinf.2009.02.011.

Sulzer, D. (2007). Multiple hit hypotheses for dopamine neuron loss in Parkinson's disease. *Trends in Neurosciences*, *30*(5), 244–250. https://doi.org/10.1016/j.tins.2007.03.009.

Tiraboschi, P., Xhani, R., Zerbi, S. M., Corso, A., Martinelli, I., Fusi, L., et al. (2021). Postinfectious neurologic complications in COVID-19: A complex case report. *Journal of Nuclear Medicine*, *62*(8), 1171–1176. https://doi.org/10.2967/jnumed.120.256099.

Tufekci, K. U., Meuwissen, R., Genc, S., & Genc, K. (2012). Inflammation in Parkinson's disease. *Advances in Protein Chemistry and Structural Biology*, *88*, 69–132. https://doi.org/10.1016/b978-0-12-398314-5.00004-0.

Tysnes, O. B., & Storstein, A. (2017). Epidemiology of Parkinson's disease. *Journal of Neural Transmission (Vienna)*, *124*(8), 901–905. https://doi.org/10.1007/s00702-017-1686-y.

Umemura, A., Oeda, T., Tomita, S., Hayashi, R., Kohsaka, M., Park, K., et al. (2014). Delirium and high fever are associated with subacute motor deterioration in Parkinson disease: A nested case-control study. *PLoS One*, *9*(6), e94944. https://doi.org/10.1371/journal.pone.0094944.

van Wamelen, D. J., Wan, Y.-M., Chaudhuri, K. R., & Jenner, P. (2020). Stress and cortisol in Parkinson's disease. *International Review of Neurobiology*, *152*, 131–156. https://doi.org/10.1016/bs.irn.2020.01.005. https://pubmed.ncbi.nlm.nih.gov/32450994/.

Wan, D., Du, T., Hong, W., Chen, L., Que, H., Lu, S., et al. (2021). Neurological complications and infection mechanism of SARS-COV-2. *Signal Transduction and Targeted Therapy*, *6*(1), 406. https://doi.org/10.1038/s41392-021-00818-7.

Williams, A., Branscome, H., Khatkar, P., Mensah, G. A., Al Sharif, S., Pinto, D. O., et al. (2021). A comprehensive review of COVID-19 biology, diagnostics, therapeutics, and disease impacting the central nervous system. *Journal of Neurovirology*, *27*(5), 667–690. https://doi.org/10.1007/s13365-021-00998-6.

Wood, H. (2020). New insights into the neurological effects of COVID-19. *Nature Reviews. Neurology*, *16*(8), 403. https://doi.org/10.1038/s41582-020-0386-7.

Worldmeter.info. (2021). *COVID-19 coronavirus pandemic*. Retrieved from https://www.worldometers.info/coronavirus/. 24/10/2021.

Wu, X., Wu, W., Pan, W., Wu, L., Liu, K., & Zhang, H.-L. (2015). Acute necrotizing encephalopathy: An underrecognized clinicoradiologic disorder. *Mediators of Inflammation*, *2015*, 792578792578–792587. https://doi.org/10.1155/2015/792578.

Xing, F., Marsili, L., & Truong, D. D. (2022). Parkinsonism in viral, paraneoplastic, and autoimmune diseases. *Journal of the Neurological Sciences*, *433*, 120014. https://doi.org/10.1016/j.jns.2021.120014.

CHAPTER FIVE

Smell deficits in COVID-19 and possible links with Parkinson's disease

Aron Emmi[a,b,c], Michele Sandre[b,c], Andrea Porzionato[a,c], and Angelo Antonini[b,c,]*

[a]Institute of Human Anatomy, Department of Neuroscience, University of Padova, Padova, Italy
[b]Parkinson and Movement Disorders Unit, Department of Neuroscience, Centre for Rare Neurological Diseases (ERN-RND), University of Padova, Padova, Italy
[c]Center for Neurodegenerative Disease Research (CESNE), University of Padova, Padova, Italy
*Corresponding author: e-mail address: angelo.antonini@unipd.it

Contents

Abstract

Olfactory impairment is a common symptom in Coronavirus Disease 2019 (COVID-19), the disease caused by Severe Acute Respiratory Syndrome—Coronavirus 2 (SARS-CoV-2) infection. While other viruses, such as influenza viruses, may affect the ability to smell, loss of olfactory function is often smoother and associated to various degrees of nasal symptoms. In COVID-19, smell loss may appear also in absence of other symptoms, frequently with a sudden onset. However, despite great clinical interest in COVID-19 olfactory alterations, very little is known concerning the mechanisms underlying these phenomena. Moreover, olfactory dysfunction is observed in neurological conditions like Parkinson's disease (PD) and can precede motor onset by many years, suggesting that viral infections, like COVID-19, and regional inflammatory responses may trigger defective protein aggregation and subsequent neurodegeneration, potentially linking COVID-19 olfactory impairment to neurodegeneration. In the following chapter, we report the neurobiological and neuropathological underpinnings of olfactory impairments encountered in COVID-19 and discuss the implications of these findings in the context of neurodegenerative disorders, with particular regard to PD and alpha-synuclein pathology.

International Review of Neurobiology, Volume 165
ISSN 0074-7742
https://doi.org/10.1016/bs.irn.2022.08.001
Copyright © 2022 Elsevier Inc.
All rights reserved.

1. Introduction

Coronavirus Disease 2019 (COVID-19), the disease caused by Severe Acute Respiratory Syndrome—Coronavirus 2 (SARS-CoV-2), is often associated with a wide spectrum of neurological manifestations (Ellul et al., 2020; Helms et al., 2020; Huang et al., 2021; Iadecola, Anrather, & Kamel, 2020; Mao et al., 2020). Intriguingly, the COVID-19 pandemic brought to the attention of clinicians an astonishing number of patients affected by mild symptoms, the most frequent being a sudden and complete loss of olfaction, which may last for a variable period of time (Guerrero et al., 2021).

Often wrongfully regarded as an ancillary sense, olfaction covers a great role in our everyday life for its involvement in lifeguarding processes (e.g., avoiding dangerous chemicals), as well as in food evaluation. As the pleasantness of gustatory and olfactory stimuli represents an important contributor to patients' quality of life, olfactory impairment is also frequently associated to anhedonia, lack of motivation and depressive symptoms in patients with COVID-19 (Athanassi, Dorado Doncel, Bath, & Mandairon, 2021; Voruz et al., 2022).

Aside of the functional impairments deriving from olfactory dysfunction, the olfactory system directly links the brain to the external environment, representing a gateway through which many viruses, such as herpesvirus-1 and 6, may access the central nervous system (CNS) (Duarte et al., 2019; Harberts et al., 2011). Also, the olfactory system is involved in the pathogenesis of different neurodegenerative diseases, like Parkinson's (PD) and Alzheimer's disease (Sulzer, 2007), due to its connections to the entorhinal cortex and limbic system, suggesting special care for possible long-term consequences of viral infections. In this context, olfactory transmucosal invasion of SARS-CoV-2, and the inflammatory correlates of COVID-19 at the level of the olfactory system, must be carefully considered, as they may represent possible factors for the establishment, or the precipitation, of protein aggregation and neurodegenerative phenomena.

2. Clinical features of COVID-19 olfactory impairment

2.1 Individual variables influencing COVID-19 olfactory impairment

During the first waves of the COVID-19 pandemic, olfactory impairment was regarded as a common clinical feature experienced by numerous patients. Early studies correlated higher propensity for acute olfactory loss with a more indolent course, but subsequent work suggested elevated prevalence of smell loss across most COVID-19 cases, regardless of severity (Brann et al., 2020). Hornuss et al. (2020) evidenced that olfactory alterations often occur unnoticed in COVID-19 patients, and that they do not represent a predictor of severe COVID-19 manifestations. Moreover, a substantial proportion of patients with previous mild-to-moderate symptomatic COVID-19 characterized by new onset of chemosensory dysfunction, still complained on altered sense of smell or taste 1 year after the onset (Brann et al., 2020). These long-term effects of COVID-19 on chemosensory functions have been investigated recently by Shelton et al. (2022), evidencing that only a restricted cluster of patients experience long-term smell loss. The authors discovered a genetic mutation in COVID-19 patients that was associated with a greater propensity for smell or taste loss. The mutation was found in two overlapping genes, called UGT2A1 and UGT2A2, encoding proteins that remove odor molecules from the nostrils after they have been detected. However, interactions between SARS-CoV-2 and the aforementioned genes remain to be clarified.

Among the main issues concerning the evaluation of individual olfactory outcomes in COVID-19, objective testing of olfactory function represents a major aspect of concern. In the case of chemical senses, two main objective tests have been developed and used over the years in the clinic, namely the Sniffin' Sticks (Hummel et al., 2009) and the University of Pennsylvania Smell Inventory Test (UPSIT) (Doty et al., 2014), while others are being developed. Their use is mandatory in order to have an objective evaluation of the level of impairment, since the subjective report is often misleading. However, they both require the direct testing of the patient, which may be difficult or impossible in the case of COVID-19 infected patients. Conversely, collecting data from patients may be of paramount importance to follow the disease, even though the utility of self-reporting about chemical sensitivity has been repeatedly questioned (Hummel et al., 2009).

2.2 SARS-CoV-2 variants and their relationship with olfactory impairment

Intriguingly, aside from intra-individual differences underlying more or less severe olfactory impairment in COVID-19, SARS-CoV-2 variants may as well affect olfactory outcome in the general populations. Coelho, Reiter, French, and Costanzo (2022) surveyed 616,318 people in the United States who had COVID-19. The authors found that, compared to those who had been infected with the original virus, people who had contracted the Alpha variant—the first variant of concern to arise—were 50% as likely to have chemosensory disruption. This probability fell to 44% for the later Delta variant, and to 17% for the latest variant, Omicron. While this clinical data appears to be in line with a less severe clinical outcome associated with newer variants, no studies to date have examined the impact of different SARS-CoV-2 variants on the olfactory mucosa, either in animal models, or in the neuropathological setting. Thus, the pathophysiology of olfactory impairment in new SARS-CoV-2 variants, such as the recently circulating Omicron variant, remain to be investigated.

3. Neuropathology of COVID-19 olfactory impairment

Other coronaviruses, such as Severe Acute Respiratory Syndrome Coronavirus 1 (SARS-CoV-1), are known to enter the brain of transgenic mice via the olfactory bulb, leading to neuronal necrosis in the absence of encephalitis (Netland, Meyerholz, Moore, Cassell, & Perlman, 2008). Similarly, high levels of viral RNA were found in the brainstem and thalamus of transgenic mice following intranasal inoculation of the Middle-Eastern Respiratory Syndrome Coronavirus (MERS-CoV) (Li et al., 2016), suggesting for a common neuroinvasive potential.

As far as SARS-CoV-2 is concerned, most studies agree on the extensive inflammatory processes occurring in the olfactory mucosa and bulb of COVID-19 patients (Emmi et al., 2021; Matschke et al., 2020; Meinhardt et al., 2021; Schwabenland et al., 2021; Thakur et al., 2021), but often conflicting findings are reported concerning SARS-CoV-2 neurotropism. Douaud et al. (2022) investigated structural brain changes before and after a SARS-CoV-2 infection in a large sample of UK Biobank participants, revealing significant reductions in gray matter thickness and tissue-contrast in the orbitofrontal cortex and parahippocampal gyrus, along with prominent alterations in brain areas functionally connected to the primary olfactory cortex. Positron emission tomography (PET) hypometabolism in long COVID patients encompassing the olfactory gyrus and the connected limbic/paralimbic regions, extended to the brainstem and the cerebellum (Guedj et al., 2021), and in the right parahippocampal gyrus and the right thalamus (Sollini et al., 2021) was also reported.

From a neuropathological perspective, several studies identified SARS-CoV-2 viral antigens and genomic sequences through RT-PCR, immunohistochemistry, in-situ hybridization and even electron microscopy in the human olfactory system (Khan et al., 2021; Meinhardt et al., 2021; Zazhytska et al., 2022). Meinhardt et al. (2021) detected SARS-CoV-2 Spike protein in primary olfactory neurons of the olfactory mucosa in a large sample of COVID-19 patients, suggesting an olfactory-transmucosal route of infection throughout the CNS, as testified also by the detection of viral RNA at the level of the olfactory bulb and the medulla oblongata.

Conversely, two recent studies found viral genomic sequences and antigens in sustentacular cells of the olfactory epithelium, but not in olfactory neurons (Khan et al., 2021; Zazhytska et al., 2022). This was associated with the reorganization of nuclear architecture and downregulation of olfactory receptors, as well as their signaling pathways, in neuronal cells of the olfactory mucosa, hinting towards a non-cell autonomous cause of anosmia (Zazhytska et al., 2022). Two studies of single nucleus RNA sequencing of brains of COVID-19 patients, focusing on the olfactory system and the choroid plexus, detected broad perturbations, with upregulation of genes involved in innate antiviral response and inflammation, microglial activation and neurodegeneration, but found no direct evidence of viral RNA in the tissue (Fullard et al., 2021; Yang et al., 2021). Similarly, authors have not detected viral proteins/ RNA through immunohistochemistry or in-situ hybridization, even though viral genomic sequences were found via RT-PCR assays (Lee et al., 2021; Solomon et al., 2020; Thakur et al., 2021). In concordance to these findings, animal model studies suggest that loss of smell is related to damage to the cilia and olfactory epithelium, but not infection of the olfactory neurons. For example, in an experiment where hamsters were nasally infected with SARS-CoV-2, the olfactory epithelium and cilia became very damaged, leading to anosmia, but no infection was observed in the olfactory neurons (Bryche et al., 2020). Furthermore, Brann et al. (2020) demonstrated that mouse, non-human primate and human olfactory mucosa express two key genes involved in SARS-CoV-2 entry, ACE2 and TMPRSS2. However, single cell sequencing revealed that ACE2 is expressed in support cells, stem cells, and perivascular cells, rather than in neurons.

Concerning neuroinflammation occurring in the olfactory system of COVID-19 patients, Schwabenland et al. (2021) performed deep spatial profiling of the local immune response through imaging mass spectrometry, revealing significant immune activation in the medulla oblongata and in the olfactory bulb, with a prominent role mediated by CD8 + T-cell—microglia crosstalk in the parenchyma. Similarly, other authors (Emmi et al., 2021; Matschke et al., 2020; Solomon et al., 2020) detected prominent astrogliosis, microglial activation and microglial nodules in the brainstem and olfactory structures of COVID-19 subjects, as seen in Fig. 1. Hence, even though the detection of SARS-CoV-2 in olfactory neuronal cells has not been consistently reproduced throughout studies, most findings support marked neuroinflammation of the olfactory system in COVID-19. Combined with recent studies indicating indirect downregulation of olfactory receptor pathways mediated by sustentacular cell infection, it appears that olfactory impairment in COVID-19 may not always be associated to direct viral invasion of olfactory neuronal cells, with subsequent olfactory-transmucosal spread of the virus throughout the CNS but can also be mediated by sustentacular cell infection and dysregulation of olfactory receptor pathways.

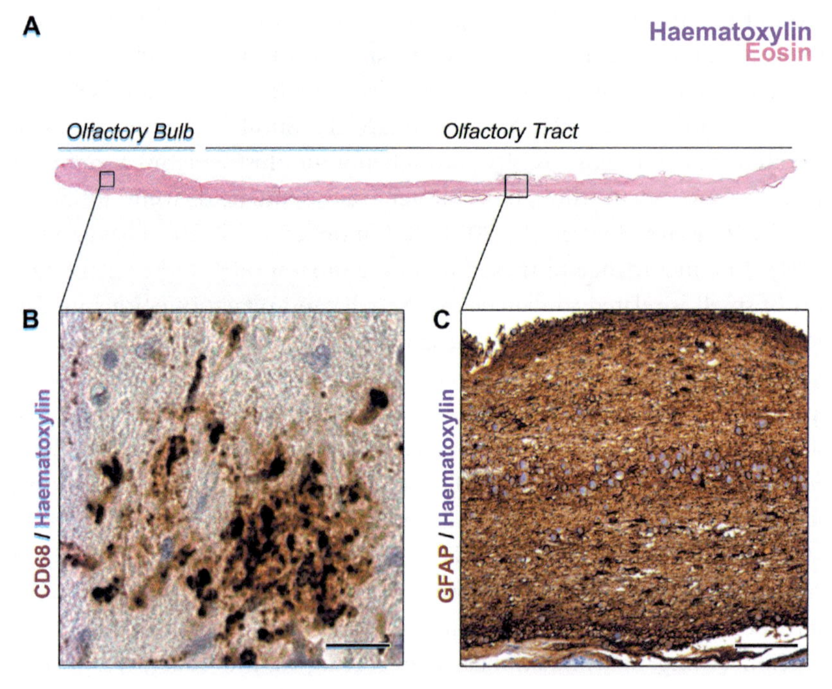

Fig. 1 (A) Olfactory bulb and tract of a person who died due to COVID-19, Hematoxylin & Eosin staining. (B) Microglial nodule in the olfactory bulb, as demonstrated by CD68 immunoperoxidase staining. (C) Prominent astrogliosis of the olfactory tract with numerous corpora amylacea representing a non-specific finding in COVID-19, often difficult to disentangle from patient comorbidities. GFAP immunoperoxidase staining. This material has been obtained from an autopsy performed at the Institute of Anatomy of the University of Padua, Italy.

4. Olfactory impairment in COVID-19 and parkinsonism

Despite being counted as a respiratory virus, SARS-CoV-2 is associated to frequent and sometimes severe neurological manifestations, posing important interrogatives on the short and long-term effects of SARS-CoV-2 infection on the CNS. Particularly concerning is the hypothesis that SARS-CoV-2 infection of the CNS may predispose, or quickly precipitate, the development of PD (Beauchamp, Finkelstein, Bush, Evans, & Barnham, 2020; Bouali-Benazzouz & Benazzouz, 2021; Brundin, Nath, & Beckham, 2020). This hypothesis arises both from the 1917 Spanish flu and von Economo's encephalitis lethargica pandemics, which have seen a surge of post-encephalitic parkinsonism following the waves of the pandemic, and the known association between viral infection and the development of transient or permanent movement disorders (Jang, Boltz, Webster, & Smeyne, 2009). In fact, pathogens, and in particular respiratory viruses, have been suggested as a potential etiopathogenic factor for PD, leading to parkinsonism in subjects over the age of 50, regardless of genetic substrate (Beauchamp et al., 2020; Tanner et al., 1999). The well-established neuropathological staging of PD, suggested by Braak & Braak, also appears to account for neurotropic pathogens that may infect the CNS either through the vagus nerve (pneumo-gastric pathway) or by accessing the brain through the olfactory systems. Both the olfactory bulb and tract, as well as the medulla oblongata where the vagal nuclei are located, represent the very first sites of early PD neuropathology, and interestingly also represent the main sites of inflammation/ infection encountered in COVID-19 (Hawkes, Del Tredici, & Braak, 2007; Klingelhoefer & Reichmann, 2015). Furthermore, viral-related inflammation might render the CNS susceptible to preceding or subsequent stressors (Sulzer, 2007), even in the absence of direct viral invasion; indeed, past history of infection was associated with a 20% higher risk of presenting PD in the future (Meng, Shen, & Ji, 2019). In the case of coronaviruses (CoV), higher antibodies titers against common CoV have been detected in the cerebrospinal fluid (CSF) of people with PD (PwP) when compared to controls, while there was evidence of post-encephalitic parkinsonism in mice infected with a CoV strain (MHV-A59) (Fazzini, Fleming, & Fahn, 1992; Fishman et al., 1985).

In a recent study, Semerdzhiev, Fakhree, Segers-Nolten, Blum, and Claessens (2022) demonstrated that in the presence of SARS-CoV-2 Nucleocapsid protein, the onset of alpha-synuclein aggregation into amyloid fibrils is strongly accelerated, indicating that N-protein facilitates the formation of a critical nucleus for aggregation. Fibril formation does not only appear to accelerate, but also proceeds in an unusual two-step process. In cells, the presence of Nucleocapsid protein changes the distribution of alpha-synuclein over different conformations, which likely represent different functions at already short time scales. Similarly, Charnley et al. (2022) identified two peptides from the SARS-CoV-2 proteome that self-assemble into amyloid assemblies. These amyloids were shown to be highly toxic to neuronal cells and are hypothesized to trigger neurological symptoms in COVID-19. The cytotoxicity and protease-resistant structure of these assemblies may result in their persistent presence in the CNS of patients, even following infection, and could partially explain the lasting neurological symptoms of COVID-19, especially those that are novel in relation to other post-viral syndromes, such as those following the original SARS-CoV-1. The outlook in relation to triggering of progressive neurodegenerative disease remains uncertain. Given the typically slow progress of neurodegenerative disease, if such a phenomenon exists, it will most probably take some time to become evident epidemiologically. In an animal model study performed by Käufer et al. (2022), microglial activation and neuronal proteinopathy persisted even beyond viral clearance. Viral protein exposure in the nasal cavity led to pronounced microglia activation in the olfactory bulb beyond viral clearance. Cortical, but not hippocampal, neurons accumulated hyperphosphorylated tau and alpha-synuclein, in the absence of overt inflammation and neurodegeneration. Importantly, not all brain regions were affected, in line with the selective vulnerability hypothesis. In this animal model, despite the absence of virus in brain, neurons developed signatures of proteinopathies that may contribute to progressive neuronal dysfunction.

This is further confirmed by available neuropathological studies on COVID-19 decedents, showing very little and often conflicting evidence on SARS-CoV-2 neurotropism. On the other hand, all available neuropathological studies are carried on subjects who died during the acute phases of the disease (Emmi et al., 2021; Matschke et al., 2020; Schwabenland et al., 2021; Solomon et al., 2020), and, as such, the neuropathological alterations occurring in cases of chronic infection or in the post-infection timeframe are yet unknown. Neurodegenerative changes induced by either direct infection or by the indirect effects mediated by neuroinflammation may not be appreciable or detectable in patients who died shortly after infection, and as most subjects in neuropathological studies already present important neurological comorbidities, it is not possible to determine which markers of neurodegeneration are directly related to COVID-19, and which are related to patient morbidity.

References

Athanassi, A., Dorado Doncel, R., Bath, K. G., & Mandairon, N. (2021). Relationship between depression and olfactory sensory function: A review. *Chemical Senses, 46*. https://doi.org/10.1093/chemse/bjab044.

Beauchamp, L. C., Finkelstein, D. I., Bush, A. I., Evans, A. H., & Barnham, K. J. (2020). Parkinsonism as a third wave of the COVID-19 pandemic? *Journal of Parkinson's Disease, 10*(4), 1343–1353. https://doi.org/10.3233/JPD-202211.

Bouali-Benazzouz, R., & Benazzouz, A. (2021). Covid-19 infection and parkinsonism: Is there a link? *Movement Disorders, 36*(8), 1737–1743. https://doi.org/10.1002/mds.28680.

Brann, D. H., Tsukahara, T., Weinreb, C., Lipovsek, M., Van den Berge, K., Gong, B., et al. (2020). Non-neuronal expression of SARS-CoV-2 entry genes in the olfactory system suggests mechanisms underlying COVID-19-associated anosmia. *Science Advances, 6*(31). https://doi.org/10.1126/sciadv.abc5801.

Brundin, P., Nath, A., & Beckham, J. D. (2020). Is COVID-19 a perfect storm for Parkinson's disease? *Trends in Neurosciences, 43*(12), 931–933. https://doi.org/10.1016/j.tins.2020.10.009.

Bryche, B., St Albin, A., Murri, S., Lacôte, S., Pulido, C., Ar Gouilh, M., et al. (2020). Massive transient damage of the olfactory epithelium associated with infection of sustentacular cells by SARS-CoV-2 in golden Syrian hamsters. *Brain, Behavior, and Immunity, 89*, 579–586. https://doi.org/10.1016/j.bbi.2020.06.032.

Charnley, M., Islam, S., Bindra, G. K., Engwirda, J., Ratcliffe, J., Zhou, J., et al. (2022). Neurotoxic amyloidogenic peptides in the proteome of SARS-COV2: Potential implications for neurological symptoms in COVID-19. *Nature Communications, 13*(1), 3387. https://doi.org/10.1038/s41467-022-30932-1.

Coelho, D. H., Reiter, E. R., French, E., & Costanzo, R. M. (2022). Decreasing incidence of chemosensory changes by COVID-19 variant. *Otolaryngology and Head and Neck Surgery*. https://doi.org/10.1177/01945998221097656. 1945998221097656.

Doty, R. L., Bayona, E. A., Leon-Ariza, D. S., Cuadros, J., Chung, I., Vazquez, B., et al. (2014). The lateralized smell test for detecting Alzheimer's disease: Failure to replicate. *Journal of the Neurological Sciences, 340*(1–2), 170–173. https://doi.org/10.1016/j.jns.2014.03.022.

Douaud, G., Lee, S., Alfaro-Almagro, F., Arthofer, C., Wang, C., McCarthy, P., et al. (2022). SARS-CoV-2 is associated with changes in brain structure in UK biobank. *Nature, 604*(7907), 697–707. https://doi.org/10.1038/s41586-022-04569-5.

Duarte, L. F., Farías, M. A., Álvarez, D. M., Bueno, S. M., Riedel, C. A., & González, P. A. (2019). Herpes simplex virus type 1 infection of the central nervous system: Insights into proposed interrelationships with neurodegenerative disorders. *Frontiers in Cellular Neuroscience, 13*. https://doi.org/10.3389/fncel.2019.00046.

Ellul, M. A., Benjamin, L., Singh, B., Lant, S., Michael, B. D., Easton, A., et al. (2020). Neurological associations of COVID-19. *Lancet Neurology, 19*(9), 767–783. https://doi.org/10.1016/s1474-4422(20)30221-0.

Emmi, A., Rizzo, S. M. R., Barzon, L., Carturan, E., Sinigaglia, A., Riccetti, S., et al. (2021). *COVID-19 neuropathology: Evidence for SARS-CoV-2 invasion of human brainstem nuclei*. https://doi.org/10.1101/2022.06.29.498117.

Fazzini, E., Fleming, J., & Fahn, S. (1992). Cerebrospinal fluid antibodies to coronavirus in patients with Parkinson's disease. *Movement Disorders*, 7(2), 153–158. https://doi.org/10.1002/mds.870070210.

Fishman, P. S., Gass, J. S., Swoveland, P. T., Lavi, E., Highkin, M. K., & Weiss, S. R. (1985). Infection of the basal ganglia by a murine coronavirus. *Science (New York, N.Y.)*, *229*(4716), 877–879. https://doi.org/10.1126/science.2992088.

Fullard, J. F., Lee, H.-C., Voloudakis, G., Suo, S., Javidfar, B., Shao, Z., et al. (2021). Single-nucleus transcriptome analysis of human brain immune response in patients with severe COVID-19. *Genome Medicine*, *13*(1), 118. https://doi.org/10.1186/s13073-021-00933-8.

Guedj, E., Campion, J. Y., Dudouet, P., Kaphan, E., Bregeon, F., Tissot-Dupont, H., et al. (2021). (18)F-FDG brain PET hypometabolism in patients with long COVID. *European Journal of Nuclear Medicine and Molecular Imaging*, *48*(9), 2823–2833. https://doi.org/10.1007/s00259-021-05215-4.

Guerrero, J. I., Barragán, L. A., Martínez, J. D., Montoya, J. P., Peña, A., Sobrino, F. E., et al. (2021). Central and peripheral nervous system involvement by COVID-19: A systematic review of the pathophysiology, clinical manifestations, neuropathology, neuroimaging, electrophysiology, and cerebrospinal fluid findings. *BMC Infectious Diseases*, *21*(1), 515. https://doi.org/10.1186/s12879-021-06185-6.

Harberts, E., Yao, K., Wohler, J. E., Maric, D., Ohayon, J., Henkin, R., et al. (2011). Human herpesvirus-6 entry into the central nervous system through the olfactory pathway. *Proceedings of the National Academy of Sciences*, *108*(33), 13734–13739. https://doi.org/10.1073/pnas.1105143108.

Hawkes, C. H., Del Tredici, K., & Braak, H. (2007). Parkinson's disease: A dual-hit hypothesis. *Neuropathology and Applied Neurobiology*, *33*(6), 599–614. https://doi.org/10.1111/j.1365-2990.2007.00874.x.

Helms, J., Kremer, S., Merdji, H., Clere-Jehl, R., Schenck, M., Kummerlen, C., et al. (2020). Neurologic features in severe SARS-CoV-2 infection. *The New England Journal of Medicine*, *382*(23), 2268–2270. https://doi.org/10.1056/NEJMc2008597.

Hornuss, D., Lange, B., Schröter, N., Rieg, S., Kern, W. V., & Wagner, D. (2020). Anosmia in COVID-19 patients. *Clinical Microbiology and Infection*, *26*(10), 1426–1427. https://doi.org/10.1016/j.cmi.2020.05.017.

Huang, C., Huang, L., Wang, Y., Li, X., Ren, L., Gu, X., et al. (2021). 6-month consequences of COVID-19 in patients discharged from hospital: A cohort study. *Lancet*, *397*(10270), 220–232. https://doi.org/10.1016/s0140-6736(20)32656-8.

Hummel, T., Rissom, K., Reden, J., Hähner, A., Weidenbecher, M., & Hüttenbrink, K.-B. (2009). Effects of olfactory training in patients with olfactory loss. *The Laryngoscope*, *119*(3), 496–499. https://doi.org/10.1002/lary.20101.

Iadecola, C., Anrather, J., & Kamel, H. (2020). Effects of COVID-19 on the nervous system. *Cell*, *183*(1), 16–27.e11. https://doi.org/10.1016/j.cell.2020.08.028.

Jang, H., Boltz, D. A., Webster, R. G., & Smeyne, R. J. (2009). Viral parkinsonism. *Biochimica et Biophysica Acta*, *1792*(7), 714–721. https://doi.org/10.1016/j.bbadis.2008.08.001.

Käufer, C., Schreiber, C. S., Hartke, A.-S., Denden, I., Stanelle-Bertram, S., Beck, S., et al. (2022). Microgliosis and neuronal proteinopathy in brain persist beyond viral clearance in SARS-CoV-2 hamster model. *eBioMedicine*, *79*, 103999. https://doi.org/10.1016/j.ebiom.2022.103999.

Khan, M., Yoo, S.-J., Clijsters, M., Backaert, W., Vanstapel, A., Speleman, K., et al. (2021). Visualizing in deceased COVID-19 patients how SARS-CoV-2 attacks the respiratory and olfactory mucosae but spares the olfactory bulb. *Cell*, *184*(24), 5932–5949.e5915. https://doi.org/10.1016/j.cell.2021.10.027.

Klingelhoefer, L., & Reichmann, H. (2015). Pathogenesis of Parkinson disease- -the gut-brain axis and environmental factors. *Nature Reviews. Neurology*, *11*(11), 625–636. https://doi.org/10.1038/nrneurol.2015.197.

Lee, M. H., Perl, D. P., Nair, G., Li, W., Maric, D., Murray, H., et al. (2021). Microvascular injury in the brains of patients with Covid-19. *The New England Journal of Medicine*, *384*(5), 481–483. https://doi.org/10.1056/NEJMc2033369.

Li, K., Wohlford-Lenane, C., Perlman, S., Zhao, J., Jewell, A. K., Reznikov, L. R., et al. (2016). Middle East respiratory syndrome coronavirus causes multiple organ damage and lethal disease in mice transgenic for human dipeptidyl peptidase 4. *The Journal of Infectious Diseases*, *213*(5), 712–722. https://doi.org/10.1093/infdis/jiv499.

Mao, L., Jin, H., Wang, M., Hu, Y., Chen, S., He, Q., et al. (2020). Neurologic manifestations of hospitalized patients with coronavirus disease 2019 in Wuhan, China. *JAMA Neurol*, 77(6), 683–690. https://doi.org/10.1001/jamaneurol.2020.1127.

Matschke, J., Lütgehetmann, M., Hagel, C., Sperhake, J. P., Schröder, A. S., Edler, C., et al. (2020). Neuropathology of patients with COVID-19 in Germany: A post-mortem case series. *The Lancet Neurology*, *19*(11), 919–929. https://doi.org/10.1016/S1474-4422(20)30308-2.

Meinhardt, J., Radke, J., Dittmayer, C., Franz, J., Thomas, C., Mothes, R., et al. (2021). Olfactory transmucosal SARS-CoV-2 invasion as a port of central nervous system entry in individuals with COVID-19. *Nature Neuroscience*, *24*(2), 168–175. https://doi.org/10.1038/s41593-020-00758-5.

Meng, L., Shen, L., & Ji, H. F. (2019). Impact of infection on risk of Parkinson's disease: A quantitative assessment of case-control and cohort studies. *Journal of Neurovirology*, *25*(2), 221–228. https://doi.org/10.1007/s13365-018-0707-4.

Netland, J., Meyerholz, D. K., Moore, S., Cassell, M., & Perlman, S. (2008). Severe acute respiratory syndrome coronavirus infection causes neuronal death in the absence of encephalitis in mice transgenic for human ACE2. *Journal of Virology*, *82*(15), 7264–7275. https://doi.org/10.1128/jvi.00737-08.

Schwabenland, M., Salié, H., Tanevski, J., Killmer, S., Lago, M. S., Schlaak, A. E., et al. (2021). Deep spatial profiling of human COVID-19 brains reveals neuroinflammation with distinct microanatomical microglia-T-cell interactions. *Immunity*, *54*(7), 1594–1610.e1511. https://doi.org/10.1016/j.immuni.2021.06.002.

Semerdzhiev, S. A., Fakhree, M. A. A., Segers-Nolten, I., Blum, C., & Claessens, M. M. A. E. (2022). Interactions between SARS-CoV-2 N-protein and α-synuclein accelerate amyloid formation. *ACS Chemical Neuroscience*, *13*(1), 143–150. https://doi.org/10.1021/acschemneuro.1c00666.

Shelton, J. F., Shastri, A. J., Fletez-Brant, K., Auton, A., Chubb, A., Fitch, A., et al. (2022). The UGT2A1/UGT2A2 locus is associated with COVID-19-related loss of smell or taste. *Nature Genetics*, *54*(2), 121–124. https://doi.org/10.1038/s41588-021-00986-w.

Sollini, M., Morbelli, S., Ciccarelli, M., Cecconi, M., Aghemo, A., Morelli, P., et al. (2021). Long COVID hallmarks on [18F]FDG-PET/CT: A case-control study. *European Journal of Nuclear Medicine and Molecular Imaging*, *48*(10), 3187–3197. https://doi.org/10.1007/s00259-021-05294-3.

Solomon, I. H., Normandin, E., Bhattacharyya, S., Mukerji, S. S., Keller, K., Ali, A. S., et al. (2020). Neuropathological features of Covid-19. *The New England Journal of Medicine*, *383*(10), 989–992. https://doi.org/10.1056/NEJMc2019373.

Sulzer, D. (2007). Multiple hit hypotheses for dopamine neuron loss in Parkinson's disease. *Trends in Neurosciences*, *30*(5), 244–250.

Tanner, C. M., Ottman, R., Goldman, S. M., Ellenberg, J., Chan, P., Mayeux, R., et al. (1999). Parkinson disease in twins: An etiologic study. *JAMA*, *281*(4), 341–346. https://doi.org/10.1001/jama.281.4.341.

Thakur, B., Dubey, P., Benitez, J., Torres, J. P., Reddy, S., Shokar, N., et al. (2021). A systematic review and meta-analysis of geographic differences in comorbidities and associated severity and mortality among individuals with COVID-19. *Scientific Reports*, *11*(1), 8562. https://doi.org/10.1038/s41598-021-88130-w.

Thakur, K. T., Miller, E. H., Glendinning, M. D., Al-Dalahmah, O., Banu, M. A., Boehme, A. K., et al. (2021). COVID-19 neuropathology at Columbia University Irving medical center/New York presbyterian hospital. *Brain*, *144*(9), 2696–2708. https://doi.org/10.1093/brain/awab148.

Voruz, P., Allali, G., Benzakour, L., Nuber-Champier, A., Thomasson, M., Jacot de Alcântara, I., et al. (2022). Long COVID neuropsychological deficits after severe, moderate, or mild infection. *Clinical and Translational Neuroscience*, *6*(2), 9. Retrieved from https://www.mdpi.com/2514-183X/6/2/9.

Yang, A. C., Kern, F., Losada, P. M., Agam, M. R., Maat, C. A., Schmartz, G. P., et al. (2021). Dysregulation of brain and choroid plexus cell types in severe COVID-19. *Nature*, *595*(7868), 565–571. https://doi.org/10.1038/s41586-021-03710-0.

Zazhytska, M., Kodra, A., Hoagland, D. A., Frere, J., Fullard, J. F., Shayya, H., et al. (2022). Non-cell-autonomous disruption of nuclear architecture as a potential cause of COVID-19-induced anosmia. *Cell*, *185*(6), 1052–1064.e1012. https://doi.org/10.1016/j.cell.2022.01.024.

CHAPTER SIX

Spotlight on non-motor symptoms and Covid-19

Silvia Rota[a,b,†], Iro Boura[a,b,c,†], Yi-Min Wan[a,b,d], Claudia Lazcano-Ocampo[b,e,f], Mayela Rodriguez-Violante[g], Angelo Antonini[h], and Kallol Ray Chaudhuri[a,b,*]

[a]Department of Basic and Clinical Neuroscience, Institute of Psychiatry, Psychology & Neuroscience, King's College London, London, United Kingdom
[b]Parkinson's Foundation Centre of Excellence, King's College Hospital NHS Foundation Trust, London, United Kingdom
[c]Medical School, University of Crete, Heraklion, Crete, Greece
[d]Department of Psychiatry, Ng Teng Fong General Hospital, Singapore, Singapore
[e]Department of Neurology, Movement Disorders Unit, Hospital Sotero del Rio, Santiago, Chile
[f]Department of Neurology, Clínica INDISA, Santiago, Chile
[g]Instituto Nacional de Neurologia y Neurocirugia, Ciudad de México, Mexico
[h]Parkinson and Movement Disorders Unit, Department of Neuroscience, Centre for Rare Neurological Diseases (ERN-RND), University of Padova, Padova, Italy
*Corresponding author: e-mail address: ray.chaudhuri@kcl.ac.uk

Contents

Abstract

The Coronavirus Disease 2019 (Covid-19) pandemic has profoundly affected the quality of life (QoL) and health of the general population globally over the past 2 years, with a clear impact on people with Parkinson's Disease (PwP, PD). Non-motor symptoms have

† These authors contributed equally to this work.

International Review of Neurobiology, Volume 165
ISSN 0074-7742
https://doi.org/10.1016/bs.irn.2022.04.001
Copyright © 2022 Elsevier Inc.
All rights reserved.

been widely acknowledged to hold a vital part in the clinical spectrum of PD, and, although often underrecognized, they significantly contribute to patients' and their caregivers' QoL. Up to now, there have been numerous reports of newly emerging or acutely deteriorating non-motor symptoms in PwP who had been infected by the Severe Acute Respiratory Syndrome Coronavirus-2 (SARS-CoV-2), while some of these symptoms, like fatigue, pain, depression, anxiety and cognitive impairment, have also been identified as part of the long-COVID syndrome due to their persistent nature. The subjacent mechanisms, mediating the appearance or progression of non-motor symptoms in the context of Covid-19, although probably multifactorial in origin, remain largely unknown. Such mechanisms might be, at least partly, related solely to the viral infection per se or the lifestyle changes imposed during the pandemic, as many of the non-motor symptoms seem to be prevalent even among Covid-19 patients without PD. Here, we summarize the available evidence and implications of Covid-19 in non-motor PD symptoms in the acute and chronic, if applicable, phase of the infection, with a special reference on studies of PwP.

1. Introduction

The Coronavirus Disease 2019 (Covid-19), caused by the Severe Acute Respiratory Syndrome Coronavirus-2 (SARS-CoV-2), has profoundly affected the global morbidity and mortality during the past 2 years with the survival of older individuals and those with chronic conditions, being particularly vulnerable (Abate, Checkol, & Mantefardo, 2021; Cascella, Rajnik, Aleem, Dulebohn, & Di Napoli, 2021). Researchers have been increasingly exploring the interaction of SARS-CoV-2 with pre-existing conditions, including Parkinson's Disease (PD), as numerous studies have revealed newly acquired or a worsening of pre-existent symptoms among people with PD (PwP) after being diagnosed with Covid-19. It is now widely acknowledged that non-motor symptoms, although often underrecognized, constitute a vital part in the multifaceted clinical spectrum of PD, significantly contributing to patients' and caregivers' disability and quality of life (QoL) (Chaudhuri, Healy, & Schapira, 2006; Martinez-Martin, Rodriguez-Blazquez, Kurtis, & Chaudhuri, 2011). Many of these symptoms, including constipation, hyposmia, rapid eye movement (REM) behavior disorder (RBD), fatigue and depression, might even precede the emergence of motor symptoms in PD by decades (Chaudhuri, Yates, & Martinez-Martin, 2005).

Our aim is to review and summarize the available evidence exploring the effect of Covid-19 on non-motor symptoms of PD both in the acute and, possibly, the chronic phase of the infection, as well as potential mechanisms that might mediate such effects on PwP (Fig. 1).

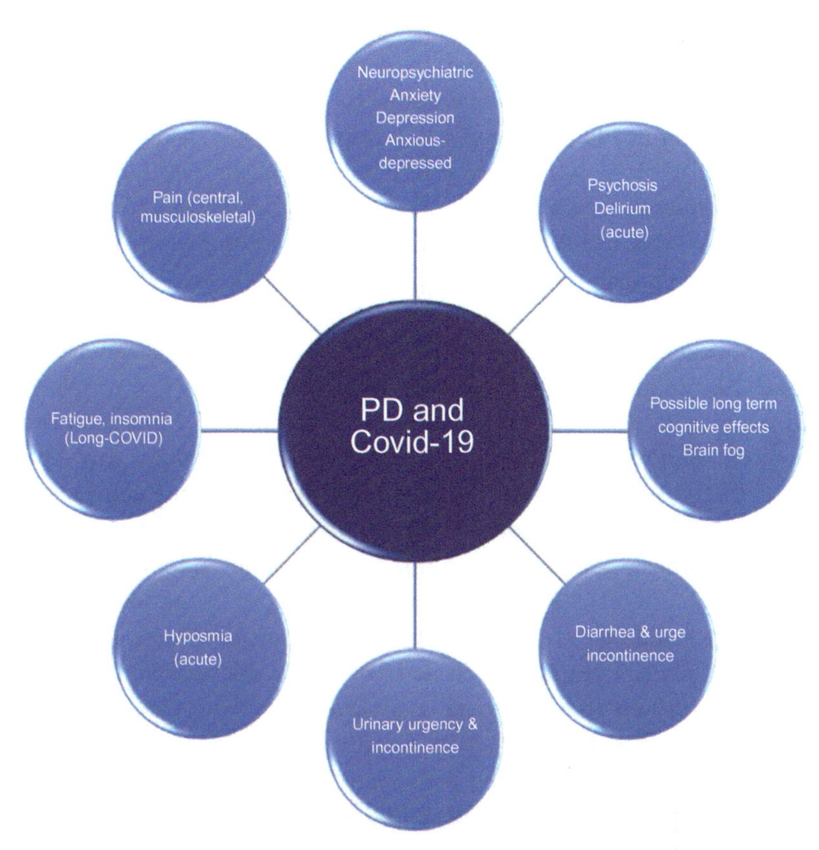

Fig. 1 Summary of the non-motor effects in PwP affected by Covid-19 both in the acute stage and in the long-term (Covid-19, Coronavirus disease 2019; PD, Parkinson's Disease; PwP, people with Parkinson's Disease).

2. Depression

Depression has classically been known to correlate with social isolation among PwP (Karlsen, Larsen, Tandberg, & Maeland, 1999). Socially isolated caregivers of PwP have also experienced loneliness, leading to depression (McRae et al., 2009), which is even more critical in the current setting of the Covid-19 pandemic (Subramanian, Farahnik, & Mischley, 2020). Studies have shown that non-motor symptoms of PD, including mood changes, have generally deteriorated during the pandemic era, presenting either as part of the Covid-19 symptomatology (Hainque & Grabli, 2020) or in the course of the illness per se (Antonini, Leta, Teo, & Chaudhuri, 2020; Xia et al., 2020), leading to decreased QoL during this period (Suzuki et al., 2021). In the online study Fox Insight, a large cohort of adults with and without PD (5429 and 1452, respectively), 71% of PwP infected with SARS-CoV-2 have experienced new-onset or worsening of pre-existing depression, while the corresponding percentage for the non-infected PwP was 36.5% (Brown et al., 2020). Interestingly, the PwP mostly affected were usually women, irrespective of their Covid-19 status, even when social factors were accounted for (Brown et al., 2020; Picillo et al., 2021; Suzuki et al., 2021; Xia et al., 2020). Among those diagnosed with Covid-19, mood outcomes have been largely similar between people with and without PD (Brown et al., 2020; Suzuki et al., 2021). In a multicenter series of 27 PwP, depression has also been identified as part of the long-COVID syndrome, the entity used to describe the long-term sequelae of Covid-19 (Covid-19-associated symptoms persisting for over 12 weeks, without another possible explanation), affecting about 7.4% of the patients (Leta et al., 2021). Moreover, a recent large longitudinal general population-based cohort, exploring the rates of suicide among PwP (Chen et al., 2021) over 11 years, has indicated that suicide risk was two times higher in PwP compared to the general population, similarly to earlier studies (Eliasen, Dalhoff, & Horwitz, 2018; Erlangsen et al., 2020; Li et al., 2018), with late-life suicide rates being particularly high in East Asia. PD severity and co-morbid depression have been highlighted as the main correlates.

Factors associated with mood decline throughout the pandemic, include interruptions in medical care delivery, in essential daily activities and in regular exercise, loss of social contacts/activities, and/or self-isolation/quarantine (Brown et al., 2020; Kitani-Morii et al., 2021; van der Heide, Meinders, Bloem, & Helmich, 2020). A cross-sectional survey ($n=156$) in Brazil has shown that lower physical activity level in PwP was associated with increased thoughts of death (Haas et al., 2022). Although the effect of the Covid-19 pandemic on mood changes is still unclear, increased awareness and screening for depression is strongly suggested among PwP, as the added impact of social distancing, limitations in exercise and PD-related complications might render patients particularly vulnerable.

3. Anxiety

It is widely acknowledged that PwP are at high risk of suffering from anxiety (Broen, Narayen, Kuijf, Dissanayaka, & Leentjens, 2016), and may be particularly susceptible to external stressors (Helmich & Bloem, 2020;

van Wamelen, Wan, Ray Chaudhuri, & Jenner, 2020). Symptoms of both PD and anxiety, as well as the stress-related responses, have been thought to be involved in the cascade of dopaminergic, serotoninergic, and adrenergic pathways (Godoy, Rossignoli, Delfino-Pereira, Garcia-Cairasco, & de Lima Umeoka, 2018; Kano et al., 2011; van Wamelen et al., 2020).

It should, thus, come as no surprise that global converging evidence has indicated an increase in anxiety and stress among PwP infected with SARS-CoV-2 compared to healthy controls (Brown et al., 2020; Cilia et al., 2020; Shalash et al., 2020; Suzuki et al., 2021), although anxiety was the only significant predictor of worsening stress in PwP during the early days of the pandemic (Salari et al., 2020). A subjective worsening of anxiety, along with cognitive symptoms (with a mean Mini Mental State Examination score (MMSE) drop of 0.5 points), have been reported by 28 PwP in Pisa, Italy, throughout the course of lockdown (Palermo et al., 2020). In Rome, another study of 162 PwP has revealed that half of non-infected patients had experienced a worsening in their motor symptoms, with 25% of them reporting augmented anxiety during the pandemic (Schirinzi et al., 2020). A subsequent case-controlled survey in Tuscany has shown that 29.6% of non-infected patients in a PD cohort ($n=733$) had experienced exacerbation of their motor symptoms with similar worsening of mood (24.7%), anxiety (25%), and poor sleep (22.2%) (Del Prete et al., 2021). An online survey exploring the impact of the pandemic on 358 PwP has shown that the pre-pandemic presence of neuropsychiatric symptoms, as well as cognitive dysfunction, was associated with increased levels of psychological distress within the pandemic (van der Heide et al., 2020).

Being home-bound and the loss of social contacts remained the commonest stressors among PwP (Zipprich, Teschner, Witte, Schonenberg, & Prell, 2020); others included reduced access to non-pharmacological treatments, such as physiotherapy and psychotherapy (Antonini et al., 2020; Song et al., 2020; Sulzer et al., 2020). Concurrent anxiety, stress, isolation, and physical inactivity have been found to be a particularly deleterious combination for PwP (Helmich & Bloem, 2020; Krzyszton et al., 2021).

In some countries with mandated lockdowns, the impact of anxiety among PwP appeared bleaker, with 25.5%–43.8% of them recording severe anxiety compared to controls, and more than half of PwP caregivers also reporting heightened anxiety (Oppo et al., 2020; Salari et al., 2020; Shalash et al., 2020). Specific contributing factors were mainly fear of the patient or the loved ones contracting SARS-CoV-2 (Krzyszton et al., 2021; Montanaro et al., 2022; Salari et al., 2020; Zipprich et al., 2020) and fear of limited drug availability (Guo et al., 2020; Montanaro et al., 2022;

Salari et al., 2020; Shalash et al., 2020; Xia et al., 2020). In contrast, during the same time period in India, a telephone survey reported that patients and caregivers were "well-informed and coping well," with only 11% of patients having experienced a worsening of their motor and non-motor symptoms (Prasad et al., 2020). This incongruity has been attributed to be due to the fact that considerably fewer people ($n = 156$) had died in India due to Covid-19 in the period immediately preceding the published article compared to other countries (World Health Organization, 2022).

A large, prospective cohort study had shown that anxiety levels before the pandemic, along with other motor and non-motor features, may predict future development of treatment-related motor complications (Kelly et al., 2019). This notion was reflected in a study taking place in March 2020 (De Micco et al., 2021), which revealed that the increased psychological impact of a 40-day quarantine in 94 PwP was significantly associated with a high pre-lockdown degree of anxiety, treatment-related PD motor complications, as well as the number of lockdown hours per day. PwP have reported markedly increased irritability, related to frequent thoughts of the lockdown, despite efforts not to think about it. There was no observed association noted between disease duration and severity with psychological well-being.

In a telephone survey of 100 individuals with advanced PD, Montanaro and colleagues have shown that anxiety and depression (as evaluated via the Hospital Anxiety and Depression Scale (HADS)) occurred in more than 30% of the participants and were significantly correlated with the type of PD treatment (Montanaro et al., 2022); patients on standard medical therapy and levodopa/carbidopa intestinal gel (LCIG) infusion exhibited the highest anxiety rates. Similarly to aforementioned studies (De Micco et al., 2021), no association was found with disease duration. The study also demonstrated that 40% and 21.7% of caregivers experienced anxiety and depression respectively (Montanaro et al., 2022).

Contrary to the majority of studies stating that the Covid-19-related lifestyle changes have led to negative outcomes in PwP, there has been some evidence that reduced external demands and activity levels may have been a relief to those that constantly had to adapt to disease-related changes or experience social pressure, thus reducing stress levels (Corti et al., 2018; HØrmann Thomsen, Wallerstedt, Winge, & Bergquist, 2021). A recent study has shown that the immediate effects of the Covid-19 period may have not been as disruptive as expected, with patients ratings and free-text descriptions suggesting that, despite the admittedly increased anxiety, the pandemic and the related changes in society have been associated with improvement in health-related QoL, improved sleep, and an overall feeling that the "pressure" is gone (HØrmann Thomsen et al., 2021). Another study in Morocco has shown that a 6-week confinement had not significantly affected the overall anxiety and depression scores of PwP ($n=50$). It has been postulated that the unexpected improvement in mood and anxiety scores of a few patients may have been due to the regular presence of family members and an increased family support during the lockdown (El Otmani et al., 2021). The discrepant findings could be due to disparities in the socio-cultural framework across the different countries. Emergent PD anxiety may be secondary to an amalgamation of environmental vulnerability, dysfunctional coping strategies, as well as predisposing personality traits (Corti et al., 2018), factors likely to come to fore within the context of a crisis, like the Covid-19 pandemic.

In this period, PwP with high perceived stress levels have experienced more anxiety, ruminations, and neuroticism, scoring lower on cognitive abilities, social support, trait resilience, optimism, and positive appraisal style (van der Heide et al., 2020). Similarly to depression, those affected by anxiety were more frequently women (Picillo et al., 2021; Suzuki et al., 2021). Understanding the coping mechanisms and determinants of resilience for PwP, which may serve as prognostic factors (Weems, Costa, Dehon, & Berman, 2004; Whitworth et al., 2013), may be crucial in surviving a crisis, such as the Covid-19 pandemic.

However, the lack of consistent data on the risk of PwP developing Covid-19 (Artusi et al., 2020) has perpetuated the feelings of doubts and uncertainty, hence, increasing the level of anxiety; indeed, recent studies have recommended the critical requisite of correct and consistent information during the outbreak (Montanaro et al., 2022; Schirinzi et al., 2020). This highlights the key need for interventional planning during the post-lockdown phase (e.g., telemedicine services), aiming at bolstering the resilience of PwP toward a rapid resolution of potential mental health issues, triggered from this particularly stressful experience (De Micco et al., 2021).

4. Cognitive impairment

Acute viral central nervous system (CNS) infections, including cytomegalovirus (CMV), Epstein-Barr virus (EBV), human immunodeficiency virus (HIV), herpes viruses, varicella zoster virus (VZV) and hepatitis C virus

(HCV), have been suggested to contribute to the subsequent development of cognitive impairment, even in previously healthy individuals, either via a direct viral insult of the nervous system or indirectly through promoting inflammation, epigenetic changes or a hypercoagulable state, which might alter brain structure and function (Damiano et al., 2021). An analysis of a large cohort of older individuals, who had been cognitively intact at baseline, has shown that acute care and critical or non-critical illness hospitalizations were associated with a higher risk of cognitive impairment during the following years (Ehlenbach et al., 2010). For older patients, who had been hospitalized due to acute systematic infections, a 1-year follow-up study has shown that premorbid dementia and delirium during hospitalization constituted risk factors for subsequent cognitive deterioration (Silva et al., 2021). Furthermore, a large recent cohort study of elderly nursing home residents has found that infection-related hospitalization was linked to an immediate and sustained cognitive decline, with older patients being more vulnerable (Gracner et al., 2021), while preliminary evidence supports that hospitalization per se due to an acute illness might precipitate cognitive impairment in older adults (Chinnappa-Quinn et al., 2020).

Recent data suggests that Covid-19, even when asymptomatic, might have both short- and long-term effects on patients' cognition (Damiano et al., 2021). A population-based study of 153 Covid-19 cases has revealed that 4% presented with a new-onset dementia-like syndrome during the acute phase of Covid-19 (Varatharaj et al., 2020). Small case series of Covid-19 patients, assessed using standardized neuropsychological tools, have revealed a decline in cognition detected about 2–4 weeks after infection with their attention being mostly impaired (Negrini et al., 2021; Zhou et al., 2020). In a cohort of 57 patients on rehabilitation after Covid-19 the vast majority has exhibited cognitive decline with attention and executive function being the most commonly affected domains (Jaywant et al., 2021). A large cross-sectional study in China has shown that Covid-19 was associated with long-term impairment in cognitive performance (assessment at 6 months) among patients older than 60, with severe Covid-19 and delirium being risk factors (Liu et al., 2021). A small sample size study has shown that younger Covid-19 patients with mild to moderate symptoms had persistent subclinical cognitive impairment, mostly affecting short-term memory, attention and concentration; these findings were not correlated to depressed mood or fatigue (Woo et al., 2020). Importantly, a recently published study, assessing brain changes in a sample of 785 UK Biobank participants (401 Covid-19-positive individuals and 384 controls), who had undergone a brain scan before and after being infected with SARS-CoV-2, has revealed a significant reduction in global brain size and in gray matter thickness and tissue-contrast in the orbitofrontal cortex and parahippocampal gyrus, along with a decrease in the average cognitive performance before and after Covid-19 (Douaud et al., 2022). Although the authors could not conclude on the reversibility of these findings, they have highlighted the neurodegenerative potential of the virus.

A recent study assessing mice models and individuals with a mild SARS-CoV-2 infection has found a sustained elevation of CCL11, a circulating chemokine which has been associated with cognitive impairment and suspension of neurogenesis (Fernández-Castañeda et al., 2022). According to the authors, humans suffering from long-COVID, including cognitive symptoms, had higher levels of CCL11 compared to those with long-COVID, but without any cognitive complaints. What is more, affected mice have exhibited a persistent impairment in hippocampal neurogenesis, plus myelin loss in subcortical white matter. These pathophysiological changes might contribute to significant neurological sequelae, even after mild Covid-19.

Cognitive dysfunction is quite prevalent in PD with up to 83% of PwP presenting with some degree of cognitive impairment at some point of the disease course, usually at the advanced stages, and might be accompanied by psychosis and hallucinations (Schapira, Chaudhuri, & Jenner, 2017). However, mild or subclinical cognitive deficits, usually presenting as executive dysfunction or visuospatial impairment, can be diagnosed even in earlier PD stages (Aarsland et al., 2021). It is also of interest that, due to nigrostriatal dopamine depletion, PwP might experience cognitive inflexibility and decreased ability to cope with new circumstances, like the new daily routine imposed during the Covid-19 pandemic, a deficit that may also be associated with a sense of loss of control and psychological stress (Douma & de Kloet, 2020; Robbins & Cools, 2014). In the Covid-19 context, cognitive impairment might result either from direct disruption and damage of the virus on the CNS and the harmful sequalae (inflammation, hypercoagulation, hypoxia) or could be a secondary effect due to depression (Rock, Roiser, Riedel, & Blackwell, 2014), anxiety (Potvin, Hudon, Dion, Grenier, & Préville, 2011) or fatigue (Neu et al., 2011), as a result of the general psycho-social stress, isolation, social deprivation and routine changes accompanying the pandemic. Depression, anxiety and fatigue are also quite common non-motor symptoms in PD, and need to be considered and properly addressed whenever cognitive deficits are revealed (Schapira et al., 2017).

In a small, multicenter, 12-week study of 27 PwP with a Covid-19 diagnosis, new-onset cognitive disturbances appeared in 22.2% of patients, including "brain fog," concentration impairment and memory deficits (Leta et al., 2021). In an online survey of 46 PwP and Covid-19, intellectual impairment has been detected in 48% of patients, although in this occasion no previous examinations had been available (Xu et al., 2021). In one community-based case-control study comparing PwP with and without Covid-19, cognitive performance has only been marginally affected (Cilia et al., 2020). In a group of 10 advanced PwP with Covid-19, cognition has been found to worsen during the infection (Antonini et al., 2020). In an Italian Movement Disorders Outpatient Clinic, 28 PwP have been diagnosed with new-onset cognitive disturbances and higher anxiety scores after the lockdown period, without being previously infected by SARS-CoV-2, thus, highlighting the effect of isolation on cognition (Palermo et al., 2020). However, these results have not been confirmed in a group of mild to moderate PD severity (Luis-Martínez et al., 2021). Researchers have attributed these new-onset clinical phenomena to a combination of factors, including the negative impact of the prolonged lockdown, the limited access to healthcare and rehabilitation services, the effect of the infection per se and the natural course of the disease. However, most of these studies have significant methodological flaws, like small sample sizes, lack of a control group or baseline assessments, so their results should be examined with caution. Interestingly, in a phone-based study of 568 Spanish PwP, it has been found that patients who had not reported a prior Covid-19 diagnosis were more likely to experience motor fluctuations (61% vs 35.7%, $P=0.052$) and hallucinations (23.4% vs 0%, $P=0.025$), and there was a tendency towards cognitive impairment and behavioral problems compared to those who had contracted a SARS-CoV-2 infection (Santos-García et al., 2020). The authors have suggested that stricter prevention measures, that might have prevented PwP from contracting Covid-19, might have also led to the above complications. However, the Covid-19-positive group in the above PD cohort was relatively small (2.6% of the total sample).

5. Psychosis

Psychosis is encountered in up to 40% of PwP, usually later in the disease course and typically manifests gradually, allowing for step-by-step approaches (Schapira et al., 2017; Simonet, Tolosa, Camara, & Valldeoriola, 2020). All dopaminergic medication can induce psychosis; levodopa in a lesser degree

(Simonet et al., 2020). On the other hand, psychotic features can emerge secondarily due to other medication, like antidepressants, pain-killers or steroids, alcohol abuse or withdrawal, systematic or CNS infections, including Covid-19, CNS lesions, acute vascular events, or they can be part of the clinical spectrum of delirium, especially in elderly hospitalized patients (Dubovsky, Arvikar, Stern, & Axelrod, 2012; Parra et al., 2020; Saldanha, Menon, Chaudari, Bhattacharya, & Guliani, 2013; Webster & Holroyd, 2000). In the Covid-19 context, incident psychosis has not been very common, although more findings have been based on case reports and case series. According to a recent review of December 2021 summarizing the known published cases, delusions were the most typical sign of new-onset Covid-19 psychosis, and most patients had a favorable outcome, although 69% required hospitalization and 33% had to be admitted in a psychiatric facility (Smith et al., 2021). However, concurrent delirium was not excluded in the majority of cases and the possibility of bias was considered high in almost one out of three patients, either due to a history of substance abuse, psychiatric disorders or other medical conditions, which might be related to secondary psychosis (Smith et al., 2021). In the case-series of 10 advanced PwP diagnosed with Covid-19, no new-onset cases of psychosis was described, but psychotic features, which had been present prior to Covid-19, were reported to worsen during the infection (Antonini et al., 2020).

6. Delirium

Of note, delirium, an often life-threatening emergency, constitutes a common complication of Covid-19, especially in critically ill patients, and has been associated with a poor outcome (Ticinesi et al., 2020). Psychotic symptoms have been found to appear in almost half of the afflicted patients (Paik, Ahn, Min, Park, & Kim, 2018; Webster & Holroyd, 2000). Hospitalization per se can be a risk factor for the elderly, as delirium has been found to complicate the clinical course of one-third of the older general medical patients (Marcantonio, 2017). For those under mechanical ventilation admitted in the ICU this percentage has been found to exceed 75% (Ely et al., 2004). Decreased functional status, cognitive impairment, depression and comorbidities are typical predisposing factors, while subjacent infections, pain, anemia, anesthesia and drugs might act as precipitating factors (Kalimisetty, Askar, Fay, & Khan, 2017; Marcantonio, 2017; Wakefield, 2002). For Covid-19 patients, lack of visitation due to safety regulations imposed by the healthcare systems has also been identified as a risk factor (Pun et al., 2021). In two small sample size case-series of PwP with Covid-19, delirium has been found in 7–10% of patients (Antonini et al., 2020; Leta et al., 2021). Evidence suggests that PwP are more vulnerable to delirium while admitted in the hospital and treating physicians should be vigilant to acknowledge and timely manage any early sings of agitation, confusion or cognitive deterioration (Vardy, Teodorczuk, & Yarnall, 2015).

7. Hyposmia

Olfactory dysfunction constitutes one of the most commonly encountered symptoms in Covid-19 (Guerrero et al., 2021), although the majority of patients are expected to recover within a month (D'Ascanio et al., 2021). It is so typically connected to Covid-19, that a recent meta-analysis has shown it could be used as a good predictor of a SARS-CoV-2 infection (Hariyanto, Rizki, & Kurniawan, 2021). The olfactory epithelium is rich in common putative molecular site entries for SARS-CoV-2, like the Angiotensin Converting Enzyme-2 (ACE2) and Neuropilin 1 (NRP1), while the olfactory bulb is not protected by the blood–brain barrier (BBB), constituting theoretically an easier target for SARS-CoV-2 or other airborne viruses (Cantuti-Castelvetri et al., 2020; Wan et al., 2021). Reverse-transcription polymerase chain reaction (RT-PCR), in situ hybridization and immunohistochemical staining techniques have, indeed, revealed an increased SARS-CoV-2 viral load in the nasal epithelium (Meinhardt et al., 2021).

Impaired sense of smell is a highly prevalent non-motor symptom in PD as well, appearing in approximately 90% of PwP in the early or premotor prodromal stage, occasionally preceding the emergence of motor symptoms by years (Haehner, Hummel, & Reichmann, 2011; Xiao, Chen, & Le, 2014). Hyposmic PwP have been found to exhibit worse motor and cognitive progression, lower QoL scores and requiring higher levodopa equivalent doses (LED) compared to normosmic ones (Gjerde et al., 2018; He et al., 2020). The high percentage of olfactory dysfunction both in the PD and the Covid-19 population has recently given rise to concerns about SARS-CoV-2 infection being the "perfect storm" of a subsequent rise in parkinsonism cases (Brundin, Nath, & Beckham, 2020; Krey, Huber, Höglinger, & Wegner, 2021), although such notions remain highly speculative.

8. Gastrointestinal dysfunction

Gastrointestinal (GI) symptoms might affect up to three out of five patients with Covid-19 (Kariyawasam, Jayarajah, Riza, Abeysuriya, & Seneviratne, 2021). They are mostly mild and self-limited, and they might be prevalent even from the early stages of the infection (Kariyawasam et al., 2021; Villapol, 2020). Common symptoms include diarrhea, affecting about 10.4–11.5% of Covid-19 patients, followed by nausea and vomiting, which range from 6.3% to 10.5%, but more generic manifestations like anorexia or abdominal discomfort are often present (Andrews, Cai, Rudd, & Sanger, 2021; D'Amico, Baumgart, Danese, & Peyrin-Biroulet, 2020; Silva et al., 2020). Although subjacent acute causes from the GI system, including inflammatory or ischemic conditions, should be excluded, SARS-CoV-2 per se has been considered to insult the digestive tract, either directly or via a local or systemic inflammatory response to the virus. More particular, ACE2 and Transmembrane Serine Protease 2 (TMPRSS2), which supposedly constitute major cell entry receptors for SARS-CoV-2 (Hoffmann et al., 2020), are highly expressed in the intestinal endothelial cells of ileum and colon (Kariyawasam et al., 2021), where the virus is thought to disrupt the intestinal mucosal barrier and promote inflammation (Ye, Wang, Zhang, Xu, & Shang, 2020). SARS-CoV-2 RNA has, indeed, been detected in biopsy specimens of esophagus, stomach, duodenum and rectum from Covid-19 patients with clinically confirmed GI symptoms (Lin et al., 2020), while histopathological studies have revealed varying degrees of degeneration, necrosis and inflammation of the GI mucosa in post-mortem Covid-19 cases (Deshmukh, Motwani, Kumar, Kumari, & Raza, 2021). All the above indications have fueled speculations that the GI tract is a potential entry site and target organ for the virus.

During the last two decades a growing body of evidence has highlighted the role of the GI system in PD (Menozzi, Macnaughtan, & Schapira, 2021). GI dysfunction is quite common among PwP and all levels of the digestive tract can be affected, causing a range of symptoms, such as drooling, dysphagia, nausea, reduced gastric motility, constipation and impaired defecation (Schapira et al., 2017). The pathology underlying these manifestations is not quite clear yet, although α-synuclein aggregates, the pathological trademark of PD (Poewe et al., 2017), have been detected in the myenteric and submucosal plexuses of the enteric nervous system with a distinct rostrocaudal pattern of deposition (Fasano, Visanji, Liu, Lang, & Pfeiffer, 2015).

These neurons originate centrally in the dorsal motor nucleus of the vagus nerve, which is also anatomically connected to the respiratory system. This has led researchers to assume that the dorsal vagal complex of the brainstem, which is highly enriched in ACE2, can be reached by SARS-CoV-2 through the vagus nerve from the periphery, allowing an entrance of the virus to the CNS (Rangon, Krantic, Moyse, & Fougère, 2020). Past experiments in mice have confirmed the transvagal transmission of Influenza A virus from the respiratory mucosa to the basal ganglia (Matsuda et al., 2004), while researchers have confirmed the potential interaction of the SARS-CoV-2 proteins with the intestine microstructure based on mice models (Mönkemüller, Fry, & Rickes, 2020). Interestingly, according to the dual-hit hypothesis of PD pathogenesis, a neurotropic pathogen, like a virus, has been speculated to enter the CNS through the nasal or gastric pathway, both of which appear to constitute sites of early pathology in PD (Hawkes, Del Tredici, & Braak, 2007; Klingelhoefer & Reichmann, 2015).

The emergence of GI symptoms in PwP inflicted by SARS-CoV-2 can overlap with already existing non-motor symptoms of the PD spectrum, complicating the patients' management, causing distress and worsening their QoL (Lubomski, Davis, & Sue, 2020; Menozzi et al., 2021). Diarrhea or vomiting in the setting of Covid-19 might impair the pharmacokinetics of orally administered dopaminergic medication, leading to reduced absorption of the drug and, consequently, an increased frequency and duration of OFF episodes. In a community-based case-control study comparing PwP with and without Covid-19 no statistically significant changes have been found considering the GI domains of the Non-Motor Symptoms Scale (NMSS) (Cilia et al., 2020). However, diarrhea has been found to exert a devasting effect on the pharmacokinetics of dopamine replacement therapy (particularly levodopa) and has been linked to an aggravation of motor PD symptoms (Cilia et al., 2020). Furthermore, dehydration due to diarrhea or reduced water intake in case of anorexia or vomiting, could precipitate or worsen already existent orthostatic hypotension, another common feature among PwP with dysautonomic features (Schapira et al., 2017; Simonet et al., 2020). Treating physicians should be aware that apart from SARS-CoV-2 per se, newly introduced medication in the context of Covid-19, like empirically used antibiotics, especially macrolides and quinolones, or the antiviral agent remdesivir, carry a significant risk of producing typical GI side effects, like nausea, vomiting, abdominal pain and diarrhea (Aleem & Kothadia, 2021; Patel & Hashmi, 2021; Yan & Bryant, 2021). Third-generation cephalosporines and quinolones have also been associated with a risk of pseudomembranous colitis, mediated by *Clostridium difficile*, a condition that requires special treatment and might further complicate the absorption and distribution of orally administered dopaminergic drugs (Arumugham, Gujarathi, & Cascella, 2021; Yan & Bryant, 2021).

Recent studies have suggested that patients with altered gut microbiota might experience more severe Covid-19 symptoms (Kim, 2021), while imbalances in the normal gut microbiota have been speculated to be involved in PD pathophysiology (Elfil, Kamel, Kandil, Koo, & Schaefer, 2020). A recent meta-analysis has confirmed consistent gut microbiome alterations among the population of PwP, which might be related to a pre-inflammatory status linked to the GI symptoms in PD (Romano et al., 2021). A small intestinal bacterial overgrowth has been observed in more than 60% of PwP and has been significantly associated with worse motor outcomes (Tan et al., 2014). Whether this parameter of gut dysbiosis might mediate part of the aggravation in motor and non-motor symptoms observed during PwP and Covid-19 needs to be further investigated.

Interestingly, recent reviews analyzing global data have shown that GI symptoms among Covid-19 survivors, including nausea, vomiting, diarrhea, anorexia, abdominal pain, acid reflux and constipation, might persist after their discharge or recovery (Silva Andrade et al., 2021; Yusuf et al., 2021). In a small, multicenter study with a small sample size of 27 PwP, followed for more than 12 weeks after a Covid-19 diagnosis, the prevalence of GI symptoms was low (7.4% for nausea and 3.7% for reduced appetite) (Leta et al., 2021). These prolonged manifestations, often referred to as long-COVID, might have an impact on patients' QoL and their pathogenesis remains to be clarified (Yusuf et al., 2021).

9. Dysautonomia

Orthostatic hypotension is an uncommon symptom in Covid-19 patients, with a frequency of 2.2%, being more prevalent in older individuals, those with concomitant high blood pressure and among beta blocker users (de Freitas et al., 2021). Despite the startling presentation, orthostatic hypotension has not been associated with a more severe outcome for Covid-19 (Oates et al., 2020).

Various pathophysiological mechanisms associated with Covid-19 and autonomic impairment have been described. The infection of the CNS nucleus tractus solitarius through the vagus nerve may affect the respiratory rhythm, heart rate regulation and blood pressure control. Likewise, the impairment of vagus nerve functioning, which regulates anti-inflammatory responses and the sympathetic overstimulation, may result in an increased release of pro-inflammatory cytokines (Del Rio, Marcus, & Inestrosa, 2020). Invasion of the vagus nucleus, specifically the nucleus ambiguous, could result in respiratory dysfunction, while excessive sympathetic activity can lead to cardiac arrhythmias and cathecolaminergic toxicity (Hassani, Fathi Jouzdani, Motarjem, Ranjbar, & Khansari, 2021). Postural orthostatic tachycardia syndrome (POTS), has been previously related to viral infections, such as influenza and EBV, and mainly appears as a post-infectious condition. In a case series of 20 patients with a prior Covid-19 diagnosis and evidence of orthostatic intolerance after examination (tilt table test), 75% has been diagnosed with POTS, which, interestingly, has persisted for 6–8 months after the infection. As POTS has been likely considered an autoimmune process, it is possible that SARS-CoV-2 might induce cross-reaction antibodies with autonomic related components, such as autonomic ganglia and nerve fibers, and cardiovascular receptors (Blitshteyn & Whitelaw, 2021; Díaz, Toledo, Andrade, Marcus, & Del Rio, 2020). Post-Covid-19 syndrome, as result of an autoimmune response after the initial SARS-CoV-2 infection, could be a trendsetting topic for future research with a possible benefit from inmunomodulatory therapies.

Early reports have suggested a higher frequency of orthostatic hypotension in PD and Covid-19 (Antonini et al., 2020), while the aforementioned online study Fox Insight has reported a worsening of dysautonomia in 38% of participants with Covid-19 and PD (Brown et al., 2020). Urge/incontinence, urine retention and nocturia in PD may be exacerbated after Covid-19 (Cilia et al., 2020; Xu et al., 2021); this phenomenon could be attributed to an increased consumption of over the counter drugs, such as antihistamine medication with anticholinergic properties used for Covid-19 mild symptoms, as they can aggravate constipation and urinary retention (Elbeddini, To, Tayefehchamani, & Wen, 2020). Motor fluctuations due to impaired absorption of dopaminergic drugs during Covid-19 could also have a deleterious impact on urinary symptoms (Cilia et al., 2020).

10. Pain

Pain is a common feature in both PD and Covid-19. It is present in around 40% of Covid-19 patients, with a higher prevalence in the female population (Lechien et al., 2020). In a survey of 46 PwP, who had been diagnosed with Covid-19 at the Columbia University Irving Medical Center, 40% described a worsening of pain and 4% noticed new-onset pain following the infection (Xu et al., 2021). In a similar trend, pain has been reported by 76% of PwP infected with SARS-CoV-2 in a cohort of 51 participants in the Fox Insight online study (Brown et al., 2020).

The mechanisms underlying the Covid-19-induced pain are still under investigation. Weng and colleagues have described different pathways associated with pain in Covid-19 (Weng, Su, & Wang, 2021). The systemic inflammation, which often accompanies Covid-19, implies an elevated release of cytokines, prostaglandins E1, E2 and bradykinin, and an up-regulation of interleukin-6 and could be associated with headache and polineuropathy, sore throat and myalgia. Likewise, elevated inflammatory markers may induce myocardial damage and retrosternal pain (Weng et al., 2021), while the increased ACE2 expression in the GI tract could potentially cause abdominal pain and diarrhea through mucosa inflammation (Lin et al., 2020).

The Covid-19 pandemic has had a negative impact on regular dopaminergic drugs intake, especially in non-white and low income individuals (Brown et al., 2020). Consequently, an increase in OFF periods and non-motor fluctuations, including pain, is expected (Rukavina et al., 2019). Systemic illness and inflammation could impair the pharmacodynamics of dopaminergic drugs and, therefore, newly induce or increase pre-existent pain. Finally, van der Heide and colleagues have found an aggravation of fatigue and pain in PwP, especially in women, as a result of psychological distress induced by the pandemic, while anxiety and increased rigidity might also have contributed (van der Heide et al., 2020). Different forms of pain have also been described as part of the long-COVID in PwP, including arthralgia (11.1%), muscular pain (7.4%), headache (18.5%) and peripheral neuropathy symptoms (11.1%). Less access to rehabilitation services and health centers during the pandemia, along with social isolation and worsening of motor and non-motor PD symptoms due to Covid-19, could be related to these long-term sequelae among PwP (Leta et al., 2021). Regular follow-up assessments of these patients could further enlight us about the impact of Covid-19 on the natural history of PD.

11. Fatigue

Fatigue is one of the most frequently reported symptoms among Covid-19 patients, affecting 38.1–63% of individuals with mild to moderate disease (Bhidayasiri, Virameteekul, Kim, Pal, & Chung, 2020;

Lechien et al., 2020), while it usually persists for more than 12 weeks (El Sayed, Shokry, & Gomaa, 2021). Fatigue, whose differential diagnosis and management is at least challenging, constitutes a highly prevalent feature in PD with a negative impact on QoL of both patients and caregivers (Lazcano-Ocampo et al., 2020). Several studies have successively reported fatigue as one of the most frequent, either newly emerging or worsening, non-motor PD symptoms during a SARS-CoV-2 infection, particularly affecting women and elderly individuals (Cilia et al., 2020; El-Qushayri et al., 2021; Lechien et al., 2020). In a small cohort of 27 PwP and Covid-19 fatigue was one of the most prevalent symptom (47%) identified in the spectrum of long-COVID, along with motor worsening, cognitive impairment and sleep disturbances (Leta et al., 2021).

Fatigue in PD is a complex symptom with central and peripheral components. Central fatigue has been associated with cytokines released during the inflammatory response, affecting the basal ganglia and dopamine function (Felger & Miller, 2012). Moreover, it has been related to high levels of IL-6 and C-Reactive Protein (CRP) in the cerebrospinal fluid (CSF) and an elevated production of IL-1β (Lazcano-Ocampo et al., 2020). Interestingly, high levels of IL-6, IL-1β and CRP have also been detected in Covid-19 patients, as part of the inflammatory response to the virus, supporting the theory that elevated inflammatory interleukines might cause fatigue, not only in neurodegenerative diseases, but also in other cronic conditions (Sulzer et al., 2020). Physical fatigue has been found significantly higher in a cohort of Covid-19 patients in Turkey and has been associated with higher levels of CRP and lactate dehydrogenase (LDH) (Tuzun, Keles, Okutan, Yildiran, & Palamar, 2021). LDH has been linked to raised lactate levels in muscles, which might induce hypoxia and lead to muscle pain (Tuzun et al., 2021). Psychological distress and social isolation due to the pandemic have also been related with the exacerbation of some non-motor symptoms, such as fatigue and anxiety (van der Heide et al., 2020). Finally, many researchers would argue that fatigue could be entirely precipitated by the SARS-CoV-2 per se, and be fully irrelevant to non-motor PD symptoms in PD (Cilia et al., 2020).

12. Sleep impairment

The effect of the Covid-19 pandemic on sleep quality has not been unanimous among different studies and subpopulations. According to two recent systematic reviews and meta-analyses, with the larger one involving more than 345,000 participants from 39 countries, disruption of sleep appears to be quite prevalent during the ongoing pandemic with most of these problems having been associated with increased psychological distress, while patients with an active SARS-CoV-2 infection have been found particularly vulnerable (corrected pooled estimated prevalence of 18% in the general population and 57% among Covid-19 patients, $P < 0.05$) (Alimoradi et al., 2021; Jahrami et al., 2021). A great share of such disturbances of the sleep pattern have been attributed to the confinement regulations imposed during the pandemic. More specifically, a national cross-sectional study in Italy ($n = 1515$), the first European country applying a national lockdown for the Covid-19 pandemic, has revealed that 42.2% of the participants have been experiencing sleep problem during the pandemic, with 17.4% reporting a moderate to severe insomnia (Gualano, Lo Moro, Voglino, Bert, & Siliquini, 2020). In a telephone-based survey in Hawai, assessing 367 participants with chronic neurological conditions, including PD (3.5%), a percentage of 37.4% has reported sleep disturbances during the Covid-19 pandemic, irrespective of whether the responders had contracted SARS-CoV-2 or not (Crocker et al., 2022). On the other hand, researchers argue that the impact of Covid-19 on sleep patterns also depends on the pre-pandemic quality of sleep, with 25% of participants in an online survey in the Netherlands mentioning a meaningful amelioration in their sleep quality during the pandemic and 20% reporting the opposite (Kocevska, Blanken, Van Someren, & Rösler, 2020). According to the authors, the former group consisted mostly of individuals with pre-pandemic severe insomnia symptoms, while in the latter participants had mentioned no sleep complaints prior to the pandemic, thus, pointing out the highly heterogeneous nature of sleep problems. Such a trend has also been confirmed in a Danish/ Swedish cohort of 67 PwP with the majority of them (88%) reporting less sleep disturbances during the pandemic compared to the pre-Covid-19 period, leading the authors to suggest that the perceived alleviation in "pressure" and activity level might have been experienced as a relief for some patients, reflecting in their quality of sleep (HØrmann Thomsen et al., 2021).

Sleep impairment constitutes a prevalent feature in PD with a gradually rising frequency as the disease progresses (Schapira et al., 2017). PwP might be troubled by different kinds of sleep problems, which can be related both to the disease per se or the relative medication, and include insomnia, RBD, periodic limb movements, restless leg syndrome (RLS) and akathisia. A potential aggravation of motor symptoms, which can be expected in PwP during the pandemic, especially if they have contracted SARS-CoV-2 (Antonini et al., 2020; Brown et al., 2020; Cilia et al., 2020; Santos-García et al., 2020), has also been associated with exacerbations in sleep disturbances due to wearing off phenomena (Schapira et al., 2017). On the other hand, insomnia has been closely associated to depression and anxiety, which, as we have already mentioned, are quite prevalent both in PD and Covid-19, with those conditions often found to overlap and reinforce each other (Oh, Kim, Na, Cho, & Chu, 2019; Taylor, Lichstein, Durrence, Reidel, & Bush, 2005).

A cross-sectional, questionnaire-based survey in China, including 119 PwP and a control group of 169 sex- and age-matched healthy individuals, has shown that PwP had a higher prevalence of sleep problems (69.9% vs 44.4%, $P<0.001$), which were independently associated to an aggravation of PD symptoms and anxiety, with female patients being particularly vulnerable (Xia et al., 2020). In an another cross-section, questionnaire-based study of PwP in India ($n=832$), new-onset or an aggravation of pre-existing sleep problems has been mentioned by 23.9% of participants, with insomnia being the most prevalent issue (51.5%), followed by RLS worsening (24.7%) and RBD (22.7%) (Kumar et al., 2021). These problems, which have been related to worst QoL and an overall exacerbation of motor and non-motor symptoms, were more prevalent among those with advanced PD and those who had undergone longer periods of home isolation (>60 days) or have been lacking proper family support during confinement. Physical activity and new hobbies during the lockdown have been found to act as protective factors on sleep impairment. In the online study of Fox Insight 4.5% and 32% of Covid-19-negative PwP have reported new-onset or worsening of sleep problems respectively (including insomnia, fatigue, excessive sleepiness, RBD), with this trend being more potent among female patients or those who had experienced interruptions in their PD-related medical care (Brown et al., 2020). Moreover, in a multicenter, international case series of 27 PwP, 22% has developed sleep disturbances as part of the long-COVID syndrome, suggesting that long-term sleep-related sequelae might also be expected (Leta et al., 2021). An online questionnaire of 417 PwP in Canada, assessing the physical and mental wellbeing, the daily activities and their PD symptoms, has revealed an aggravation of sleep disorders among those residing in Quebec, but not in other areas, leading the authors to suggest that some subpopulations might be more sensitive to social isolation and changes in their routine (de Rus Jacquet et al., 2021). This localization of results might be related to distinct rates of contracting a SARS-CoV-2 infection and different socio-cultural characteristics among separate regions even in the same country. Conclusively, it seems that sleep problems might be encountered during the Covid-19 pandemic in the population of PwP, however, a significant portion of these disturbances might be mediated by indirect factors accompanying the pandemic, and not the SARS-CoV-2 infection per se. More focused studies with larger sample size are needed to further explore this association.

13. Conclusion

The Covid-19 pandemic has had a clear impact on PwP and non-motor symptoms have been specifically affected, as they are both disease-specific and circumstantial. The unfortunate drawback of Covid-19-related public health measures, such as the imposed lockdown and isolation regulations, has had negative effects related to confinement and physical inactivity, which have been found to lead to worse QoL, stress, anxiety and depression (Pantell et al., 2013; Shalash et al., 2020) and many patients have witnessed a worsening of both motor and non-motor symptoms (Fabbri et al., 2021; Helmich & Bloem, 2020). Awareness of this constellation of non-motor symptoms affected by Covid-19 in PwP is required to improve and personalize patients' management.

References

Aarsland, D., Batzu, L., Halliday, G. M., Geurtsen, G. J., Ballard, C., Ray Chaudhuri, K., et al. (2021). Parkinson disease-associated cognitive impairment. *Nature Reviews. Disease Primers*, 7(1), 47. https://doi.org/10.1038/s41572-021-00280-3.

Abate, S. M., Checkol, Y. A., & Mantefardo, B. (2021). Global prevalence and determinants of mortality among patients with COVID-19: A systematic review and meta-analysis. *Annals of Medicine and Surgery*, *64*, 102204. https://doi.org/10.1016/j.amsu.2021.102204.

Aleem, A., & Kothadia, J. P. (2021). Remdesivir. In *StatPearls*. Treasure Island (FL): StatPearls Publishing (Copyright © 2021, StatPearls Publishing LLC).

Alimoradi, Z., Broström, A., Tsang, H. W. H., Griffiths, M. D., Haghayegh, S., Ohayon, M. M., et al. (2021). Sleep problems during COVID-19 pandemic and its' association to psychological distress: A systematic review and meta-analysis. *EClinicalMedicine*, *36*, 100916. https://doi.org/10.1016/j.eclinm.2021.100916.

Andrews, P. L. R., Cai, W., Rudd, J. A., & Sanger, G. J. (2021). COVID-19, nausea, and vomiting. *Journal of Gastroenterology and Hepatology*, *36*(3), 646–656. https://doi.org/10.1111/jgh.15261.

Antonini, A., Leta, V., Teo, J., & Chaudhuri, K. R. (2020). Outcome of Parkinson's disease patients affected by COVID-19. *Movement Disorders: Official Journal of the Movement Disorder Society*, *35*(6), 905–908. https://doi.org/10.1002/mds.28104.

Artusi, C. A., Romagnolo, A., Imbalzano, G., Marchet, A., Zibetti, M., Rizzone, M. G., et al. (2020). COVID-19 in Parkinson's disease: Report on prevalence and outcome. *Parkinsonism & Related Disorders*, *80*, 7–9. https://doi.org/10.1016/j.parkreldis.2020.09.008.

Arumugham, V. B., Gujarathi, R., & Cascella, M. (2021). *Third generation cephalosporins. In StatPearls*. Treasure Island (FL): StatPearls Publishing (Copyright © 2021, StatPearls Publishing LLC).

Bhidayasiri, R., Virameteekul, S., Kim, J. M., Pal, P. K., & Chung, S. J. (2020). COVID-19: An early review of its global impact and considerations for Parkinson's disease patient care. *The Journal of Movement Disorders*, *13*(2), 105–114. https://doi.org/10.14802/jmd.20042.

Blitshteyn, S., & Whitelaw, S. (2021). Postural orthostatic tachycardia syndrome (POTS) and other autonomic disorders after COVID-19 infection: A case series of 20 patients. *Immunologic Research*, *69*(2), 205–211. https://doi.org/10.1007/s12026-021-09185-5.

Broen, M. P., Narayen, N. E., Kuijf, M. L., Dissanayaka, N. N., & Leentjens, A. F. (2016). Prevalence of anxiety in Parkinson's disease: A systematic review and meta-analysis. *Movement Disorders*, *31*(8), 1125–1133. https://doi.org/10.1002/mds.26643.

Brown, E. G., Chahine, L. M., Goldman, S. M., Korell, M., Mann, E., Kinel, D. R., et al. (2020). The effect of the COVID-19 pandemic on people with Parkinson's disease. *Journal of Parkinson's Disease*, *10*(4), 1365–1377. https://doi.org/10.3233/jpd-202249.

Brundin, P., Nath, A., & Beckham, J. D. (2020). Is COVID-19 a perfect storm for Parkinson's disease? *Trends in Neurosciences*, *43*(12), 931–933. https://doi.org/10.1016/j.tins.2020.10.009.

Cantuti-Castelvetri, L., Ojha, R., Pedro, L. D., Djannatian, M., Franz, J., Kuivanen, S., et al. (2020). Neuropilin-1 facilitates SARS-CoV-2 cell entry and infectivity. *Science (New York, N.Y.)*, *370*(6518), 856–860. https://doi.org/10.1126/science.abd2985.

Cascella, M., Rajnik, M., Aleem, A., Dulebohn, S. C., & Di Napoli, R. (2021). Features, evaluation, and treatment of coronavirus (COVID-19). In *StatPearls*. Treasure Island (FL): StatPearls Publishing (Copyright © 2021, StatPearls Publishing LLC).

Chaudhuri, K. R., Healy, D. G., & Schapira, A. H. (2006). Non-motor symptoms of Parkinson's disease: Diagnosis and management. *Lancet Neurology*, *5*(3), 235–245. https://doi.org/10.1016/s1474-4422(06)70373-8.

Chaudhuri, K. R., Yates, L., & Martinez-Martin, P. (2005). The non-motor symptom complex of Parkinson's disease: A comprehensive assessment is essential. *Current Neurology and Neuroscience Reports*, *5*(4), 275–283. https://doi.org/10.1007/s11910-005-0072-6.

Chen, Y.-Y., Yu, S., Hu, Y.-H., Li, C.-Y., Artaud, F., Carcaillon-Bentata, L., et al. (2021). Risk of suicide among patients with Parkinson disease. *JAMA Psychiatry*, *78*(3), 293–301. https://doi.org/10.1001/jamapsychiatry.2020.4001.

Chinnappa-Quinn, L., Makkar, S. R., Bennett, M., Lam, B. C. P., Lo, J. W., Kochan, N. A., et al. (2020). Is hospitalization a risk factor for cognitive decline in older age adults? *International Psychogeriatrics*, *1-18*. https://doi.org/10.1017/s1041610220001763.

Cilia, R., Bonvegna, S., Straccia, G., Andreasi, N. G., Elia, A. E., Romito, L. M., et al. (2020). Effects of COVID-19 on Parkinson's disease clinical features: A community-based case-control study. *Movement Disorders*, *35*(8), 1287–1292. https://doi.org/10.1002/mds.28170.

Corti, E. J., Johnson, A. R., Gasson, N., Bucks, R. S., Thomas, M. G., & Loftus, A. M. (2018). Factor structure of the ways of coping questionnaire in Parkinson's disease. *Parkinsons Disease*, *2018*, 7128069. https://doi.org/10.1155/2018/7128069.

Crocker, J., Liu, K., Smith, M., Nakamoto, M., Mitchell, C., Zhu, E., et al. (2022). Early impact of the COVID-19 pandemic on outpatient neurologic Care in Hawai'i. *Hawai'i Journal of Health & Social Welfare*, *81*(1), 6–12 (Retrieved from https://pubmed.ncbi.nlm.nih.gov/35028589. https://www.ncbi.nlm.nih.gov/pmc/articles/PMC8742305/).

Damiano, R. F., Guedes, B. F., de Rocca, C. C., de Pádua Serafim, A., Castro, L. H. M., Munhoz, C. D., et al. (2021). Cognitive decline following acute viral infections: Literature review and projections for post-COVID-19. *European Archives of Psychiatry and Clinical Neuroscience, 1-16.* https://doi.org/10.1007/s00406-021-01286-4.

D'Amico, F., Baumgart, D. C., Danese, S., & Peyrin-Biroulet, L. (2020). Diarrhea during COVID-19 infection: Pathogenesis, epidemiology, prevention, and management. *Clinical Gastroenterology and Hepatology, 18*(8), 1663–1672. https://doi.org/10.1016/j.cgh.2020.04.001.

D'Ascanio, L., Pandolfini, M., Cingolani, C., Latini, G., Gradoni, P., Capalbo, M., et al. (2021). Olfactory dysfunction in COVID-19 patients: Prevalence and prognosis for recovering sense of smell. *Otolaryngology and Head and Neck Surgery, 164*(1), 82–86. https://doi.org/10.1177/0194599820943530.

de Freitas, R. F., Torres, S. C., Martín-Sánchez, F. J., Carbó, A. V., Lauria, G., & Nunes, J. P. L. (2021). Syncope and COVID-19 disease - a systematic review. *Autonomic Neuroscience, 235*, 102872. https://doi.org/10.1016/j.autneu.2021.102872.

De Micco, R., Siciliano, M., Sant'Elia, V., Giordano, A., Russo, A., Tedeschi, G., et al. (2021). Correlates of psychological distress in patients with Parkinson's disease during the COVID-19 outbreak. *Movement Disorders Clinical Practice, 8*(1), 60–68. https://doi.org/10.1002/mdc3.13108.

de Rus Jacquet, A., Bogard, S., Normandeau, C. P., Degroot, C., Postuma, R. B., Dupré, N., et al. (2021). Clinical perception and management of Parkinson's disease during the COVID-19 pandemic: A Canadian experience. *Parkinsonism & Related Disorders, 91*, 66–76. https://doi.org/10.1016/j.parkreldis.2021.08.018.

Del Prete, E., Francesconi, A., Palermo, G., Mazzucchi, S., Frosini, D., Morganti, R., et al. (2021). Prevalence and impact of COVID-19 in Parkinson's disease: Evidence from a multi-center survey in Tuscany region. *Journal of Neurology, 268*(4), 1179–1187. https://doi.org/10.1007/s00415-020-10002-6.

Del Rio, R., Marcus, N. J., & Inestrosa, N. C. (2020). Potential role of autonomic dysfunction in Covid-19 morbidity and mortality. *Frontiers in Physiology, 11*, 561749. https://doi.org/10.3389/fphys.2020.561749.

Deshmukh, V., Motwani, R., Kumar, A., Kumari, C., & Raza, K. (2021). Histopathological observations in COVID-19: A systematic review. *Journal of Clinical Pathology, 74*(2), 76–83. https://doi.org/10.1136/jclinpath-2020-206995.

Díaz, H. S., Toledo, C., Andrade, D. C., Marcus, N. J., & Del Rio, R. (2020). Neuroinflammation in heart failure: New insights for an old disease. *The Journal of Physiology, 598*(1), 33–59. https://doi.org/10.1113/jp278864.

Douaud, G., Lee, S., Alfaro-Almagro, F., Arthofer, C., Wang, C., McCarthy, P., et al. (2022). SARS-CoV-2 is associated with changes in brain structure in UK biobank. *Nature*. https://doi.org/10.1038/s41586-022-04569-5.

Douma, E. H., & de Kloet, E. R. (2020). Stress-induced plasticity and functioning of ventral tegmental dopamine neurons. *Neuroscience and Biobehavioral Reviews, 108*, 48–77. https://doi.org/10.1016/j.neubiorev.2019.10.015.

Dubovsky, A. N., Arvikar, S., Stern, T. A., & Axelrod, L. (2012). The neuropsychiatric complications of glucocorticoid use: Steroid psychosis revisited. *Psychosomatics, 53*(2), 103–115. https://doi.org/10.1016/j.psym.2011.12.007.

Ehlenbach, W. J., Hough, C. L., Crane, P. K., Haneuse, S. J., Carson, S. S., Curtis, J. R., et al. (2010). Association between acute care and critical illness hospitalization and cognitive function in older adults. *JAMA, 303*(8), 763–770. https://doi.org/10.1001/jama.2010.167.

El Otmani, H., El Bidaoui, Z., Amzil, R., Bellakhdar, S., El Moutawakil, B., & Abdoh Rafai, M. (2021). No impact of confinement during COVID-19 pandemic on anxiety and depression in parkinsonian patients. *Revue Neurologique (Paris), 177*(3), 272–274. https://doi.org/10.1016/j.neurol.2021.01.005.

El Sayed, S., Shokry, D., & Gomaa, S. M. (2021). Post-COVID-19 fatigue and anhedonia: A cross-sectional study and their correlation to post-recovery period. *Neuropsychopharmacology Reports*, *41*(1), 50–55. https://doi.org/10.1002/npr2.12154.

Elbeddini, A., To, A., Tayefehchamani, Y., & Wen, C. (2020). Potential impact and challenges associated with Parkinson's disease patient care amidst the COVID-19 global pandemic. *Journal of Clinical Movement Disorders*, 7, 7. https://doi.org/10.1186/s40734-020-00089-4.

Elfil, M., Kamel, S., Kandil, M., Koo, B. B., & Schaefer, S. M. (2020). Implications of the gut microbiome in Parkinson's disease. *Movement Disorders*, *35*(6), 921–933. https://doi.org/10.1002/mds.28004.

Eliasen, A., Dalhoff, K. P., & Horwitz, H. (2018). Neurological diseases and risk of suicide attempt: A case-control study. *Journal of Neurology*, *265*(6), 1303–1309. https://doi.org/10.1007/s00415-018-8837-4.

El-Qushayri, A. E., Ghozy, S., Reda, A., Kamel, A. M. A., Abbas, A. S., & Dmytriw, A. A. (2021). The impact of Parkinson's disease on manifestations and outcomes of Covid-19 patients: A systematic review and meta-analysis. *Reviews in Medical Virology*, *e2278*. https://doi.org/10.1002/rmv.2278.

Ely, E. W., Shintani, A., Truman, B., Speroff, T., Gordon, S. M., Harrell, F. E., Jr., et al. (2004). Delirium as a predictor of mortality in mechanically ventilated patients in the intensive care unit. *JAMA*, *291*(14), 1753–1762. https://doi.org/10.1001/jama.291.14.1753.

Erlangsen, A., Stenager, E., Conwell, Y., Andersen, P. K., Hawton, K., Benros, M. E., et al. (2020). Association between neurological disorders and death by suicide in Denmark. *JAMA*, *323*(5), 444–454. https://doi.org/10.1001/jama.2019.21834.

Fabbri, M., Leung, C., Baille, G., Béreau, M., Brefel Courbon, C., Castelnovo, G., et al. (2021). A French survey on the lockdown consequences of COVID-19 pandemic in Parkinson's disease. The ERCOPARK study. *Parkinsonism & Related Disorders*, *89*, 128–133. https://doi.org/10.1016/j.parkreldis.2021.07.013.

Fasano, A., Visanji, N. P., Liu, L. W., Lang, A. E., & Pfeiffer, R. F. (2015). Gastrointestinal dysfunction in Parkinson's disease. *Lancet Neurology*, *14*(6), 625–639. https://doi.org/10.1016/s1474-4422(15)00007-1.

Felger, J. C., & Miller, A. H. (2012). Cytokine effects on the basal ganglia and dopamine function: The subcortical source of inflammatory malaise. *Frontiers in Neuroendocrinology*, *33*(3), 315–327. https://doi.org/10.1016/j.yfrne.2012.09.003.

Fernández-Castañeda, A., Lu, P., Geraghty, A. C., Song, E., Lee, M.-H., Wood, J., et al. (2022). Mild respiratory SARS-CoV-2 infection can cause multi-lineage cellular dysregulation and myelin loss in the brain. *bioRxiv*. https://doi.org/10.1101/2022.01.07.475453. 2022.2001.2007.475453.

Gjerde, K. V., Müller, B., Skeie, G. O., Assmus, J., Alves, G., & Tysnes, O. B. (2018). Hyposmia in a simple smell test is associated with accelerated cognitive decline in early Parkinson's disease. *Acta Neurologica Scandinavica*, *138*(6), 508–514. https://doi.org/10.1111/ane.13003.

Godoy, L. D., Rossignoli, M. T., Delfino-Pereira, P., Garcia-Cairasco, N., & de Lima Umeoka, E. H. (2018). A comprehensive overview on stress neurobiology: Basic concepts and clinical implications. *Frontiers in Behavioral Neuroscience*, *12*, 127. https://doi.org/10.3389/fnbeh.2018.00127.

Gracner, T., Agarwal, M., Murali, K. P., Stone, P. W., Larson, E. L., Furuya, E. Y., et al. (2021). Association of Infection-Related Hospitalization with Cognitive Impairment among Nursing Home Residents. *JAMA Network Open*, *4*(4), e217528. https://doi.org/10.1001/jamanetworkopen.2021.7528.

Gualano, M. R., Lo Moro, G., Voglino, G., Bert, F., & Siliquini, R. (2020). Effects of Covid-19 lockdown on mental health and sleep disturbances in Italy. *International Journal of Environmental Research and Public Health*, *17*(13). https://doi.org/10.3390/ijerph17134779.

Guerrero, J. I., Barragán, L. A., Martínez, J. D., Montoya, J. P., Peña, A., Sobrino, F. E., et al. (2021). Central and peripheral nervous system involvement by COVID-19: A systematic review of the pathophysiology, clinical manifestations, neuropathology, neuroimaging, electrophysiology, and cerebrospinal fluid findings. *BMC Infectious Diseases*, *21*(1), 515. https://doi.org/10.1186/s12879-021-06185-6.

Guo, D., Han, B., Lu, Y., Lv, C., Fang, X., Zhang, Z., et al. (2020). Influence of the COVID-19 pandemic on quality of life of patients with Parkinson's disease. *Parkinsons Disease*, *2020*, 1216568. https://doi.org/10.1155/2020/1216568.

Haas, A. N., Passos-Monteiro, E., Delabary, M. D. S., Moratelli, J., Schuch, F. B., Correa, C. L., et al. (2022). Association between mental health and physical activity levels in people with Parkinson's disease during the COVID-19 pandemic: An observational cross-sectional survey in Brazil. *Sport Science for Health*, *1*-7. https://doi.org/10.1007/s11332-021-00868-y.

Haehner, A., Hummel, T., & Reichmann, H. (2011). Olfactory loss in Parkinson's disease. *Parkinson's Disease*, *2011*, 450939. https://doi.org/10.4061/2011/450939.

Hainque, E., & Grabli, D. (2020). Rapid worsening in Parkinson's disease may hide COVID-19 infection. *Parkinsonism & Related Disorders*, *75*, 126–127. https://doi.org/10.1016/j.parkreldis.2020.05.008.

Hariyanto, T. I., Rizki, N. A., & Kurniawan, A. (2021). Anosmia/hyposmia is a good predictor of coronavirus disease 2019 (COVID-19) infection: A meta-analysis. *International Archives of Otorhinolaryngology*, *25*(1), e170–e174. https://doi.org/10.1055/s-0040-1719120.

Hassani, M., Fathi Jouzdani, A., Motarjem, S., Ranjbar, A., & Khansari, N. (2021). How COVID-19 can cause autonomic dysfunctions and postural orthostatic syndrome? A review of mechanisms and evidence. *Neurology and Clinical Neuroscience*, *9*(6), 434–442. https://doi.org/10.1111/ncn3.12548.

Hawkes, C. H., Del Tredici, K., & Braak, H. (2007). Parkinson's disease: A dual-hit hypothesis. *Neuropathology and Applied Neurobiology*, *33*(6), 599–614. https://doi.org/10.1111/j.1365-2990.2007.00874.x.

He, R., Zhao, Y., He, Y., Zhou, Y., Yang, J., Zhou, X., et al. (2020). Olfactory dysfunction predicts disease progression in Parkinson's disease: A longitudinal study. *Frontiers in Neuroscience*, *14*, 569777. https://doi.org/10.3389/fnins.2020.569777.

Helmich, R. C., & Bloem, B. R. (2020). The impact of the COVID-19 pandemic on Parkinson's disease: Hidden sorrows and emerging opportunities. *Journal of Parkinson's Disease*, *10*(2), 351–354. https://doi.org/10.3233/JPD-202038.

Hoffmann, M., Kleine-Weber, H., Schroeder, S., Krüger, N., Herrler, T., Erichsen, S., et al. (2020). SARS-CoV-2 cell entry depends on ACE2 and TMPRSS2 and is blocked by a clinically proven protease inhibitor. *Cell*, *181*(2), 271–280. e278 https://doi.org/10.1016/j.cell.2020.02.052.

HØrmann Thomsen, T., Wallerstedt, S. M., Winge, K., & Bergquist, F. (2021). Life with Parkinson's disease during the COVID-19 pandemic: The pressure is "OFF". *Journal of Parkinson's Disease*, *11*(2), 491–495. https://doi.org/10.3233/jpd-202342.

Jahrami, H., BaHammam, A. S., Bragazzi, N. L., Saif, Z., Faris, M., & Vitiello, M. V. (2021). Sleep problems during the COVID-19 pandemic by population: A systematic review and meta-analysis. *Journal of Clinical Sleep Medicine*, *17*(2), 299–313. https://doi.org/10.5664/jcsm.8930.

Jaywant, A., Vanderlind, W. M., Alexopoulos, G. S., Fridman, C. B., Perlis, R. H., & Gunning, F. M. (2021). Frequency and profile of objective cognitive deficits in hospitalized patients recovering from COVID-19. *Neuropsychopharmacology: Official Publication of the American College of Neuropsychopharmacology*, *46*(13), 2235–2240. https://doi.org/10.1038/s41386-021-00978-8.

Kalimisetty, S., Askar, W., Fay, B., & Khan, A. (2017). Models for predicting incident delirium in hospitalized older adults: A systematic review. *Journal of Patient-Centered Research and Reviews*, *4*(2), 69–77. https://doi.org/10.17294/2330-0698.1414.

Kano, O., Ikeda, K., Cridebring, D., Takazawa, T., Yoshii, Y., & Iwasaki, Y. (2011). Neurobiology of depression and anxiety in Parkinson's disease. *Parkinsons Disease, 2011*, 143547. https://doi.org/10.4061/2011/143547.

Kariyawasam, J. C., Jayarajah, U., Riza, R., Abeysuriya, V., & Seneviratne, S. L. (2021). Gastrointestinal manifestations in COVID-19. *Transactions of the Royal Society of Tropical Medicine and Hygiene, 115*(12), 1362–1388. https://doi.org/10.1093/trstmh/trab042.

Karlsen, K. H., Larsen, J. P., Tandberg, E., & Maeland, J. G. (1999). Influence of clinical and demographic variables on quality of life in patients with Parkinson's disease. *Journal of Neurology, Neurosurgery, and Psychiatry, 66*(4), 431–435. https://doi.org/10.1136/jnnp.66.4.431.

Kelly, M. J., Lawton, M. A., Baig, F., Ruffmann, C., Barber, T. R., Lo, C., et al. (2019). Predictors of motor complications in early Parkinson's disease: A prospective cohort study. *Movement Disorders, 34*(8), 1174–1183. https://doi.org/10.1002/mds.27783.

Kim, H. S. (2021). Do an altered gut microbiota and an associated leaky gut affect COVID-19 severity? *MBio, 12*(1), e03022-03020. https://doi.org/10.1128/mBio.03022-20.

Kitani-Morii, F., Kasai, T., Horiguchi, G., Teramukai, S., Ohmichi, T., Shinomoto, M., et al. (2021). Risk factors for neuropsychiatric symptoms in patients with Parkinson's disease during COVID-19 pandemic in Japan. *PLoS One, 16*(1), e0245864. https://doi.org/10.1371/journal.pone.0245864.

Klingelhoefer, L., & Reichmann, H. (2015). Pathogenesis of Parkinson disease—The gut-brain axis and environmental factors. *Nature Reviews. Neurology, 11*(11), 625–636. https://doi.org/10.1038/nrneurol.2015.197.

Kocevska, D., Blanken, T. F., Van Someren, E. J. W., & Rösler, L. (2020). Sleep quality during the COVID-19 pandemic: Not one size fits all. *Sleep Medicine, 76*, 86–88. https://doi.org/10.1016/j.sleep.2020.09.029.

Krey, L., Huber, M. K., Höglinger, G. U., & Wegner, F. (2021). Can SARS-CoV-2 infection Lead to neurodegeneration and Parkinson's disease? *Brain Sciences, 11*(12). https://doi.org/10.3390/brainsci11121654.

Krzyszton, K., Mielanczuk-Lubecka, B., Stolarski, J., Poznanska, A., Kepczynska, K., Zdrowowicz, A., et al. (2021). Secondary impact of COVID-19 pandemic on people with Parkinson's disease-results of a polish online survey. *Brain Sciences, 12*(1). https://doi.org/10.3390/brainsci12010026.

Kumar, N., Gupta, R., Kumar, H., Mehta, S., Rajan, R., Kumar, D., et al. (2021). Impact of home confinement during COVID-19 pandemic on sleep parameters in Parkinson's disease. *Sleep Medicine, 77*, 15–22. https://doi.org/10.1016/j.sleep.2020.11.021.

Lazcano-Ocampo, C., Wan, Y. M., van Wamelen, D. J., Batzu, L., Boura, I., Titova, N., et al. (2020). Identifying and responding to fatigue and apathy in Parkinson's disease: A review of current practice. *Expert Review of Neurotherapeutics, 20*(5), 477–495. https://doi.org/10.1080/14737175.2020.1752669.

Lechien, J. R., Chiesa-Estomba, C. M., Place, S., Van Laethem, Y., Cabaraux, P., Mat, Q., et al. (2020). Clinical and epidemiological characteristics of 1420 European patients with mild-to-moderate coronavirus disease 2019. *Journal of Internal Medicine, 288*(3), 335–344. https://doi.org/10.1111/joim.13089.

Leta, V., Rodríguez-Violante, M., Abundes, A., Rukavina, K., Teo, J. T., Falup-Pecurariu, C., et al. (2021). Parkinson's disease and post-COVID-19 syndrome: The Parkinson's long-COVID Spectrum. *Movement Disorders, 36*(6), 1287–1289. https://doi.org/10.1002/mds.28622.

Li, W., Abbas, M. M., Acharyya, S., Ng, H. L., Tay, K. Y., Au, W. L., et al. (2018). Suicide in Parkinson's disease. *Movement Disorders Clinical Practice, 5*(2), 177–182. https://doi.org/10.1002/mdc3.12599.

Lin, L., Jiang, X., Zhang, Z., Huang, S., Zhang, Z., Fang, Z., et al. (2020). Gastrointestinal symptoms of 95 cases with SARS-CoV-2 infection. *Gut, 69*(6), 997–1001. https://doi.org/10.1136/gutjnl-2020-321013.

Liu, Y. H., Wang, Y. R., Wang, Q. H., Chen, Y., Chen, X., Li, Y., et al. (2021). Post-infection cognitive impairments in a cohort of elderly patients with COVID-19. *Molecular Neurodegeneration, 16*(1), 48. https://doi.org/10.1186/s13024-021-00469-w.

Lubomski, M., Davis, R. L., & Sue, C. M. (2020). Gastrointestinal dysfunction in Parkinson's disease. *Journal of Neurology, 267*(5), 1377–1388. https://doi.org/10.1007/s00415-020-09723-5.

Luis-Martínez, R., Di Marco, R., Weis, L., Cianci, V., Pistonesi, F., Baba, A., et al. (2021). Impact of social and mobility restrictions in Parkinson's disease during COVID-19 lockdown. *BMC Neurology, 21*(1), 332. https://doi.org/10.1186/s12883-021-02364-9.

Marcantonio, E. R. (2017). Delirium in Hospitalized Older Adults. *The New England Journal of Medicine, 377*(15), 1456–1466. https://doi.org/10.1056/NEJMcp1605501.

Martinez-Martin, P., Rodriguez-Blazquez, C., Kurtis, M. M., & Chaudhuri, K. R. (2011). The impact of non-motor symptoms on health-related quality of life of patients with Parkinson's disease. *Movement Disorders, 26*(3), 399–406. https://doi.org/10.1002/mds.23462.

Matsuda, K., Park, C. H., Sunden, Y., Kimura, T., Ochiai, K., Kida, H., et al. (2004). The vagus nerve is one route of transneural invasion for intranasally inoculated influenza a virus in mice. *Veterinary Pathology, 41*(2), 101–107. https://doi.org/10.1354/vp.41-2-101.

McRae, C., Fazio, E., Hartsock, G., Kelley, L., Urbanski, S., & Russell, D. (2009). Predictors of loneliness in caregivers of persons with Parkinson's disease. *Parkinsonism & Related Disorders, 15*(8), 554–557. https://doi.org/10.1016/j.parkreldis.2009.01.007.

Meinhardt, J., Radke, J., Dittmayer, C., Franz, J., Thomas, C., Mothes, R., et al. (2021). Olfactory transmucosal SARS-CoV-2 invasion as a port of central nervous system entry in individuals with COVID-19. *Nature Neuroscience, 24*(2), 168–175. https://doi.org/10.1038/s41593-020-00758-5.

Menozzi, E., Macnaughtan, J., & Schapira, A. H. V. (2021). The gut-brain axis and Parkinson disease: Clinical and pathogenetic relevance. *Annals of Medicine, 53*(1), 611–625. https://doi.org/10.1080/07853890.2021.1890330.

Mönkemüller, K., Fry, L., & Rickes, S. (2020). COVID-19, coronavirus, SARS-CoV-2 and the small bowel. *Revista Española de Enfermedades Digestivas, 112*(5), 383–388. https://doi.org/10.17235/reed.2020.7137/2020.

Montanaro, E., Artusi, C. A., Rosano, C., Boschetto, C., Imbalzano, G., Romagnolo, A., et al. (2022). Anxiety, depression, and worries in advanced Parkinson disease during COVID-19 pandemic. *Neurological Sciences, 43*(1), 341–348. https://doi.org/10.1007/s10072-021-05286-z.

Negrini, F., Ferrario, I., Mazziotti, D., Berchicci, M., Bonazzi, M., de Sire, A., et al. (2021). Neuropsychological features of severe hospitalized coronavirus disease 2019 patients at clinical stability and clues for Postacute rehabilitation. *Archives of Physical Medicine and Rehabilitation, 102*(1), 155–158. https://doi.org/10.1016/j.apmr.2020.09.376.

Neu, D., Kajosch, H., Peigneux, P., Verbanck, P., Linkowski, P., & Le Bon, O. (2011). Cognitive impairment in fatigue and sleepiness associated conditions. *Psychiatry Research, 189*(1), 128–134. https://doi.org/10.1016/j.psychres.2010.12.005.

Oates, C. P., Turagam, M. K., Musikantow, D., Chu, E., Shivamurthy, P., Lampert, J., et al. (2020). Syncope and presyncope in patients with COVID-19. *Pacing and Clinical Electrophysiology, 43*(10), 1139–1148. https://doi.org/10.1111/pace.14047.

Oh, C. M., Kim, H. Y., Na, H. K., Cho, K. H., & Chu, M. K. (2019). The effect of anxiety and depression on sleep quality of individuals with high risk for insomnia: A population-based study. *Frontiers in Neurology, 10*, 849. https://doi.org/10.3389/fneur.2019.00849.

Oppo, V., Serra, G., Fenu, G., Murgia, D., Ricciardi, L., Melis, M., et al. (2020). Parkinson's disease symptoms have a distinct impact on Caregivers' and Patients' stress: A study assessing the consequences of the COVID-19 lockdown. *Movement Disorders Clinical Practice*, 7(7), 865–867. https://doi.org/10.1002/mdc3.13030.

Paik, S. H., Ahn, J. S., Min, S., Park, K. C., & Kim, M. H. (2018). Impact of psychotic symptoms on clinical outcomes in delirium. *PLoS One*, *13*(7), e0200538. https://doi.org/10.1371/journal.pone.0200538.

Palermo, G., Tommasini, L., Baldacci, F., Del Prete, E., Siciliano, G., & Ceravolo, R. (2020). Impact of coronavirus disease 2019 pandemic on cognition in Parkinson's disease. *Movement Disorders*, *35*(10), 1717–1718. https://doi.org/10.1002/mds.28254.

Pantell, M., Rehkopf, D., Jutte, D., Syme, S. L., Balmes, J., & Adler, N. (2013). Social isolation: A predictor of mortality comparable to traditional clinical risk factors. *American Journal of Public Health*, *103*(11), 2056–2062. https://doi.org/10.2105/ajph.2013.301261.

Parra, A., Juanes, A., Losada, C. P., Álvarez-Sesmero, S., Santana, V. D., Martí, I., et al. (2020). Psychotic symptoms in COVID-19 patients. A retrospective descriptive study. *Psychiatry Research*, *291*, 113254. https://doi.org/10.1016/j.psychres.2020.113254.

Patel, P. H., & Hashmi, M. F. (2021). Macrolides. In *StatPearls*. Treasure Island (FL): StatPearls Publishing (Copyright © 2021, StatPearls Publishing LLC).

Picillo, M., Palladino, R., Erro, R., Alfano, R., Colosimo, C., Marconi, R., et al. (2021). The PRIAMO study: Age- and sex-related relationship between prodromal constipation and disease phenotype in early Parkinson's disease. *Journal of Neurology*, *268*(2), 448–454. https://doi.org/10.1007/s00415-020-10156-3.

Poewe, W., Seppi, K., Tanner, C. M., Halliday, G. M., Brundin, P., Volkmann, J., et al. (2017). Parkinson disease. *Nature Reviews. Disease Primers*, *3*(1), 17013. https://doi.org/10.1038/nrdp.2017.13.

Potvin, O., Hudon, C., Dion, M., Grenier, S., & Préville, M. (2011). Anxiety disorders, depressive episodes and cognitive impairment no dementia in community-dwelling older men and women. *International Journal of Geriatric Psychiatry*, *26*(10), 1080–1088. https://doi.org/10.1002/gps.2647.

Prasad, S., Holla, V. V., Neeraja, K., Surisetti, B. K., Kamble, N., Yadav, R., et al. (2020). Parkinson's disease and COVID-19: Perceptions and implications in patients and caregivers. *Movement Disorders*, *35*(6), 912–914. https://doi.org/10.1002/mds.28088.

Pun, B. T., Badenes, R., Heras La Calle, G., Orun, O. M., Chen, W., Raman, R., et al. (2021). Prevalence and risk factors for delirium in critically ill patients with COVID-19 (COVID-D): A multicentre cohort study. *The Lancet. Respiratory Medicine*, *9*(3), 239–250. https://doi.org/10.1016/S2213-2600(20)30552-X.

Rangon, C.-M., Krantic, S., Moyse, E., & Fougère, B. (2020). The vagal autonomic pathway of COVID-19 at the crossroad of Alzheimer's disease and aging: A review of knowledge. *Journal of Alzheimer's Disease Reports*, *4*(1), 537–551. https://doi.org/10.3233/ADR-200273.

Robbins, T. W., & Cools, R. (2014). Cognitive deficits in Parkinson's disease: A cognitive neuroscience perspective. *Movement Disorders*, *29*(5), 597–607. https://doi.org/10.1002/mds.25853.

Rock, P. L., Roiser, J. P., Riedel, W. J., & Blackwell, A. D. (2014). Cognitive impairment in depression: A systematic review and meta-analysis. *Psychological Medicine*, *44*(10), 2029–2040. https://doi.org/10.1017/s0033291713002535.

Romano, S., Savva, G. M., Bedarf, J. R., Charles, I. G., Hildebrand, F., & Narbad, A. (2021). Meta-analysis of the Parkinson's disease gut microbiome suggests alterations linked to intestinal inflammation. *Npj. Parkinson's Disease*, 7(1), 27. https://doi.org/10.1038/s41531-021-00156-z.

Rukavina, K., Leta, V., Sportelli, C., Buhidma, Y., Duty, S., Malcangio, M., et al. (2019). Pain in Parkinson's disease: New concepts in pathogenesis and treatment. *Current Opinion in Neurology*, *32*(4), 579–588. https://doi.org/10.1097/wco.0000000000000711.

Salari, M., Zali, A., Ashrafi, F., Etemadifar, M., Sharma, S., Hajizadeh, N., et al. (2020). Incidence of anxiety in Parkinson's disease during the coronavirus disease (COVID-19) pandemic. *Movement Disorders*, *35*(7), 1095–1096. https://doi.org/10.1002/mds.28116.

Saldanha, D., Menon, P., Chaudari, B., Bhattacharya, L., & Guliani, S. (2013). Acute psychosis: A neuropsychiatric dilemma. *Industrial Psychiatry Journal*, *22*(2), 157–158. https://doi.org/10.4103/0972-6748.132933.

Santos-García, D., Oreiro, M., Pérez, P., Fanjul, G., Paz González, J. M., Feal Painceiras, M. J., et al. (2020). Impact of coronavirus disease 2019 pandemic on Parkinson's disease: A cross-sectional survey of 568 Spanish patients. *Movement Disorders*, *35*(10), 1712–1716. https://doi.org/10.1002/mds.28261.

Schapira, A. H. V., Chaudhuri, K. R., & Jenner, P. (2017). Non-motor features of Parkinson disease. *Nature Reviews. Neuroscience*, *18*(8), 509. https://doi.org/10.1038/nrn.2017.91.

Schirinzi, T., Cerroni, R., Di Lazzaro, G., Liguori, C., Scalise, S., Bovenzi, R., et al. (2020). Self-reported needs of patients with Parkinson's disease during COVID-19 emergency in Italy. *Neurological Sciences*, *41*(6), 1373–1375. https://doi.org/10.1007/s10072-020-04442-1.

Shalash, A., Roushdy, T., Essam, M., Fathy, M., Dawood, N. L., Abushady, E. M., et al. (2020). Mental health, physical activity, and quality of life in Parkinson's disease during COVID-19 pandemic. *Movement Disorders*, *35*(7), 1097–1099. https://doi.org/10.1002/mds.28134.

Silva Andrade, B., Siqueira, S., de Assis Soares, W. R., de Souza Rangel, F., Santos, N. O., Dos Santos Freitas, A., et al. (2021). Long-COVID and post-COVID health complications: An up-to-date review on clinical conditions and their possible molecular mechanisms. *Viruses*, *13*(4), 700. https://doi.org/10.3390/v13040700.

Silva, F., Brito, B. B., Santos, M. L. C., Marques, H. S., Silva Júnior, R. T. D., Carvalho, L. S., et al. (2020). COVID-19 gastrointestinal manifestations: A systematic review. *Revista da Sociedade Brasileira de Medicina Tropical*, *53*, e20200714. https://doi.org/10.1590/0037-8682-0714-2020.

Silva, A. R., Regueira, P., Cardoso, A. L., Baldeiras, I., Santana, I., & Cerejeira, J. (2021). Cognitive trajectories following acute infection in older patients with and without cognitive impairment: An 1-year follow-up study. *Frontiers in Psychiatry*, *12*, 754489. https://doi.org/10.3389/fpsyt.2021.754489.

Simonet, C., Tolosa, E., Camara, A., & Valldeoriola, F. (2020). Emergencies and critical issues in Parkinson's disease. *Practical Neurology*, *20*(1), 15. https://doi.org/10.1136/practneurol-2018-002075.

Smith, C. M., Gilbert, E. B., Riordan, P. A., Helmke, N., von Isenburg, M., Kincaid, B. R., et al. (2021). COVID-19-associated psychosis: A systematic review of case reports. *General Hospital Psychiatry*, *73*, 84–100. https://doi.org/10.1016/j.genhosppsych.2021.10.003.

Song, J., Ahn, J. H., Choi, I., Mun, J. K., Cho, J. W., & Youn, J. (2020). The changes of exercise pattern and clinical symptoms in patients with Parkinson's disease in the era of COVID-19 pandemic. *Parkinsonism & Related Disorders*, *80*, 148–151. https://doi.org/10.1016/j.parkreldis.2020.09.034.

Subramanian, I., Farahnik, J., & Mischley, L. K. (2020). Synergy of pandemics-social isolation is associated with worsened Parkinson severity and quality of life. *NPJ Parkinsons Disease*, *6*, 28. https://doi.org/10.1038/s41531-020-00128-9.

Sulzer, D., Antonini, A., Leta, V., Nordvig, A., Smeyne, R. J., Goldman, J. E., et al. (2020). COVID-19 and possible links with Parkinson's disease and parkinsonism: From bench to bedside. *NPJ Parkinsons Disease*, *6*, 18. https://doi.org/10.1038/s41531-020-00123-0.

Suzuki, K., Numao, A., Komagamine, T., Haruyama, Y., Kawasaki, A., Funakoshi, K., et al. (2021). Impact of the COVID-19 pandemic on the quality of life of patients with Parkinson's disease and their caregivers: A single-center survey in Tochigi prefecture. *Journal of Parkinson's Disease, 11*(3), 1047–1056. https://doi.org/10.3233/JPD-212560.

Tan, A. H., Mahadeva, S., Thalha, A. M., Gibson, P. R., Kiew, C. K., Yeat, C. M., et al. (2014). Small intestinal bacterial overgrowth in Parkinson's disease. *Parkinsonism & Related Disorders, 20*(5), 535–540. https://doi.org/10.1016/j.parkreldis.2014.02.019.

Taylor, D. J., Lichstein, K. L., Durrence, H. H., Reidel, B. W., & Bush, A. J. (2005). Epidemiology of insomnia, depression, and anxiety. *Sleep, 28*(11), 1457–1464. https://doi.org/10.1093/sleep/28.11.1457.

Ticinesi, A., Cerundolo, N., Parise, A., Nouvenne, A., Prati, B., Guerra, A., et al. (2020). Delirium in COVID-19: Epidemiology and clinical correlations in a large group of patients admitted to an academic hospital. *Aging Clinical and Experimental Research, 32*(10), 2159–2166. https://doi.org/10.1007/s40520-020-01699-6.

Tuzun, S., Keles, A., Okutan, D., Yildiran, T., & Palamar, D. (2021). Assessment of musculoskeletal pain, fatigue and grip strength in hospitalized patients with COVID-19. *European Journal of Physical and Rehabilitation Medicine, 57*(4), 653–662. https://doi.org/10.23736/s1973-9087.20.06563-6.

van der Heide, A., Meinders, M. J., Bloem, B. R., & Helmich, R. C. (2020). The impact of the COVID-19 pandemic on psychological distress, physical activity, and symptom severity in Parkinson's disease. *Journal of Parkinson's Disease, 10*(4), 1355–1364. https://doi.org/10.3233/jpd-202251.

van Wamelen, D. J., Wan, Y. M., Ray Chaudhuri, K., & Jenner, P. (2020). Stress and cortisol in Parkinson's disease. *International Review of Neurobiology, 152*, 131–156. https://doi.org/10.1016/bs.irn.2020.01.005.

Varatharaj, A., Thomas, N., Ellul, M. A., Davies, N. W. S., Pollak, T. A., Tenorio, E. L., et al. (2020). Neurological and neuropsychiatric complications of COVID-19 in 153 patients: A UK-wide surveillance study. *The Lancet. Psychiatry*, 7(10), 875–882. https://doi.org/10.1016/S2215-0366(20)30287-X.

Vardy, E. R., Teodorczuk, A., & Yarnall, A. J. (2015). Review of delirium in patients with Parkinson's disease. *Journal of Neurology, 262*(11), 2401–2410. https://doi.org/10.1007/s00415-015-7760-1.

Villapol, S. (2020). Gastrointestinal symptoms associated with COVID-19: Impact on the gut microbiome. *Translational Research, 226*, 57–69. https://doi.org/10.1016/j.trsl.2020.08.004.

Wakefield, B. J. (2002). Risk for acute confusion on hospital admission. *Clinical Nursing Research, 11*(2), 153–172. https://doi.org/10.1177/105477380201100205.

Wan, D., Du, T., Hong, W., Chen, L., Que, H., Lu, S., et al. (2021). Neurological complications and infection mechanism of SARS-COV-2. *Signal Transduction and Targeted Therapy, 6*(1), 406. https://doi.org/10.1038/s41392-021-00818-7.

Webster, R., & Holroyd, S. (2000). Prevalence of psychotic symptoms in delirium. *Psychosomatics, 41*(6), 519–522. https://doi.org/10.1176/appi.psy.41.6.519.

Weems, C. F., Costa, N. M., Dehon, C., & Berman, S. L. (2004). Paul Tillich's theory of existential anxiety: A preliminary conceptual and empirical examination. *Anxiety, Stress, and Coping, 17*(4), 383–399. https://doi.org/10.1080/10615800412331318616.

Weng, L. M., Su, X., & Wang, X. Q. (2021). Pain symptoms in patients with coronavirus disease (COVID-19): A literature review. *Journal of Pain Research, 14*, 147–159. https://doi.org/10.2147/jpr.S269206.

Whitworth, S. R., Loftus, A. M., Skinner, T. C., Gasson, N., Barker, R. A., Bucks, R. S., et al. (2013). Personality affects aspects of health-related quality of life in Parkinson's disease via psychological coping strategies. *Journal of Parkinson's Disease, 3*(1), 45–53. https://doi.org/10.3233/JPD-120149.

Woo, M. S., Malsy, J., Pöttgen, J., Seddiq Zai, S., Ufer, F., Hadjilaou, A., et al. (2020). Frequent neurocognitive deficits after recovery from mild COVID-19. *Brain Communications*, *2*(2), fcaa205. https://doi.org/10.1093/braincomms/fcaa205.

World Health Organization. (2022). *WHO coronavirus (COVID-19) dashboard*. Retrieved February 4, 2022, from. https://covid19.who.int/region/searo/country/in.

Xia, Y., Kou, L., Zhang, G., Han, C., Hu, J., Wan, F., et al. (2020). Investigation on sleep and mental health of patients with Parkinson's disease during the coronavirus disease 2019 pandemic. *Sleep Medicine*, *75*, 428–433. https://doi.org/10.1016/j.sleep.2020.09.011.

Xiao, Q., Chen, S., & Le, W. (2014). Hyposmia: A possible biomarker of Parkinson's disease. *Neuroscience Bulletin*, *30*(1), 134–140. https://doi.org/10.1007/s12264-013-1390-3.

Xu, Y., Surface, M., Chan, A. K., Halpern, J., Vanegas-Arroyave, N., Ford, B., et al. (2021). COVID-19 manifestations in people with Parkinson's disease: A USA cohort. *Journal of Neurology*, *1*-7. https://doi.org/10.1007/s00415-021-10784-3.

Yan, A., & Bryant, E. E. (2021). Quinolones. In *StatPearls*. Treasure Island (FL): StatPearls Publishing (Copyright © 2021, StatPearls Publishing LLC).

Ye, Q., Wang, B., Zhang, T., Xu, J., & Shang, S. (2020). The mechanism and treatment of gastrointestinal symptoms in patients with COVID-19. *American Journal of Physiology. Gastrointestinal and Liver Physiology*, *319*(2), G245–G252. https://doi.org/10.1152/ajpgi.00148.2020.

Yusuf, F., Fahriani, M., Mamada, S. S., Frediansyah, A., Abubakar, A., Maghfirah, D., et al. (2021). Global prevalence of prolonged gastrointestinal symptoms in COVID-19 survivors and potential pathogenesis: A systematic review and meta-analysis. *F1000Res*, *10*, 301. https://doi.org/10.12688/f1000research.52216.1.

Zhou, H., Lu, S., Chen, J., Wei, N., Wang, D., Lyu, H., et al. (2020). The landscape of cognitive function in recovered COVID-19 patients. *Journal of Psychiatric Research*, *129*, 98–102. https://doi.org/10.1016/j.jpsychires.2020.06.022.

Zipprich, H. M., Teschner, U., Witte, O. W., Schonenberg, A., & Prell, T. (2020). Knowledge, attitudes, practices, and burden during the COVID-19 pandemic in people with Parkinson's disease in Germany. *Journal of Clinical Medicine*, *9*(6). https://doi.org/10.3390/jcm9061643.

CHAPTER SEVEN

Treatment paradigms in Parkinson's Disease and Covid-19

Iro Boura[a,b,c,*], Lucia Batzu[b,c], Espen Dietrichs[d], and Kallol Ray Chaudhuri[b,c]

[a]Medical School, University of Crete, Heraklion, Crete, Greece
[b]Department of Basic and Clinical Neuroscience, Institute of Psychiatry, Psychology & Neuroscience, King's College London, London, United Kingdom
[c]Parkinson's Foundation Centre of Excellence, King's College Hospital NHS Foundation Trust, London, United Kingdom
[d]Department of Neurology, Oslo University Hospital and University of Oslo, Oslo, Norway
*Corresponding author: e-mail addresses: boura.iro@gmail.com; iro.boura@kcl.ac.uk

Contents

Abstract

People with Parkinson's Disease (PwP) may be at higher risk for complications from the Coronavirus Disease 2019 (Covid-19) due to older age and to the multi-faceted nature of Parkinson's Disease (PD) per se, presenting with a variety of motor and non-motor symptoms. Those on advanced therapies may be particularly vulnerable. Taking the above into consideration, along with the potential multi-systemic impact of Covid-19

International Review of Neurobiology, Volume 165
ISSN 0074-7742
https://doi.org/10.1016/bs.irn.2022.03.002
Copyright © 2022 Elsevier Inc.
All rights reserved.

on affected patients and the complications of hospitalization, we are providing an evidence-based guidance to ensure a high standard of care for PwP affected by Covid-19 with varying severity of the condition. Adherence to the dopaminergic medication of PwP, without abrupt modifications in dosage and frequency, is of utmost importance, while potential interactions with newly introduced drugs should always be considered. Treating physicians should be cautious to acknowledge and timely address any potential complications, while consultation by a neurologist, preferably with special knowledge on movement disorders, is advised for patients admitted in non-neurological wards. Non-pharmacological approaches, including the patient's mobilization, falls prevention, good sleep hygiene, emotional support, and adequate nutritional and fluid intake, are essential and the role of telemedicine services should be strengthened and encouraged.

1. Introduction

Over the past 2 years the epidemic disease caused by Severe Acute Respiratory Syndrome Coronavirus-2 (SARS-CoV-2) has spread worldwide and profoundly affected the global morbidity and mortality (Abate, Checkol, & Mantefardo, 2021; Piroth et al., 2021). Older age (over 60), along with coexistent comorbidities, were found to increase the risk for severe or critical Coronavirus disease 2019 (Covid-19) (Cascella, Rajnik, Aleem, Dulebohn, & Di Napoli, 2021). More specifically, a large study of confirmed Covid-19 cases reported to the Center for Disease Control and Prevention (CDC) has shown that infected patients are six and 12 times more likely to be hospitalized (45.4 vs 7.6%) or die (19.5 vs 1.6%) respectively, if they have an underlying medical condition, including neurological or neurodevelopmental disabilities, compared to those with an unremarkable medical history (Zhu et al., 2020).

Parkinson's disease (PD) appears with a frequency of 2%–3% among those over the age of 65 and is nowadays acknowledged as the second most common neurodegenerative disorder worldwide (Poewe et al., 2017). Thus, it would not be an uncommon clinical scenario for healthcare professionals to encounter people with PD (PwP) affected with SARS-CoV-2. Through the prism of modern neurology, PD is no longer treated solely as a motor impairment disorder, but affects various systems, exhibiting both motor and non-motor manifestations (Chaudhuri, Healy, & Schapira, 2006). This constellation of symptoms might make the simultaneous management of these two conditions, PD and Covid-19, quite challenging, especially as these patients tend to be admitted in non-neurological wards.

Despite the recent outbreak of SARS-CoV-2, humanity has long been acquainted with the coronavirus (CoV) family with numerous reported outbreaks of CoV species highly pathogenic to humans (Cui, Li, & Shi, 2019). It has been estimated that the percentage of healthy carriers of a CoV strain in the general population is up to 2%, plus these viruses are responsible for about 5-10% of acute respiratory infections in general (Cascella et al., 2021). With these in mind, previous data derived from the management of PwP with respiratory infections or in situations requiring hospitalization, including mechanical ventilation, can prove particularly valuable in the era of the Covid-19 pandemic.

Although PD is more prevalent among the elderly and age might be an aggravating factor in the outcome of Covid-19 patients, no clear associations have been described so far between the two conditions, as the former does not weaken the immune system per se. The difficulty in the management of PwP infected with SARS-CoV-2 lays not only in the nature of PD symptoms, but also in the fact that PwP follow complicated therapeutic schemes, especially in the advanced stages of the disease. Introduction of new drugs, often required in the setting of Covid-19, might predispose to drug-to-drug interactions. Furthermore, the clinical and metabolic changes accompanying Covid-19 might affect the intensity of the parkinsonian symptoms and impose changes in the standard anti-PD treatment. A recent systematic review and meta-analysis showed that the hospitalization and mortality rates among PwP infected with SARS-CoV-2 are 49% and 12% respectively (Khoshnood et al., 2021). Small case series of older PwP with advanced disease state have depicted an even higher mortality rate (40%), especially for those on advanced therapies (50%) due to older age and comorbidities (Antonini, Leta, Teo, & Chaudhuri, 2020), rendering this population particularly vulnerable to the Covid-19 impact.

2. Management of PwP according to Covid-19 severity

According to the National Institute of Health (NIH) there are five categories of Covid-19 severity (Table 1).

Covid-19 usually progresses in two phases (Cascella et al., 2021): an initial early period which coincides or precedes the onset of symptoms when the virus vastly replicates, and a later phase, which is dominated by inflammation induced by cytokines release and activation of the coagulation system, leading to a prothrombotic state. In the former phase, antiviral- and

Table 1 Types of severity of Covid-19.

	Suggestive symptoms[a]	Shortness of breath or lung infiltrates < 50%	$SatO_2$ < 94% or PaO_2/FiO_2 < 300 and RR > 30 or lung infiltrates > 50%	Acute respiratory failure or septic shock or multiple organ dysfunction
Asymptomatic				
Mild				
Moderate				
Severe				
Critical				

[a]Fever, cough, sore throat, malaise, headache, muscle pain, nausea, vomiting, diarrhea, anosmia, dysgeusia.

PaO_2/FiO_2: ratio of partial pressure of arterial oxygen to fraction of inspired oxygen; RR: respiratory ration (breaths/min); $SatO_2$: oxygen saturation.

antibody-based medication might effectively control the clinical symptoms, while in the later period anti-inflammatory drugs, namely corticosteroids and/or immunomodulating agents, are more efficient.

3. Asymptomatic or mild Covid-19 infection

In asymptomatic or mild Covid-19 infection, no hospital admission is required, and patients are advised to isolate themselves and monitor their symptoms. PwP, especially the elderly, are strongly advised to inform their treating neurologist about their symptoms and be closely monitored until full clinical recovery. Since regular follow-up visits in the Outpatients Clinics have become less frequent (or even completely canceled) during the pandemic due to lockdown regulations imposed by numerous countries, the role of telemedicine has been substantially promoted, covering a wide range of services, such as medical visits, psychotherapy, and physiotherapy, with outcomes comparable to in-person approaches (Cubo, Hassan, Bloem, & Mari, 2020). Although telemedicine services have not been globally applied (some treating physicians would remain in contact with their patients solely via phone calls or emails) (Fasano et al., 2020), such initiatives lay the foundations for future innovations, as the pandemic persists.

3.1 Anti-parkinsonian treatment: Role in Covid-19

It is essential for PwP, not only to adhere to their routine anti-parkinsonian medication during Covid-19, but also to undergo frequent reviews of their regimen, whether they are hospitalized or not, as numerous studies emphasize that increases in dopaminergic therapy during an infection are often required due to aggravation of motor and non-motor symptoms (Antonini et al., 2020; Cilia et al., 2020). It has been suggested that this aggravation is partly mediated by the subsequent systemic inflammatory response triggered by SARS-CoV-2 (Brugger et al., 2015). Another reason would be the emergence of diarrhea, a common finding among Covid-19 cases (D'Amico, Baumgart, Danese, & Peyrin-Biroulet, 2020), which might exert a devastating effect on the pharmacokinetics of orally administered anti-parkinsonian medications, especially levodopa, leading to reduced absorption of the drug and, thus, an increased frequency and duration of OFF episodes (Cilia et al., 2020).

Recent evidence has suggested that some anti-parkinsonian medication might play a beneficial role in the pathophysiology of Covid-19. A potential protective role of dopamine has been suggested, as a dopamine D1 receptor agonist was found to suppress endotoxin-induced pulmonary inflammation in mice (Bone, Liu, Pittet, & Zmijewski, 2017). On the other hand, L-Dopa decarboxylase (DDC), an essential enzyme in the biosynthesis of both dopamine and serotonin, seems to be the most significantly co-expressed and co-regulated gene with Angiotensin-Converting Enzyme 2 (ACE2) in non-neuronal cell types, thus, affecting the dopamine blood levels (Nataf, 2020). ACE2 has been identified as a major cell entry receptor for SARS-CoV-2 (Zhang, Penninger, Li, Zhong, & Slutsky, 2020), while levodopa in PwP is always administered with a DDC inhibitor, such as carbidopa or benserazide, to prevent levodopa from peripherally converting to dopamine. Researchers showed that DDC levels rose in asymptomatic or mild severity Covid-19 patients, while an inverse relationship was noted between SARS-CoV-2 RNA levels and DDC expression, leading them to assume a detrimental effect of DDC inhibitors on Covid-19 course (Fasano et al., 2020; Mpekoulis et al., 2021). Amantadine, a drug with mild anti-parkinsonian properties, which is classically used to manage dyskinesias in PD, has been found to exhibit anti-viral properties, while growing evidence has shown that its use might mitigate, or even prevent, the effects of Covid-19 (Abreu, Aguilar, Covarrubias, & Durán, 2020; Cortés-Borra & Aranda-Abreu, 2021; Kamel et al., 2021). Finally, an analysis of SARS-CoV-2 and human proteins revealed the Catechol-*O*-methyltransferase (COMT) inhibitor entacapone as a potential antiviral therapeutic agent against SARS-CoV-2 (Gordon et al., 2020). Despite the above findings, Fasano and colleagues, who conducted a multi-center study using a cohort of PwP with Covid-19, commented that no clear associations have arisen between anti-parkinsonian drugs and Covid-19 outcome, although larger cohorts are needed to further explore such effects (Fasano, Elia, et al., 2020).

3.1.1 Potential complications from disruptions in dopaminergic treatment

According to a survey conducted by the Movement Disorders Society (MDS) Epidemiology group, 45.4% of PwP worldwide encountered difficulties getting their regular prescriptions for dopamine replacement therapy during the pandemic with more devastating effects observed in lower income countries (Cheong et al., 2020). An abrupt cessation or a significant reduction of anti-parkinsonian medication might predispose to the so-called parkinsonism-hyperpyrexia syndrome, a rare, but life-threatening condition, caused by acute deficiency of iatrogenic dopamine, which presents with hyperthermia, severe rigidity, impaired consciousness, stupor, autonomic dysfunction (including respiratory failure) and high serum creatinine kinase levels (Newman, Grosset, & Kennedy, 2009). It might also be precipitated by respiratory, gastrointestinal or urinary infections (Simonet, Tolosa, Camara, & Valldeoriola, 2020). Parkinsonism-hyperpyrexia syndrome is the modern term, referring to the previously used "neuroleptic malignant-like syndrome." It is an urgent condition, requiring hospitalization, occasionally in the Intensive Care Unit (ICU), supportive care, intravenous fluids and, most importantly, that the patient resumes dopaminergic treatment as soon as possible, although symptoms might be resistant to levodopa (Newman et al., 2009; Serrano-Dueñas, 2003). If treatment with levodopa fails, administration of subcutaneous apomorphine or transdermal rotigotine or intravenous amantadine, if available, might be of use (Simonet et al., 2020). A pulse steroid trial could also be used, although evidence on effectiveness is scarce (Clarke, 2004). Administration of dantrolene sodium and intragastric bromocriptine have also been reported in the literature as useful treatment options (Takubo et al., 2003). Parkinsonism-hyperpyrexia syndrome has a mortality rate of up to 4%, but about 30% of patients are left with permanent deficits (Newman et al., 2009).

Abrupt cessation or even tapering of dopamine agonists might lead to dopamine agonist withdrawal syndrome, which is characterized by prominent psychiatric features (anxiety, agitation, panic attacks, depression, suicidal ideation), dysautonomia and generalized pain. Symptoms are often refractory to levodopa and psychotropic medication and respond to resuming dopamine agonists therapy (Chaudhuri et al., 2015; Nirenberg, 2013).

Symptoms might be self-limited, although the severity varies greatly among different patients, with higher daily doses of dopamine agonists and total dose of dopaminergic medication being the most significant risk factors (Yu & Fernandez, 2017).

In case of worsening of parkinsonism, PwP should contact their treating neurologist for an urgent consultation. The above syndromes are severe, although rare. Aggravation of symptoms might be attributed to the natural course of PD per se, especially if patients have not visited their treating neurologist for a while. Covid-19 could also contribute to aggravation of both motor and non-motor PD symptoms (Fearon & Fasano, 2021). Finally, other metabolic causes, like electrolyte imbalances or co-infections, should be excluded (Brugger et al., 2015).

3.1.2 Management of advanced therapies in PD and Covid-19

As far as PD advanced therapies are concerned, management of Deep Brain Stimulation (DBS) parameters might be particularly challenging during the pandemic. Minor changes can be implemented by patients or caregivers themselves, following instructions through telemedicine services or phone consultations (Miocinovic et al., 2020; Sharma et al., 2021). It should be emphasized to both patients and caregivers to pay close attention to the battery status of non-rechargeable stimulators, and to ensure that rechargeable stimulators are always kept charged. Potential cessation of DBS functioning might lead to the rare, but life-threatening situation of DBS-withdrawal syndrome, characterized by a severe akinetic crisis, which might not respond to dopamine replacement therapy (Reuter et al., 2018). Those with a PD diagnosis for more than 15 years, who are efficiently managed with bilateral subthalamic nuclei stimulation are more at risk of developing this urgent condition (Cartella, Terranova, Rizzo, Quartarone, & Girlanda, 2021). Infections of the DBS hardware might also lead to disruption of treatment (Reuter et al., 2018). Patients should be instructed by their treating neurologist to always keep a stock of orally administered levodopa to substitute for any unexpected DBS malfunctioning. If patients notice any worsening of their PD symptoms, they should immediately contact their treating neurologist and get proper instructions, as decision making under these circumstances is personalized for each case, depending on the severity of symptoms, the battery status and the patient's tolerance to a potential DBS interruption (Miocinovic et al., 2020). Subjacent metabolic causes should also be excluded. Even during the pandemic, there is still room for urgent procedures, if the patient's safety due to DBS malfunction is compromised. In case of new, non-urgent DBS implementations, delays might be expected, although such interventions are still scheduled with respect to the different countries healthcare system regulations with priority given to the most debilitated patients (Siddiqui et al., 2021).

Dose adjustments in PwP under infusion therapies might also be challenging, as patients might be asked to make such modifications by themselves under the remote guidance of their treating neurologist, assuming that the pump is set in the non-locked mode. The role of an attentive carer is extremely valuable under these circumstances, as patients' manual dexterity or cognitive performance might be impaired. Video consultations could help the treating neurologists get a more objective evaluation of the clinical status of the patient in these circumstances. However, such evaluations are often conducted through phone calls, posing risks for the aptness of dose changes. Furthermore, allowing patients such an access to dose modifications might render them vulnerable at presenting with dopamine dysregulation syndrome, especially if they use an apomorphine pump or have a history of impulse control disorders (ICDs) (Fasano, Antonini, et al., 2020). In case patients are not properly educated to make such adjustments themselves, some pump manufacturers provide the possibility of a specialized nurse to make home visits, depending on the geography of the patient's residency, following strictly defined safety protocols (Fasano, Antonini, et al., 2020). Manufacturers can also ship new equipment to replace the hardware of the pump in case of damage. It is very important for patients under infusion therapies to be instructed to keep a stock of orally administered levodopa (preferably dispersible or liquid formulations of levodopa) for "rescue" therapy, should this be required in case of pump malfunction or failure due to the emergent risk of parkinsonism-hyperpyrexia syndrome (Cartella et al., 2021). In case of PD symptoms worsening, patients should contact their treating neurologist, as personalized arrangements might be needed. Catheter blockage is a common cause of pump malfunction and can be usually managed by the patient or caregiver without further complications. Subjacent metabolic causes, including skin or hardware or *Helicobacter pylori* infections, or other gastrointestinal complications, like a peptic ulcer, should be excluded (Fasano, Antonini, et al., 2020; Tan et al., 2015). Delays in placement of new infusion therapies are to be expected during the pandemic, although the majority of these procedures are eventually scheduled in due time according to the different healthcare systems regulations and safety protocols and with priority given to more frail patients (Richter et al., 2021).

3.2 Pharmacological therapy in Covid-19 and potential complications in PwP

3.2.1 Monoclonal antibodies

According to NIH guidelines, SARS-CoV-2 neutralizing monoclonal antibodies, such as REGN_COV2 (casirivimab and imdevimab) or bamlanivimab/etesevimab or sotrovimab can be considered for outpatients with mild severity of Covid-19, who are thought to be at risk of disease progression with a low threshold for hospitalization (Cascella et al., 2021), especially for those over 65 years old (Nathan et al., 2021; Weinreich et al., 2021). No absolute contraindications have been reported for the above medications and they could be safely administered to PwP weighing over 40 kg (Gupta et al., 2021; Nathan et al., 2021; Weinreich et al., 2021). NIH recommends against the administration of dexamethasone at this stage (Cascella et al., 2021).

3.2.2 Common cold medications

There have been reports about people who tend to self-medicate with easily accessible analgesics and drugs used to relieve common cold symptoms, some of them sold over the counter (Quincho-Lopez, Benites-Ibarra, Hilario-Gomez, Quijano-Escate, & Taype-Rondan, 2021). Concomitant administration of selective monoamine oxidase (MAO) inhibitors, such as selegiline and rasagiline, with sympathomimetic substances, such as those used to relieve nasal congestion and other rhinitis symptoms (e.g., budesonide, triamcinolone and xylometazoline), might lead to hypertensive crisis. Other common cold medication, like ephedrine, pseudoephedrine, and dextromethorphan, increase the possibility of serotonin syndrome when co-administered with MAO inhibitors, as both drug categories raise serotonin levels. PwP should be properly and timely informed by their treating neurologist that co-administration of the above medication is strictly contraindicated or consult their pharmacist before purchasing such products. Occurrence of serotonin syndrome might be even more likely in patients who are treated simultaneously with both MAO inhibitors and selective serotonin reuptake inhibitors (SSRIs), as depression is, indeed, a prevalent condition among PwP, commonly treated with SSRIs (Seppi et al., 2019). This combination is usually well-tolerated; however, cases of serotonin syndrome have been described in the literature (Aboukarr & Giudice, 2018), therefore, extra caution is needed. Serotonin syndrome, a potentially life-threatening condition, bears similarities to the neuroleptic malignant syndrome, but, apart from agitation and confusion, it is characterized by a mixture of neuromuscular abnormalities (akathisia, tremor, myoclonus, hyperreflexia, rigidity) and autonomic hyperactivity (Volpi-Abadie, Kaye, & Kaye, 2013). Symptoms usually appear acutely, about 6–12h after the introduction or the dose increase of the causative agent (Simonet et al., 2020). It is a diagnosis of exclusion and it is of the utmost importance to early identify and withhold any drugs that might contribute to the hyper-serotonergic situation, while supportive care needs to be promptly provided (Volpi-Abadie et al., 2013). Non-selective serotonin receptor antagonists, like cyproheptadine and methysergide, might also be of help (Simonet et al., 2020). Under these circumstances, the temporary and prophylactic cessation of MAO inhibitors is advised. These do not necessarily need to be replaced by other antiparkinsonian drugs; however, if such need arises, amantadine could be a reasonable option. Amantadine is assumed to both block dopamine reuptake in presynaptic vesicles and increase dopamine release, therefore imitating the action of MAO inhibitors, which reduce dopamine metabolism and prolong dopamine availability in the synaptic cleft. An alternative option in case of sub-treatment would be an up-titration of the total levodopa dose.

4. Moderate Covid-19 infection

4.1 Hospitalization in PwP with Covid-19 and potential complications

Patients with clinically moderate Covid-19 are usually admitted to the hospital for closer monitoring, especially if other comorbidities are present (Cascella et al., 2021). There have been numerous reports describing the dissatisfaction of PwP and their caregivers in relation to the care received during hospitalization. Most complaints refer to delays, or even omissions in their regular anti-parkinsonian regimen, leading to worsening of motor performance, increased rate of falls and longer admission periods due to complications, such as co-infections (Buetow, Henshaw, Bryant, & O'Sullivan, 2010; Gerlach, Broen, van Domburg, Vermeij, & Weber, 2012; Magdalinou, Martin, & Kessel, 2007).

There is a sub-population of almost 40% among PwP with respiratory dysfunction in the absence of lung or cardiovascular disease (Baille et al., 2019). Respiratory abnormalities in PD, including both upper and lower obstructive or restrictive patterns, have been well-recognized, and attributed to dysautonomia, camptocormia and kyphoscoliosis (Arrigo et al., 2020). An impaired brainstem ventilatory control might also contribute to the respiratory dysfunction, even in premotor stages of PD, affecting the central drive of breathing (Vijayan, Singh, Ghosh, Stell, & Mastaglia, 2020). Since a number of studies have found that respiratory impairment might worsen during OFF periods, most researchers strongly support the protective role of anti-parkinsonian medication against respiratory failure (Arrigo et al., 2020). It is, thus, of the utmost importance for PwP to receive their medication in an appropriate and timely manner, especially in the advanced stage of the disease, when patients usually follow complicated and personalized treatment schemes. Consulting neurologists should highlight this necessity when PwP with Covid-19 are admitted in non-neurological wards. A detailed and thorough medical history should be taken by treating physicians prior to admission, so that the personnel who is unfamiliar with PD understands the significance of adhering to defined dose intervals. Under special circumstances, patients could be allowed to receive their medications by themselves. Special caution is advised when considering the use of anticholinergic drugs in PwP, which are commonly used for obstructive pulmonary disorders, as they have been associated with cognitive impairment and delirium, especially in advanced stages of PD, when cognition might be already compromised (Arrigo et al., 2020).

4.2 Pharmacological therapy in Covid-19 and potential complications in PwP

4.2.1 Dexamethasone

Dexamethasone remains the cornerstone of treatment for Covid-19 patients who require hospitalization, either alone or in combination with remdesivir (Cascella et al., 2021). Administration of dexamethasone is not contraindicated for PwP, although special care should be taken to avoid any adverse events. Since an inflammatory component has long been presumed in the pathogenesis of PD (Tufekci, Meuwissen, Genc, & Genc, 2012), dexamethasone was found to exert a protective effect on dopamine neurons and was associated with a better motor performance in experimental parkinsonian mice models (Joshi & Singh, 2018; Tentillier et al., 2016).

The use of corticosteroids has been connected to a potential neuropsychiatric impairment with about 2-60% of patients presenting with mood swings, depression, mania, suicidality, anxiety, confusion, and behavior changes, along with impairment of their sleep pattern (Chen et al., 2021; Dubovsky, Arvikar, Stern, & Axelrod, 2012; Noreen, Maqbool, & Madni, 2021). The term "steroid psychosis," which is often used in the literature to describe such phenomena, is considered by many rather misleading, as neuropsychiatric impairment exhibits a vast diversity of symptoms. In about 6% of patients who receive steroids, the neuropsychiatric effects are expected to be rather severe with high doses considered as the most significant precipitating factor (Dubovsky et al., 2012). The use of dexamethasone in critically ill patients was also found to aggravate the occurrence of delirium (Wu, Li, Liao, & Wang, 2021).

In PwP, anxiety, psychosis and clinically significant depression are encountered in up to 60%, 40% and 35% respectively (Schapira, Chaudhuri, & Jenner, 2017). It is therefore of particular importance to differentiate whether such symptoms might be steroid-induced or merely exacerbations of non-motor symptoms of PD. If such phenomena occur during OFF periods, it is preferable to adjust the anti-parkinsonian regimen before introducing any further modifications. It is also crucial to exclude other potential metabolic causes, such as electrolyte imbalances, hypo- or hyperglycemia, intoxication, side effects of other medications or additional infections (Janes, Kuster, Goldson, & Forjuoh, 2019). SARS-CoV-2 infection per se may exacerbate or trigger psychiatric symptoms, even in the absence of steroidal treatment, most of which are self-remitting (Ferrando et al., 2020; Varatharaj et al., 2020). In a large meta-analysis from Asia, North America and Europe, steroid-induced mania and psychosis were only reported in 13 (0.7%) of 1744 cases in the acute stage, although other neuropsychiatric symptoms, like confusion, depression, anxiety, impaired memory and insomnia were quite common, all of them presenting in more than 30% of inflicted patients (Rogers et al., 2020).

Dexamethasone has a 36- to 72-h period of action and neuropsychiatric symptoms usually appear within 2 weeks after administration; a subacute onset of up to 12 weeks has also been described (Wada et al., 2001). No evidence-based guidelines are available concerning the treatment of corticosteroid-induced neuropsychiatric impairment, as most information is based on case reports and case series. However, such phenomena are expected to typically resolve in more than 90% of cases with corticosteroids tapering at a dose of less than the equivalent of 40 mg/day of prednisone and, eventually, with the complete cessation of steroids (Dubovsky et al., 2012; Warrington & Bostwick, 2006). In most cases, any associated delirium commonly resolves within days and psychosis within a week, though depression or manic symptoms may last up to 6 weeks after discontinuation of steroids (Janes et al., 2019).

Treating physicians need to be vigilant in identifying such phenomena, as treatment modifications might be required in PwP, while corticosteroids tapering might not be an option, especially in more severe cases of Covid-19. Dopamine agonists might need to be gradually discontinued if psychotic features appear and replaced with the equivalent dose of levodopa. In patients with mania, several publications support the efficient use of carbamazepine, valproate, lamotrigine, and phenytoin as mood stabilizers (Dubovsky et al., 2012; Wada et al., 2001). For PwP treated with pergolide, the use of valproate and phenytoin is contraindicated, as these drugs compete for protein-binding sites resulting in toxic concentrations of either or both of them (Dalvi & Ford, 1998). Moreover, the use of carbamazepine should be avoided in PwP receiving clozapine, as concomitant administration might lead to development of neuroleptic malignant syndrome (Dalvi & Ford, 1998). To timely address manic symptoms, antipsychotics can be co-administered or given alone. Although phenothiazines, butyrophenones and zotepine have been effectively used in steroid psychosis (many support that a good therapeutic response to low doses of haloperidol or olanzapine constitutes a characteristic of corticosteroid-induced psychosis (Wada et al., 2001)), these drugs are not suitable for PwP, as they tend to exacerbate parkinsonian symptoms. The only antipsychotics that can be safely used with appropriate monitoring in PwP are clozapine, quetiapine and pimavanserin (Seppi et al., 2019), although no reports on their effects on corticosteroids-induced psychosis are currently available.

Previous reports suggested that antidepressants, like tricyclic antidepressants (TCAs), should be avoided in steroid psychosis, partly due to their anticholinergic effects which might exacerbate delirium. However, newer data suggests that the indication for antidepressants should be re-examined (Wada et al., 2001), as fluoxetine, sertraline and venlafaxine have been successfully used in various case reports (Dubovsky et al., 2012). Fluoxetine, in particular, has been shown to be a promising drug in the context of Covid-19 due to reported anti-inflammatory properties (Dąbrowska, Galińska-Skok, & Waszkiewicz, 2021). TCAs like nortriptyline and desipramine have been labeled as "likely efficacious" and "clinically useful" to treat depression in PwP (Seppi et al., 2019) and could be considered also for anxiety in PD, although they should be avoided in patients with suicidal ideation or cardiovascular problems (Prediger, Matheus, Schwarzbold, Lima, & Vital, 2012). Venlafaxine is considered 'efficacious' and 'clinically useful' in alleviating depressive symptoms in PD (Seppi et al., 2019). Its role in relieving anxiety symptoms is controversial and it should be avoided in patients with cardiac disease, electrolyte imbalance and hypertension (Prediger et al., 2012), all of which are often encountered in the Covid-19 setting (Azer, 2020). Fluoxetine and sertraline are thought to be 'possibly clinically useful' in treating depression in PD (Seppi et al., 2019), although their role in anxiety is also controversial and some clinicians prefer them only when depression is also present (Prediger et al., 2012). Benzodiazepines do not constitute a good therapeutic option for anxiety in PD, especially in the elderly, as they have been associated with sedation, cognitive impairment, confusion and falls, while the risk of abuse and dependence, as well as the possibility of withdrawal syndrome, should be considered (Prediger et al., 2012).

4.2.2 Remdesivir

Remdesivir, a broad-spectrum anti-viral agent, along with dexamethasone, might be considered for patients who require supplemental oxygen (Cascella et al., 2021), although evidence is conflicting (Pan et al., 2021). Remdesivir is administered intravenously at a weight-based dose (5- or 10-day scheme) and has been approved for adults weighing a minimum of 40 kg (Aleem & Kothadia, 2021). Although there is evidence that it inhibits numerous cytochrome P450 (CYP450) enzymes in vitro, no such indications have arisen in vivo, as there are no reports of severe drug-to-drug interactions up to now, and it is generally considered a well-tolerated treatment (Aleem & Kothadia, 2021). Remdesivir can be safely administered to PwP, as no interactions with anti-parkinsonian medications have been reported up to now, although the pharmacokinetics of the drug have not been extensively studied for those over 65 years old. PwP and their treating physician need to by particularly alert to potential adverse events of the drug, including hypotension, arrhythmias, nausea and vomiting, constipation and gastroparesis (Aleem & Kothadia, 2021), as dysautonomia might be a prominent feature in PwP (Chaudhuri, 2021). Close monitoring with a holistic approach to the patients' symptoms is advised.

4.2.3 Prophylactic thromboembolic agents

Thromboembolic prophylaxis with appropriate anticoagulation is also indicated in moderate Covid-19 (Cascella et al., 2021). Anticoagulation of any kind does not constitute a contraindication in PwP, even in those with advanced therapies, including DBS or infusion therapies.

4.2.4 Antibiotics

In case of a potential bacterial co-infection in Covid-19 patients, empirical antibiotic treatment should be initiated (Cascella et al., 2021). For PwP, in particular, increased awareness is advised, as they might present with swallowing difficulties, especially in advanced stages (Suttrup & Warnecke, 2016), while the cough reflex might be impaired, even from the early stages of the disease, rendering the removal of secretions problematic (Ebihara et al., 2003) and predisposing to aspiration pneumonia, one of the leading death causes among PwP (Won, Byun, Oh, Park, & Seo, 2021).

Considering the choice of antibiotics, most physicians adhere to local community-acquired pneumonia guidelines (Bendala Estrada et al., 2021). In a meta-analysis of 13,932 Covid-19 patients, beta-lactams, especially third-generation cephalosporines, were the most commonly administered agents (72.0%), followed by macrolides (60.2%) and fluoroquinolones (13.3%) (Bendala Estrada et al., 2021). In a different meta-analysis of 375 patients, fluoroquinolones were the most commonly used antibiotics (56.8%), followed by ceftriaxone (39.5%) and azithromycin (29.1%) (Chedid et al., 2021). Other broad-spectrum antibiotics, like linezolid, have also been used (Miranda et al., 2020). In a large meta-analysis of 3338 Covid-19 patients, bacterial co-infection and secondary infection were identified in 3.5% and 14.3% of them respectively, though the majority of patients (71.8%) did receive antibiotics (Langford et al., 2020). Although most antibiotics are generally well-tolerated by patients, including those with PD, the general rule is that physicians should weigh benefits and potential risks and be extra-cautious to avoid initiating such medications when unnecessary. Early de-escalation of empirical antibiotic treatment is strongly encouraged, as longer treatments might predispose to development of antimicrobial resistance and opportunistic pathogens superinfections, which are associated with increased morbidity and mortality in vulnerable Covid-19 patients (Chedid et al., 2021).

In regard to antibiotics used in PwP, there have been reports about potential neuroprotective effects in PD animal models (Yadav, Thakur, Shekhar, & Ayushi., 2021), with ceftriaxone found to offer several advantages in improving clinical aspects of parkinsonism due to a long-term upregulation of glutamate transporter GLT-1 and removal of synaptic glutamate (Kelsey & Neville, 2014). On the other hand, a significantly increased PD prevalence was found among patients using narrow-spectrum penicillin and penicillinase-resistant penicillin, which was attributed to gut microbial imbalance (dysbiosis) in PwP and inflammation, promoted by consumption of antibiotics (Ternák, Kuti, & Kovács, 2020). Since alterations in the gut microbiome of PwP have been confirmed (Romano et al., 2021) and recent studies suggest that patients with altered gut microbiota might experience more severe Covid-19 symptoms (Kim, 2021), empirical treatment for bacterial pneumonia in PwP should only be initiated when clinical suspicion is high.

Most antibiotics carry a significant risk of typical gastrointestinal adverse events, like nausea, vomiting, abdominal pain and diarrhea, especially macrolides and quinolones (Patel & Hashmi, 2021; Yan & Bryant, 2021). PwP should be made aware of such side effects, as these symptoms might overlap with typical non-motor symptoms of PD or interfere with the absorption of orally administered parkinsonian medication. Treating physicians should also be aware that third-generation cephalosporines, and more commonly quinolones, have been associated with a risk of pseudomembranous colitis triggered by *Clostridium difficile* in order to timely differentiate and treat this condition in PwP (Arumugham, Gujarathi, & Cascella, 2021; Yan & Bryant, 2021).

Third-generation cephalosporines have known neurotoxic effects, including epileptogenic activity, new-onset movement disorders (asterixis, myoclonus) and encephalopathy characterized by limb numbness or weakness, behavior impairment and cognitive deficits. Symptoms typically present acutely within 1 week of antibiotic therapy initiation and subside within 1 week after treatment cessation (Arumugham et al., 2021).

Macrolides had been widely used in Covid-19 treatment during the first months of the pandemic due to their antiviral features (Bendala Estrada et al., 2021). Initial studies had shown that azithromycin, either alone or in combination with hydroxychloroquine, was effective against SARS-CoV-2 (Gautret et al., 2020; Touret et al., 2020), while macrolides use was associated to a higher survival ratio in Covid-19 patients (Bendala Estrada et al., 2021). Although generally well-tolerated, macrolides exhibit a variety of side effects which are particularly relevant for PwP. More specifically, they have been associated with QT interval prolongation in the cardiac cycle, predisposing to ventricular tachycardia, ventricular fibrillation and Torsades de Pointes syndrome, which are severe, life-threatening arrhythmias, especially in the presence of electrolyte abnormalities, like hypokalemia (Patel & Hashmi, 2021). Many anti-parkinsonian drugs, including apomorphine or medication commonly used in PD comorbidities, like antidepressants (citalopram, escitalopram, venlafaxine, nortriptyline) or antipsychotics (quetiapine, clozapine), have also been linked to QT prolongation (Tisdale, 2016). Thus, concomitant use of macrolides is not recommended for these patients. If macrolides administration is deemed necessary, electrolytes blood levels, including calcium, potassium and magnesium, should be measured and the patient should be carefully monitored with serial electrocardiograms (ECG), while the treating physicians should also consider cessation of the interacting agents (Patel & Hashmi, 2021). Patients using apomorphine, either as a pen or as a pump, might need to discontinue this medication before introducing macrolides. While apomorphine injections can be abruptly discontinued, the apomorphine pump infusion needs to be gradually withdrawn. In the former case, extra doses of levodopa could substitute the apomorphine shots in order to manage morning or unexpected OFF state. Replacement of an apomorphine pump with the levodopa equivalent dose can be more challenging, as gradually increasing doses of multiple levodopa administrations will be needed, following a personalized scheme and a close follow-up of the patient. The same principles apply to quinolones, which have also been associated with QT interval prolongation (Yan & Bryant, 2021).

Macrolides are also thought to inhibit the liver metabolism of bromocriptine, thus leading to accumulation of the drug and manifestation of ergotism (Dalvi & Ford, 1998). Moreover, the macrolide spiramycin might form a non-absorbable complex with carbidopa, which affects the pharmacokinetics of levodopa when the two compounds are co-administered (Dalvi & Ford, 1998).

Arthralgias are commonly encountered in patients treated with quinolones and self-resolve after treatment cessation (Yan & Bryant, 2021). Pain in PD is a prominent non-motor feature (Schapira et al., 2017), which might affect a patient's mobility, thus, such side effects might cause extra discomfort in a PwP. Moreover, tendon rupture, another typical side effect of quinolones, is more common among people over the age of 60 and might also jeopardize the vulnerable mobility of PwP (Yan & Bryant, 2021). Since the quinolone ciprofloxacin strongly inhibits CYP1A2, it may increase the bioavailability of ropinirole (Kaye & Nicholls, 2000). Although their co-administration was found well-tolerated in vivo, treating physicians should consider a dosage adjustment in PwP treated with ropinirole, if ciprofloxacin is initiated (Kaye & Nicholls, 2000).

Linezolid is considered a broad spectrum antibiotic with a reversible MAO inhibitor potential (Quinn & Stern, 2009). Concomitant use of linezolid and serotonergic agents (SSRIs, serotonin, and norepinephrine reuptake inhibitors (SNRIs), TCAs, rasagiline/selegiline, bromocriptine) is contraindicated. If linezolid is to be administered, it is recommended that the serotonergic drug is discontinued with a washout period of at least 2 weeks, while monitoring the patient for development of serotonin syndrome (Hisham, Sivakumar, Nandakumar, & Lakshmikanthcharan, 2016). Although no specific case reports have been published about serotonin syndrome among patients treated with linezolid and levodopa-carbidopa this combination should also be avoided (Pettit et al., 2016).

5. Severe Covid-19 infection

5.1 Pharmacological therapy in Covid-19 and potential complications in PwP

Patients with severe Covid-19 need to be hospitalized. The same principles mentioned above for prophylactic anticoagulation and empiric antibiotic treatment in case of bacterial infection also apply for severe Covid-19. A stronger indication for dexamethasone administration applies on such cases, usually in combination with remdesivir or baricitinib or tocilizumab, if patients require oxygen supplementation with high flow nasal cannula (HFNC) or non-invasive ventilation (Cascella et al., 2021). In case dexamethasone is contraindicated, the combination of baricitinib plus remdesivir can be of use (Cascella et al., 2021).

5.1.1 Baricitinib

Baricitinib is a Janus kinase (JAK) inhibitor, which can be used in combination with remdesivir or dexamethasone (Cascella et al., 2021). Although it bears a small risk of thromboembolic events and secondary infections, it is unlikely to cause any serious complications when administered for short time periods like in the Covid-19 setting (Hsu, Mao, Liu, & Lai, 2021). Baricitinib can be safely administered to PwP with severe Covid-19, and no drug-to-drug interactions have been described with antiparkinsonian medication.

5.1.2 Tocilizumab

Tocilizumab is an anti-interleukin (IL)-6 receptor alpha monoclonal antibody, which is effectively used in Covid-19 patients who manifest rapid respiratory deterioration (Cascella et al., 2021). Although a number of adverse events have been reported to be associated with tocilizumab use, it has not been associated with any known serious drug-to-drug interactions or any contraindications for PwP (Hsu et al., 2021). Caution is advised with concurrent use of myelotoxic medication or immunosuppressive drugs due to the risk of toxicity or secondary infections respectively (Hsu et al., 2021).

5.1.3 Norepinephrine

Norepinephrine is a first-line vasopressor aiming to maintain a mean arterial pressure of 60–65 mmHg in cases of critically ill Covid-19 patients with hypotension that do not respond to volume resuscitation with intravenous administration of fluids (Cascella et al., 2021; Smith & Maani, 2021). There are no absolute contraindications in norepinephrine use (Smith & Maani, 2021). Special caution is advised for patients who are treated with MAO inhibitors or those on amitriptyline or imipramine-type antidepressants, which might be the case in PwP, as this combination might lead to resistant hypertension (Smith & Maani, 2021).

5.1.4 Plasma transfusion

Food and Drug Administration (FDA) has approved the use of convalescent plasma in patients with severe life-threatening Covid-19, although reported results are mixed (Cascella et al., 2021). There are no known contraindications for plasma transfusion in PwP or interactions with anti-parkinsonian medication (Liumbruno et al., 2009).

5.2 Intubation, ICU admission, and potential complications in PwP and Covid-19

In severe Covid-19 cases, intubation and mechanical ventilation, along with further support of the patient in the ICU, might be deemed necessary. The published evidence commenting on the duration of ICU admission and ventilation among Covid-19 patients with and without PD are conflicting, with a large German study supporting that figures did not significantly differ between the two populations (Huber et al., 2021), while others found opposite results (Zhang, Schultz, Aldridge, Simmering, & Narayanan, 2020). In case of prolonged endotracheal intubation, it is preferable to proceed early in performing a tracheostomy (Gerlach, Winogrodzka, & Weber, 2011).

Adhering to the established dopaminergic medications dosing scheme in intubated PwP with Covid-19 is particularly important to prevent the development of rigidity with contractures and impairment of the respiratory function with low vital capacity and peak expiratory flow and aggravation of PD symptoms (Fasano, Antonini, et al., 2020). Discontinuation of dopaminergic medication in ICU might result in chest wall rigidity, thus impairing the ventilator management (Freeman et al., 2007). Interestingly, levodopa has been found to strengthen diaphragm contraction in anesthetized dogs and patients with chronic obstructive pulmonary disease (COPD) and acute respiratory dysfunction (Aubier et al., 1989; Fujii, 2006). Maintaining the standard regimen would also lower the risk of parkinsonism-hyperpyrexia syndrome in case of abrupt cessation or decrease in levodopa equivalent dose; insertion of a nasogastric tube is highly recommended to allow administration of antiparkinsonian medication (Roberts & Lewis, 2018). Of note, enteral nutrition with high protein-enriched supplements, commonly used in the ICU, was found to constitute an independent risk factor for parkinsonism-hyperpyrexia syndrome, as dietary protein might compromise the absorption of levodopa (Bonnici, Ruiner, St-Laurent, & Hornstein, 2010). Proper levodopa dosing is also expected to minimize the risk of rigidity and distress on emergence of anesthesia (Roberts & Lewis, 2018). For PwP under infusion therapies, it is advised for these therapies to be continued (Fasano, Antonini, et al., 2020). Alternatively, the levodopa equivalent dose could be orally administered, although it would be better if infusion therapies were not abruptly discontinued. If required, the levodopa tablets can be administered in a crushed form through a nasogastric tube to achieve faster absorption and action, or identical doses of a soluble levodopa preparation can be used. Replacement of an apomorphine pump with the levodopa equivalent dose can be challenging, as gradually increasing doses of multiple levodopa administrations will be needed, following a personalized scheme and a close follow-up of the PwP. However, if the oral administration of levodopa is not feasible for any reason, subcutaneous administration of apomorphine is recommended, even in PwP without any prior exposure to apomorphine (Fasano, Antonini, et al., 2020). The rotigotine patch would also be a reasonable alternative under these circumstances, although it is considered a less potent anti-parkinsonian drug compared to levodopa or apomorphine (Raeder et al., 2021). It could either replace or even complement the basic antiparkinsonian treatment, in cases when escalation is required. Intravenous administration of amantadine is another option, although it is considered less efficacious than levodopa and with more adverse events, and it is not available in all countries (Fasano, Antonini, et al., 2020).

5.2.1 General anesthesia

In severe cases where intubation and mechanical ventilation might be deemed necessary, general anesthesia medication might also cause, mostly minor, issues in PwP infected with SARS-CoV-2. There are no overall guidelines considering the care of PwP admitted in ICU and most information concerning the effect of general anesthetics and analgesics on PwP is based on case reports of PwP going through an operation. More specifically, the concomitant use of MAO inhibitors and pethidine is contraindicated due to inhibition of serotonin reuptake and risk of serotonin syndrome (Zornberg, Bodkin, & Cohen, 1991). Halothane renders the heart more vulnerable to the action of catecholamines and may predispose to arrhythmias in patients treated with levodopa. Isoflurane, sevoflurane and enflurane have been reported as safer alternatives (Roberts & Lewis, 2018). Although propofol has been typically considered a safe agent to induce and maintain general anesthesia in PwP, bearing a favorable pharmacokinetic profile, it has been associated with a risk of triggering of dyskinetic and dystonic movements. Such phenomena, although unusual, have been effectively managed in published cases with the use of dexmedetomidine (Roberts & Lewis, 2018). Other anesthetics which have been associated with aggravation of extrapyramidal signs are fentanyl (Buxton, Gauthier, Kinshella, & Godwin, 2018; Zesiewicz et al., 2009), alfentanil and thiopental (Shaikh & Verma, 2011). Low doses of ketamine were reported to facilitate problematic airway management and tremor in PwP perioperatively (Wright, Goodnight, & McEvoy, 2009).

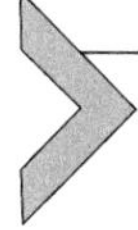

6. Other pharmacological interventions in Covid-19 and possible complications in PwP

Chloroquine and its derivative hydroxychloroquine with the combination of azithromycin used to be a first-line treatment in Covid-19 serious infections during the first months of the pandemic (Touret & de Lamballerie, 2020). However, their efficacy was not confirmed in randomized control trials, while their potential cardiotoxic effect (QT prolongation) renders their clinical use problematic (Ho et al., 2021).

A recent meta-analysis of non-randomized cohort studies has shown that anakinra, a recombinant IL-1 receptor antagonist might be related to reduced mortality and a lower risk of mechanical ventilation when administered in patients with severe Covid-19 (Pasin et al., 2021). It has a safe administration profile with no special contraindications for PwP or reported interactions with anti-parkinsonian medication (Anakinra. LiverTox: Clinical and Research Information on Drug-Induced Liver Injury [Internet], 2012).

Ruxolitinib and tofacitinib are selective JAK inhibitors with promising results against severe Covid-19 according to ongoing research (Cascella et al., 2021). No absolute contraindications have been reported for administration in PwP or serious interactions with drugs commonly used in PD, with the exception of clozapine due to the risk of QT prolongation, along with a potential additive hematological toxicity (neutropenia) (Dhillon, 2017; Gatti, De Ponti, & Pea, 2021). Both drugs are largely metabolized by the isoenzyme of CYP3A4, therefore treating physicians should be cautious with the co-administration of drugs which are strong inducers or inhibitors of this isoenzyme, like carbamazepine or valproic acid (Dhillon, 2017; Gatti et al., 2021).

It is generally advised vitamin D levels to be assessed in PwP, as vitamin D deficiency has been reported to contribute to PD clinical course (Fullard & Duda, 2020). In a large case-controlled survey in Italy, it was found that PwP non-supplemented with vitamin D were at greater risk of being diagnosed with Covid-19 compared to those who were well-supplemented, highlighting a potentially protective role of vitamin D (Fasano et al., 2020). In addition, administration of a moderate and well-calculated vitamin D3 dosage in PwP might help mitigate the effect of Covid-19 complications (Azzam, Ghozy, & Azab, 2022).

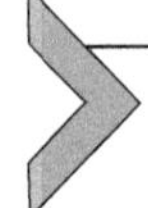

7. Usual complications in hospitalized PwP and Covid-19

7.1 Nausea

Nausea or vomiting are quite common among the constellation of Covid-19 symptoms (Andrews, Cai, Rudd, & Sanger, 2021). Metoclopramide, a dopamine D2 receptor antagonist, is an antiemetic typically used in the clinical setting to alleviate nausea in Covid-19 patients (Ai et al., 2020). However, it is generally advised for this drug to be avoided in PwP, due to its central mode of action and the potential of aggravating parkinsonism (Travagli, Browning, & Camilleri, 2020). The same applies for the dopamine antagonists in the category of phenothiazines, thioxanthenes and butyrophenones (Roberts & Lewis, 2018). Under these circumstances, the use of domperidone is preferred as, despite being a D2 receptor antagonist, it cannot cross the blood-brain barrier and exhibits a better tolerability and safety profile (Travagli et al., 2020). It should be noted, though, that domperidone can prolong the QT interval, increasing the risk of life-threatening arrhythmias; a QT interval exceeding 450 ms in men or 470 ms in women on the baseline ECG should prevent treating physicians from applying this therapy (Reddymasu, Soykan, & McCallum, 2007). However, a randomized controlled study conducted in healthy volunteers showed that this effect was absent in domperidone doses of less than

80 mg daily (Biewenga et al., 2015). Ondansetron, a selective serotonin 5-HT_3 receptor antagonist, might be a safe alternative for PwP with nausea or vomiting, although constipation could be an expected side effect (Wilde & Markham, 1996). Finally, the antihistamine cyclizine might be of use in PwP, with sedation, dizziness, confusion, palpitations, constipation, and urinary retention being the most common complications (Cyclizine, 2012; Roberts & Lewis, 2018).

7.1.1 Delirium

Delirium appears in one third of older general medical patients, with half of these cases occurring on admission and the rest during hospitalization, and can be a life-threatening emergency (Marcantonio, 2017). For patients under mechanical ventilation admitted in the ICU the percentage of delirium exceeds 75% (Ely et al., 2004). Older age, decreased functional status, cognitive impairment (even mild), depression and comorbidities constitute the most common predisposing factors for delirium, while a subjacent infection, acute or severe illness, pain, anemia, anesthesia and drugs (including sedative hypnotics and anticholinergics) are the most frequently encountered precipitating factors (Kalimisetty, Askar, Fay, & Khan, 2017; Marcantonio, 2017; Wakefield, 2002). Undernourishment and dehydration on admission might be aggravating factors (Wakefield, 2002). Delirium constitutes a common complication of Covid-19, especially in critically ill patients, and is associated with a poor outcome (Ticinesi et al., 2020). Lack of family visitation due to isolation regulations has also been identified as a potential risk factor (Pun et al., 2021).

Evidence suggests that PwP are more vulnerable to delirium when admitted to hospital and treating physicians should be vigilant to acknowledge early sings of agitation, confusion, or cognitive deterioration (Vardy, Teodorczuk, & Yarnall, 2015). In case of delirium, a number of non-pharmacological approaches could be applied. Treating physicians should check for newly introduced drugs, which might act as causative factors (such as opiates, fluoroquinolones) and could be discontinued or administered at a lower dose (Marcantonio, 2017). Electrolyte imbalances, infections, intracranial disorders (such as hemorrhages, tumors, CNS infections), urinary retention, fecal impaction and potential myocardial or pulmonary conditions (such as hypotension, hypoxia, anemia) should be properly addressed (Marcantonio, 2017). Use of eyeglasses and hearing aids, if applicable, should be encouraged to prevent decreased sensory stimulation, along with regular contact with family members, even remotely using virtual means. Patients should be mobilized during hospitalization, even with assistance, and should be monitored for pressure ulcers (Marcantonio, 2017). It is better if the use of physical restraints is withhold, as it has been associated with a more prolonged state of delirium (Inouye et al., 2007). A non-pharmacological sleep-hygiene schedule, avoiding unnecessary patients' awakenings (such as nurse rounds for evaluating vital signs) and noise, preventing sleeping during the day and providing a hospital environment with adequate lighting during the day, along with a sufficient dietary intake (with proper supplementation if necessary) would also be helpful (Marcantonio, 2017). It is also important for staff to remember to reorient patients to time, place and person at least three times per day (Marcantonio, 2017). In case of hyperactive delirium, a serial tapering of anti-parkinsonian drug categories should be considered, starting with anticholinergics and then followed by MAO inhibitors, amantadine, dopamine agonists and COMT inhibitors (Ebersbach et al., 2019), although the order of dose reduction or discontinuation can be individualized. An abrupt cessation is not recommended. Sedatives, like benzodiazepines, should better be avoided, while low doses of antipsychotics can be used, if needed. For PwP, quetiapine and clozapine are considered as the safest options to avoid motor impairment, although no randomized studies have provided clear evidence on their use (Vardy et al., 2015). Quetiapine can be initiated at 25 mg and gradually increased to 100–150 mg daily, while clozapine can be started at 6.25–12.5 mg and gradually increased to 75–100 mg daily with regular monitoring for agranulocytosis (Ebersbach et al., 2019). The successful use of the cholinesterase inhibitor rivastigmine has also been described in a case report of a PwP developing delirium, but evidence considering its use is rather rare (Dautzenberg, Wouters, Oudejans, & Samson, 2003). Administration of prophylactic medication to prevent development of delirium is not recommended (Zoremba & Coburn, 2019).

7.1.2 Psychosis

Psychosis is a common complication in PwP, usually appearing later in the disease course (Schapira et al., 2017). Although dopaminergic therapy per se can induce psychosis, subjacent systemic conditions, including infections and Covid-19 in particular, or other drugs use, like antidepressants or painkillers, could also act as precipitating factors (Parra et al., 2020; Simonet et al., 2020). Psychotic symptoms in PwP typically manifest gradually, allowing for step-by-step approaches to minimize polypharmacy (Simonet et al., 2020). Underlying metabolic causes should be excluded or properly addressed, if present. Similarly to delirium, a common strategy, which can be personalized, is to gradually taper anti-parkinsonian medication, starting from anticholinergic drugs and followed by MAO inhibitors or amantadine, dopamine agonists and COMT inhibitors (Simonet et al., 2020). An abrupt cessation is not recommended. Tricyclic antidepressants and opiates are also advised to be slowly withdrawn (Simonet et al., 2020). In case drug reduction is not feasible or psychosis persists despite the modifications, introduction of an atypical antipsychotic drug could be considered. Clozapine has been characterized as an "efficacious" and "clinically useful" option to treat PD psychosis, although specialized monitoring is required (Seppi et al., 2019). Although quetiapine has been characterized only as "possibly useful", it is more easily accessible in clinical practice due to its better tolerated profile (Seppi et al., 2019). Finally, pimavanserin has been recently labeled as "efficacious" and "clinically useful" in PD psychosis management with an acceptable safety profile (Seppi et al., 2019). The use of olanzapine, risperidone or other atypical antipsychotics is generally discouraged in PD psychosis due to exacerbation of parkinsonism, while evidence considering cholinesterase inhibitors (rivastigmine, donepezil) is conflicting and seem to be more efficacious in chronic rather than acute psychosis in PD (Goldman, Vaughan, & Goetz, 2011).

It is of interest that clozapine therapy, especially duration of therapy and not dosage, has been associated with an increased risk of Covid-19 infection (Aubignat, 2021). Following this observation, international recommendations have been drafted, stating that patients on clozapine with any Covid-19 symptoms should undergo a medical assessment and complete blood cell count (Siskind et al., 2020). In case they are diagnosed with Covid-19, the treating physician should consider reducing the dose of clozapine by half. The initial dose can be gradually resumed, starting 3 days after fever resolution (Siskind et al., 2020). No instructions have been given considering initiation of clozapine during a SARS-CoV-2 infection and all data currently available refer to patients with psychosis and not PD (Aubignat, 2021).

7.1.3 Orthostasis

Orthostatic hypotension appears in up to 58% of PwP (Schapira et al., 2017). In the setting of Covid-19, external factors like dehydration secondary to fever, diarrhea, reduced water intake due to anorexia, anemia or subjacent causes of cardiac dysfunction, including arrhythmias, should be excluded or properly addressed (Simonet et al., 2020). The patient's regimen should also be reviewed with special concern on the use of antihypertensives or α-blockers, while almost all dopaminergic medication can precipitate orthostatic hypotension. In severe orthostasis attributed to dysautonomia in the context of PD, administration of fludrocortisone (0.1 mg once to three times daily), midodrine (2.5–10 mg three times daily) or droxidopa (300 mg three times daily) could be possibly useful (Seppi et al., 2019; Simonet et al., 2020).

7.1.4 Dysphagia—Nutritional intake difficulties

Malnutrition has been linked to impairment of the immune system, thus rendering individuals more susceptible to potential infections (Bourke, Berkley, & Prendergast, 2016). Immobilization and assisted breathing for extended time periods, which are common among hospitalized patients, especially those with impaired motor performance, lead to loss of muscle mass, resulting in an even more problematic recovery (Fernández-Quintela et al., 2020). Indeed, up to 60% of acutely ill were reported to be malnourished (Felder et al., 2015). Malnutrition in the context of Covid-19 might be a result of gastrointestinal complications, hypoalbuminemia, hypermetabolism and excessive nitrogen loss (Fernández-Quintela et al., 2020). In addition to the above, up to 80% of PwP might be suffering from dysphagia at some point of their disease course, which might predispose to aspiration (Suttrup & Warnecke, 2016). It is, thus, of the utmost importance for PwP inflicted with SARS-CoV-2 to achieve a sufficient caloric intake along with proper supplementation.

For PwP with dysphagia, a soft diet of adequate caloric content, high in fibers, is recommended with the addition of a liquid thickener if necessary (Arrigo et al., 2020). Instructing some patients to swallow following the chin-down posture might also be useful (Arrigo et al., 2020). In the Covid-19 era, having a speech or swallowing therapist to visit hospitalized patients might not be allowed, however, a video-assisted swallowing therapy might be possible via telemedicine services. Especially for those patients presenting with drooling and dysphagia, which is frequently encountered in advanced stages of PD, increasing the risk of aspiration, botulinum injections in the salivary glands might be indicated (van Wamelen et al., 2020). In severe dysphagia, treating physicians should consider inserting a nasogastric tube, not only to ensure sufficient caloric intake, but also to resume levodopa administration. Levodopa/Carbidopa

Intestinal Gel (LCIG) continuous infusion could be useful under these circumstances, as the percutaneous endoscopic gastrostomy (PEG) tube allows administration of nutrition and fluids. If this is not feasible, subjacent apomorphine, transdermal rotigotine or intravenous amantadine could be reasonable options to avoid an abrupt cessation of dopamine replacement therapy.

8. Conclusion

The Covid-19 pandemic has disrupted healthcare services and patients' routine and quality of life worldwide. This is of particular importance for PwP, as they are required to balance between minimizing the risk of contracting SARS-CoV-2, while getting access to the best possible care for their PD symptoms and maintaining a healthy way of living (exercise, physiotherapy, social interactions). Providers should be aware of potential delays in care delivery, especially in relation to advanced therapies. Under these circumstances, this health crisis provides an opportunity for healthcare professionals to develop and incorporate standardized tools and services for diagnosis and management for chronic conditions, like PD, remotely, using virtual means with respect to local regulations.

The recurrent infection outbreaks have increased frustration and placed an overwhelming burden to hospitals around the globe, with the recent omicron SARS-CoV-2 variant spreading rapidly. Despite the expected high transmissibility of this new variant and the numerous reported re-infections, no alarming clinical concerns have arisen up to now, thus, therapeutic protocols have not been modified yet, although there is a possibility that monoclonal antibodies efficiency might be compromised (Karim & Karim, 2021). PwP are strongly advised to get vaccinated, as the combination of public health prevention measures and vaccination has been associated with a much lower risk of severe disease, hospitalization and death, especially among the elderly (Karim & Karim, 2021). Treating physicians should remain vigilant to acknowledge early signs of complications related either to PD or Covid-19, appropriately guide their patients and timely introduce any interventions. Especially during hospitalization of PwP affected with SARS-CoV-2, the role of a consulting neurologist with special knowledge on movement disorders is crucial and might affect the patients' outcome.

References

Abate, S. M., Checkol, Y. A., & Mantefardo, B. (2021). Global prevalence and determinants of mortality among patients with COVID-19: A systematic review and meta-analysis. *Annals of Medicine and Surgery (London)*, *64*, 102204. https://doi.org/10.1016/j.amsu.2021.102204.

Aboukarr, A., & Giudice, M. (2018). Interaction between monoamine oxidase B inhibitors and selective serotonin reuptake inhibitors. *The Canadian Journal of Hospital Pharmacy*, *71*(3), 196–207.

Abreu, G. E. A., Aguilar, M. E. H., Covarrubias, D. H., & Durán, F. R. (2020). Amantadine as a drug to mitigate the effects of COVID-19. *Medical Hypotheses*, *140*, 109755. https://doi.org/10.1016/j.mehy.2020.109755.

Ai, J.-W., Zi, H., Wang, Y., Huang, Q., Wang, N., Li, L.-Y., et al. (2020). Clinical characteristics of COVID-19 patients with gastrointestinal symptoms: An analysis of seven patients in China. *Frontiers in Medicine*, 7, 308. https://doi.org/10.3389/fmed.2020.00308.

Aleem, A., & Kothadia, J. P. (2021). Remdesivir. In *StatPearls*. Treasure Island (FL): StatPearls Publishing LLC.

Anakinra. (2012). *Anakinra. LiverTox: Clinical and research information on drug-induced liver injury [internet]*. Bethesda (MD): National Institute of Diabetes and Digestive and Kidney Diseases.

Andrews, P. L. R., Cai, W., Rudd, J. A., & Sanger, G. J. (2021). COVID-19, nausea, and vomiting. *Journal of Gastroenterology and Hepatology*, *36*(3), 646–656. https://doi.org/10.1111/jgh.15261.

Antonini, A., Leta, V., Teo, J., & Chaudhuri, K. R. (2020). Outcome of Parkinson's disease patients affected by COVID-19. *Movement Disorders: Official Journal of the Movement Disorder Society*, *35*(6), 905–908. https://doi.org/10.1002/mds.28104.

Arrigo, A., Floro, S., Bartesaghi, F., Casellato, C., Sferrazza Papa, G. F., Centanni, S., et al. (2020). Respiratory dysfunction in Parkinson's disease: a narrative review. *ERJ Open Research*, *6*(4), 00165–02020. https://doi.org/10.1183/23120541.00165-2020.

Arumugham, V. B., Gujarathi, R., & Cascella, M. (2021). Third generation cephalosporins. In *StatPearls*. Treasure Island (FL): StatPearls Publishing LLC.

Aubier, M., Murciano, D., Menu, Y., Boczkowski, J., Mal, H., & Pariente, R. (1989). Dopamine effects on diaphragmatic strength during acute respiratory failure in chronic obstructive pulmonary disease. *Annals of Internal Medicine*, *110*(1), 17–23. https://doi.org/10.7326/0003-4819-110-1-17.

Aubignat, M. (2021). Clozapine-related immunodeficiency: Implications for Parkinson's disease psychosis in the context of the COVID-19 pandemic. *Revue Neurologique*, *177*(8), 849–851. https://doi.org/10.1016/j.neurol.2021.05.002.

Azer, S. A. (2020). COVID-19: Pathophysiology, diagnosis, complications and investigational therapeutics. *New Microbes and New Infections*, *37*, 100738. https://doi.org/10.1016/j.nmni.2020.100738.

Azzam, A. Y., Ghozy, S., & Azab, M. A. (2022). Vitamin D and its' role in Parkinson's disease patients with SARS-CoV-2 infection. A review article. *Interdisciplinary neurosurgery*, *27*, 101441. https://doi.org/10.1016/j.inat.2021.101441.

Baille, G., Chenivesse, C., Perez, T., Machuron, F., Dujardin, K., Devos, D., et al. (2019). Dyspnea: An underestimated symptom in Parkinson's disease. *Parkinsonism & Related Disorders*, *60*, 162–166. https://doi.org/10.1016/j.parkreldis.2018.09.001.

Bendala Estrada, A. D., Calderón Parra, J., Fernández Carracedo, E., Muiño Míguez, A., Ramos Martínez, A., Muñez Rubio, E., et al. (2021). Inadequate use of antibiotics in the covid-19 era: Effectiveness of antibiotic therapy. *BMC Infectious Diseases*, *21*(1), 1144. https://doi.org/10.1186/s12879-021-06821-1.

Biewenga, J., Keung, C., Solanki, B., Natarajan, J., Leitz, G., Deleu, S., et al. (2015). Absence of QTc prolongation with domperidone: A randomized, double-blind, placebo- and positive-controlled thorough QT/QTc study in healthy volunteers. *Clinical Pharmacology in Drug Development*, *4*(1), 41–48. https://doi.org/10.1002/cpdd.126.

Bone, N. B., Liu, Z., Pittet, J.-F., & Zmijewski, J. W. (2017). Frontline Science: D1 dopaminergic receptor signaling activates the AMPK-bioenergetic pathway in macrophages and alveolar epithelial cells and reduces endotoxin-induced ALI. *Journal of Leukocyte Biology*, *101*(2), 357–365. https://doi.org/10.1189/jlb.3HI0216-068RR.

Bonnici, A., Ruiner, C. E., St-Laurent, L., & Hornstein, D. (2010). An interaction between levodopa and enteral nutrition resulting in neuroleptic malignant-like syndrome and prolonged ICU stay. *The Annals of Pharmacotherapy*, *44*(9), 1504–1507. https://doi.org/10.1345/aph.1P242.

Bourke, C. D., Berkley, J. A., & Prendergast, A. J. (2016). Immune dysfunction as a cause and consequence of malnutrition. *Trends in Immunology*, *37*(6), 386–398. https://doi.org/10.1016/j.it.2016.04.003.

Brugger, F., Erro, R., Balint, B., Kägi, G., Barone, P., & Bhatia, K. P. (2015). Why is there motor deterioration in Parkinson's disease during systemic infections-a hypothetical view. *NPJ Parkinson's Disease*, *1*, 15014. https://doi.org/10.1038/npjparkd.2015.14.

Buetow, S., Henshaw, J., Bryant, L., & O'Sullivan, D. (2010). Medication timing errors for Parkinson's disease: Perspectives held by caregivers and people with Parkinson's in New Zealand. *Parkinsons Disease*, *2010*, 432983. https://doi.org/10.4061/2010/432983.

Buxton, J. A., Gauthier, T., Kinshella, M. W., & Godwin, J. (2018). A 52-year-old man with fentanyl-induced muscle rigidity. *CMAJ*, *190*(17), E539–e541. https://doi.org/10.1503/cmaj.171468.

Cartella, S. M., Terranova, C., Rizzo, V., Quartarone, A., & Girlanda, P. (2021). Covid-19 and Parkinson's disease: An overview. *Journal of Neurology*, *268*(12), 4415–4421. https://doi.org/10.1007/s00415-021-10721-4.

Cascella, M., Rajnik, M., Aleem, A., Dulebohn, S. C., & Di Napoli, R. (2021). Features, evaluation, and treatment of coronavirus (COVID-19). In *StatPearls*. Treasure Island (FL): StatPearls Publishing.

Chaudhuri, K. R. (2021). Thirty years of research on autonomic dysfunction, non-motor features, and endophenotypes in Parkinson disease. *Clinical Autonomic Research*, *31*(1), 37–39. https://doi.org/10.1007/s10286-021-00771-z.

Chaudhuri, K. R., Healy, D. G., & Schapira, A. H. (2006). Non-motor symptoms of Parkinson's disease: Diagnosis and management. *Lancet Neurology*, *5*(3), 235–245. https://doi.org/10.1016/s1474-4422(06)70373-8.

Chaudhuri, K. R., Todorova, A., Nirenberg, M. J., Parry, M., Martin, A., Martinez-Martin, P., et al. (2015). A pilot prospective, multicenter observational study of dopamine agonist withdrawal syndrome in Parkinson's disease. *Movement Disorders Clinical Practice*, *2*(2), 170–174. https://doi.org/10.1002/mdc3.12141.

Chedid, M., Waked, R., Haddad, E., Chetata, N., Saliba, G., & Choucair, J. (2021). Antibiotics in treatment of COVID-19 complications: A review of frequency, indications, and efficacy. *Journal of Infection and Public Health*, *14*(5), 570–576. https://doi.org/10.1016/j.jiph.2021.02.001.

Chen, F., Hao, L., Zhu, S., Yang, X., Shi, W., Zheng, K., et al. (2021). Potential adverse effects of dexamethasone therapy on COVID-19 patients: Review and recommendations. *Infectious Diseases and Therapy*, *10*(4), 1907–1931. https://doi.org/10.1007/s40121-021-00500-z.

Cheong, J. L., Goh, Z. H. K., Marras, C., Tanner, C. M., Kasten, M., & Noyce, A. J. (2020). The impact of COVID-19 on access to Parkinson's disease medication. *Movement Disorders*, *35*(12), 2129–2133. https://doi.org/10.1002/mds.28293.

Cilia, R., Bonvegna, S., Straccia, G., Andreasi, N. G., Elia, A. E., Romito, L. M., et al. (2020). Effects of COVID-19 on Parkinson's disease clinical features: A community-based case-control study. *Movement Disorders*, *35*(8), 1287–1292. https://doi.org/10.1002/mds.28170.

Clarke, C. E. (2004). Efficacy of methylprednisolone pulse therapy on neuroleptic malignant syndrome in Parkinson's disease. *Journal of Neurology, Neurosurgery, and Psychiatry*, *75*(3), 510–511 (author reply 511).

Cortés-Borra, A., & Aranda-Abreu, G. E. (2021). Amantadine in the prevention of clinical symptoms caused by SARS-CoV-2. *Pharmacological Reports*, *73*(3), 962–965. https://doi.org/10.1007/s43440-021-00231-5.

Cubo, E., Hassan, A., Bloem, B. R., & Mari, Z. (2020). Implementation of telemedicine for urgent and ongoing healthcare for patients with Parkinson's disease during the COVID-19 pandemic: New expectations for the future. *Journal of Parkinson's Disease*, *10*(3), 911–913. https://doi.org/10.3233/jpd-202108.

Cui, J., Li, F., & Shi, Z.-L. (2019). Origin and evolution of pathogenic coronaviruses. *Nature Reviews. Microbiology*, *17*(3), 181–192. https://doi.org/10.1038/s41579-018-0118-9.

Cyclizine. (2012). *In LiverTox: Clinical and research information on drug-induced liver injury*. Bethesda (MD): National Institute of Diabetes and Digestive and Kidney Diseases.

Dąbrowska, E., Galińska-Skok, B., & Waszkiewicz, N. (2021). Depressive and neurocognitive disorders in the context of the inflammatory background of COVID-19. *Life (Basel)*, *11*(10). https://doi.org/10.3390/life11101056.

Dalvi, A., & Ford, B. (1998). Antiparkinsonian agents: Clinically significant drug interactions and adverse effects, and their management. *CNS Drugs*, *9*(4), 291–310. https://doi.org/10.2165/00023210-199809040-00005.

D'Amico, F., Baumgart, D. C., Danese, S., & Peyrin-Biroulet, L. (2020). Diarrhea during COVID-19 infection: Pathogenesis, epidemiology, prevention, and management. *Clinical Gastroenterology and Hepatology*, *18*(8), 1663–1672. https://doi.org/10.1016/j.cgh.2020.04.001.

Dautzenberg, P. L., Wouters, C. J., Oudejans, I., & Samson, M. M. (2003). Rivastigmine in prevention of delirium in a 65 years old man with Parkinson's disease. *International Journal of Geriatric Psychiatry*, *18*(6), 555–556. https://doi.org/10.1002/gps.867.

Dhillon, S. (2017). Tofacitinib: A review in rheumatoid arthritis. *Drugs*, 77(18), 1987–2001. https://doi.org/10.1007/s40265-017-0835-9.

Dubovsky, A. N., Arvikar, S., Stern, T. A., & Axelrod, L. (2012). The neuropsychiatric complications of glucocorticoid use: Steroid psychosis revisited. *Psychosomatics*, *53*(2), 103–115. https://doi.org/10.1016/j.psym.2011.12.007.

Ebersbach, G., Ip, C. W., Klebe, S., Koschel, J., Lorenzl, S., Schrader, C., et al. (2019). Management of delirium in Parkinson's disease. *Journal of Neural Transmission (Vienna)*, *126*(7), 905–912. https://doi.org/10.1007/s00702-019-01980-7.

Ebihara, S., Saito, H., Kanda, A., Nakajoh, M., Takahashi, H., Arai, H., et al. (2003). Impaired efficacy of cough in patients with Parkinson disease. *Chest*, *124*(3), 1009–1015. https://doi.org/10.1378/chest.124.3.1009.

Ely, E. W., Shintani, A., Truman, B., Speroff, T., Gordon, S. M., Harrell, F. E., Jr., et al. (2004). Delirium as a predictor of mortality in mechanically ventilated patients in the intensive care unit. *JAMA*, *291*(14), 1753–1762. https://doi.org/10.1001/jama.291.14.1753.

Fasano, A., Antonini, A., Katzenschlager, R., Krack, P., Odin, P., Evans, A. H., et al. (2020). Management of advanced therapies in Parkinson's disease patients in times of humanitarian crisis: The COVID-19 experience. *Movement Disorders Clinical Practice*, 7(4), 361–372. https://doi.org/10.1002/mdc3.12965.

Fasano, A., Cereda, E., Barichella, M., Cassani, E., Ferri, V., Zecchinelli, A. L., et al. (2020). COVID-19 in Parkinson's disease patients living in Lombardy, Italy. *Movement Disorders: Official Journal of the Movement Disorder Society, 35*(7), 1089–1093. https://doi.org/10.1002/mds.28176.

Fasano, A., Elia, A. E., Dallocchio, C., Canesi, M., Alimonti, D., Sorbera, C., et al. (2020). Predictors of COVID-19 outcome in Parkinson's disease. *Parkinsonism & Related Disorders, 78*, 134–137. https://doi.org/10.1016/j.parkreldis.2020.08.012.

Fearon, C., & Fasano, A. (2021). Parkinson's disease and the COVID-19 pandemic. *Journal of Parkinson's Disease, 11*(2), 431–444. https://doi.org/10.3233/jpd-202320.

Felder, S., Lechtenboehmer, C., Bally, M., Fehr, R., Deiss, M., Faessler, L., et al. (2015). Association of nutritional risk and adverse medical outcomes across different medical inpatient populations. *Nutrition, 31*(11 – 12), 1385–1393. https://doi.org/10.1016/j.nut.2015.06.007.

Fernández-Quintela, A., Milton-Laskibar, I., Trepiana, J., Gómez-Zorita, S., Kajarabille, N., Léniz, A., et al. (2020). Key aspects in nutritional management of COVID-19 patients. *Journal of Clinical Medicine, 9*(8). https://doi.org/10.3390/jcm9082589.

Ferrando, S. J., Klepacz, L., Lynch, S., Tavakkoli, M., Dornbush, R., Baharani, R., et al. (2020). COVID-19 psychosis: A potential new neuropsychiatric condition triggered by novel coronavirus infection and the inflammatory response? *Psychosomatics, 61*(5), 551–555. https://doi.org/10.1016/j.psym.2020.05.012.

Freeman, W. D., Tan, K. M., Glass, G. A., Linos, K., Foot, C., & Ziegenfuss, M. (2007). ICU management of patients with Parkinson's disease or Parkinsonism. *Current Anaesthesia & Critical Care, 18*(5), 227–236. https://doi.org/10.1016/j.cacc.2007.09.007.

Fujii, Y. (2006). Olprinone/dopamine combination for improving diaphragmatic fatigue in pentobarbital-anesthetized dogs. *Current Therapeutic Research, Clinical and Experimental, 67*(3), 204–213. https://doi.org/10.1016/j.curtheres.2006.06.003.

Fullard, M. E., & Duda, J. E. (2020). A review of the relationship between vitamin D and Parkinson disease symptoms. *Frontiers in Neurology, 11*, 454. https://doi.org/10.3389/fneur.2020.00454.

Gatti, M., De Ponti, F., & Pea, F. (2021). Clinically significant drug interactions between psychotropic agents and repurposed COVID-19 therapies. *CNS Drugs, 35*(4), 345–384. https://doi.org/10.1007/s40263-021-00811-2.

Gautret, P., Lagier, J. C., Parola, P., Hoang, V. T., Meddeb, L., Mailhe, M., et al. (2020). Hydroxychloroquine and azithromycin as a treatment of COVID-19: Results of an open-label non-randomized clinical trial. *International Journal of Antimicrobial Agents, 56*(1), 105949. https://doi.org/10.1016/j.ijantimicag.2020.105949.

Gerlach, O. H. H., Broen, M. P. G., van Domburg, P. H. M. F., Vermeij, A. J., & Weber, W. E. J. (2012). Deterioration of Parkinson's disease during hospitalization: Survey of 684 patients. *BMC Neurology, 12*, 13. https://doi.org/10.1186/1471-2377-12-13.

Gerlach, O. H. H., Winogrodzka, A., & Weber, W. E. J. (2011). Clinical problems in the hospitalized Parkinson's disease patient: Systematic review. *Movement Disorders: Official Journal of the Movement Disorder Society, 26*(2), 197–208. https://doi.org/10.1002/mds.23449.

Goldman, J. G., Vaughan, C. L., & Goetz, C. G. (2011). An update expert opinion on management and research strategies in Parkinson's disease psychosis. *Expert Opinion on Pharmacotherapy, 12*(13), 2009–2024. https://doi.org/10.1517/14656566.2011.587122.

Gordon, D. E., Jang, G. M., Bouhaddou, M., Xu, J., Obernier, K., O'Meara, M. J., et al. (2020). A SARS-CoV-2-human protein-protein interaction map reveals drug targets and potential drug-repurposing. *bioRxiv*. https://doi.org/10.1101/2020.03.22.002386.

Gupta, A., Gonzalez-Rojas, Y., Juarez, E., Crespo Casal, M., Moya, J., Falci, D. R., et al. (2021). Early treatment for Covid-19 with SARS-CoV-2 neutralizing antibody sotrovimab. *The New England Journal of Medicine, 385*(21), 1941–1950. https://doi.org/10.1056/NEJMoa2107934.

Hisham, M., Sivakumar, M. N., Nandakumar, V., & Lakshmikanthcharan, S. (2016). Linezolid and Rasagiline—A culprit for serotonin syndrome. *Indian Journal of Pharmacology, 48*(1), 91–92. https://doi.org/10.4103/0253-7613.174573.

Ho, T. C., Wang, Y. H., Chen, Y. L., Tsai, W. C., Lee, C. H., Chuang, K. P., et al. (2021). Chloroquine and hydroxychloroquine: Efficacy in the treatment of the COVID-19. *Pathogens (Basel, Switzerland), 10*(2). https://doi.org/10.3390/pathogens10020217.

Hsu, J.-Y., Mao, Y.-C., Liu, P.-Y., & Lai, K.-L. (2021). Pharmacology and adverse events of emergency-use authorized medication in moderate to severe COVID-19. *Pharmaceuticals (Basel, Switzerland), 14*(10), 955. https://doi.org/10.3390/ph14100955.

Huber, M. K., Raichle, C., Lingor, P., Synofzik, M., Borgmann, S., Erber, J., et al. (2021). Outcomes of SARS-CoV-2 infections in patients with neurodegenerative diseases in the LEOSS cohort. *Movement Disorders: Official Journal of the Movement Disorder Society, 36*(4), 791–793. https://doi.org/10.1002/mds.28554.

Inouye, S. K., Zhang, Y., Jones, R. N., Kiely, D. K., Yang, F., & Marcantonio, E. R. (2007). Risk factors for delirium at discharge: Development and validation of a predictive model. *Archives of Internal Medicine, 167*(13), 1406–1413. https://doi.org/10.1001/archinte.167.13.1406.

Janes, M., Kuster, S., Goldson, T. M., & Forjuoh, S. N. (2019). Steroid-induced psychosis. *Proceedings (Baylor University. Medical Center), 32*(4), 614–615. https://doi.org/10.1080/08998280.2019.1629223.

Joshi, N., & Singh, S. (2018). Updates on immunity and inflammation in Parkinson disease pathology. *Journal of Neuroscience Research, 96*(3), 379–390. https://doi.org/10.1002/jnr.24185.

Kalimisetty, S., Askar, W., Fay, B., & Khan, A. (2017). Models for predicting incident delirium in hospitalized older adults: A systematic review. *Journal of Patient-Centered Research and Reviews, 4*(2), 69–77. https://doi.org/10.17294/2330-0698.1414.

Kamel, W. A., Kamel, M. I., Alhasawi, A., Elmasry, S., AlHamdan, F., & Al-Hashel, J. Y. (2021). Effect of pre-exposure use of amantadine on COVID-19 infection: A hospital-based cohort study in patients with Parkinson's disease or multiple sclerosis. *Frontiers in Neurology, 12*, 704186. https://doi.org/10.3389/fneur.2021.704186.

Karim, S. S. A., & Karim, Q. A. (2021). Omicron SARS-CoV-2 variant: A new chapter in the COVID-19 pandemic. *Lancet, 398*(10317), 2126–2128. https://doi.org/10.1016/s0140-6736(21)02758-6.

Kaye, C. M., & Nicholls, B. (2000). Clinical pharmacokinetics of ropinirole. *Clinical Pharmacokinetics, 39*(4), 243–254. https://doi.org/10.2165/00003088-200039040-00001.

Kelsey, J. E., & Neville, C. (2014). The effects of the β-lactam antibiotic, ceftriaxone, on forepaw stepping and L-DOPA-induced dyskinesia in a rodent model of Parkinson's disease. *Psychopharmacology, 231*(12), 2405–2415. https://doi.org/10.1007/s00213-013-3400-6.

Khoshnood, R. J., Zali, A., Tafreshinejad, A., Ghajarzadeh, M., Ebrahimi, N., Safari, S., et al. (2021). Parkinson's disease and COVID-19: A systematic review and meta-analysis. *Neurological Sciences: Official Journal of the Italian Neurological Society and of the Italian Society of Clinical Neurophysiology, 1-9*. https://doi.org/10.1007/s10072-021-05756-4.

Kim, H. S. (2021). Do an altered gut microbiota and an associated leaky gut affect COVID-19 severity? *mBio, 12*(1), e03022-03020. https://doi.org/10.1128/mBio.03022-20.

Langford, B. J., So, M., Raybardhan, S., Leung, V., Westwood, D., MacFadden, D. R., et al. (2020). Bacterial co-infection and secondary infection in patients with COVID-19: A living rapid review and meta-analysis. *Clinical Microbiology and Infection: The Official Publication of the European Society of Clinical Microbiology and Infectious Diseases*, *26*(12), 1622–1629. https://doi.org/10.1016/j.cmi.2020.07.016.

Liumbruno, G., Bennardello, F., Lattanzio, A., Piccoli, P., Rossetti, G., & Italian Society of Transfusion Medicine and Immunohaematology (SIMTI) Work Group. (2009). Recommendations for the transfusion of plasma and platelets. *Blood Transfusion = Trasfusione del sangue*, 7(2), 132–150. https://doi.org/10.2450/2009.0005-09.

Magdalinou, K. N., Martin, A., & Kessel, B. (2007). Prescribing medications in Parkinson's disease (PD) patients during acute admissions to a District General Hospital. *Parkinsonism & Related Disorders*, *13*(8), 539–540. https://doi.org/10.1016/j.parkreldis.2006.11.006.

Marcantonio, E. R. (2017). Delirium in Hospitalized Older Adults. *The New England Journal of Medicine*, *377*(15), 1456–1466. https://doi.org/10.1056/NEJMcp1605501.

Miocinovic, S., Ostrem, J. L., Okun, M. S., Bullinger, K. L., Riva-Posse, P., Gross, R. E., et al. (2020). Recommendations for deep brain stimulation device management during a pandemic. *Journal of Parkinson's Disease*, *10*(3), 903–910. https://doi.org/10.3233/jpd-202072.

Miranda, C., Silva, V., Capita, R., Alonso-Calleja, C., Igrejas, G., & Poeta, P. (2020). Implications of antibiotics use during the COVID-19 pandemic: Present and future. *The Journal of Antimicrobial Chemotherapy*, *75*(12), 3413–3416. https://doi.org/10.1093/jac/dkaa350.

Mpekoulis, G., Frakolaki, E., Taka, S., Ioannidis, A., Vassiliou, A. G., Kalliampakou, K. I., et al. (2021). Alteration of L-Dopa decarboxylase expression in SARS-CoV-2 infection and its association with the interferon-inducible ACE2 isoform. *PLoS One*, *16*(6), e0253458. https://doi.org/10.1371/journal.pone.0253458.

Nataf, S. (2020). An alteration of the dopamine synthetic pathway is possibly involved in the pathophysiology of COVID-19. *Journal of Medical Virology*, *92*(10), 1743–1744. https://doi.org/10.1002/jmv.25826.

Nathan, R., Shawa, I., De La Torre, I., Pustizzi, J. M., Haustrup, N., Patel, D. R., et al. (2021). A narrative review of the clinical practicalities of bamlanivimab and etesevimab antibody therapies for SARS-CoV-2. *Infectious Diseases and Therapy*, *10*(4), 1933–1947. https://doi.org/10.1007/s40121-021-00515-6.

Newman, E. J., Grosset, D. G., & Kennedy, P. G. (2009). The parkinsonism-hyperpyrexia syndrome. *Neurocritical Care*, *10*(1), 136–140. https://doi.org/10.1007/s12028-008-9125-4.

Nirenberg, M. J. (2013). Dopamine agonist withdrawal syndrome: Implications for patient care. *Drugs & Aging*, *30*(8), 587–592. https://doi.org/10.1007/s40266-013-0090-z.

Noreen, S., Maqbool, I., & Madni, A. (2021). Dexamethasone: Therapeutic potential, risks, and future projection during COVID-19 pandemic. *European Journal of Pharmacology*, *894*, 173854. https://doi.org/10.1016/j.ejphar.2021.173854.

Pan, H., Peto, R., Henao-Restrepo, A. M., Preziosi, M. P., Sathiyamoorthy, V., Abdool Karim, Q., et al. (2021). Repurposed antiviral drugs for Covid-19—Interim WHO solidarity trial results. *The New England Journal of Medicine*, *384*(6), 497–511. https://doi.org/10.1056/NEJMoa2023184.

Parra, A., Juanes, A., Losada, C. P., Álvarez-Sesmero, S., Santana, V. D., Martí, I., et al. (2020). Psychotic symptoms in COVID-19 patients. A retrospective descriptive study. *Psychiatry Research*, *291*, 113254. https://doi.org/10.1016/j.psychres.2020.113254.

Pasin, L., Cavalli, G., Navalesi, P., Sella, N., Landoni, G., Yavorovskiy, A. G., et al. (2021). Anakinra for patients with COVID-19: A meta-analysis of non-randomized cohort studies. *European Journal of Internal Medicine, 86*, 34–40. https://doi.org/10.1016/j.ejim.2021.01.016.

Patel, P. H., & Hashmi, M. F. (2021). Macrolides. In *StatPearls*. Treasure Island (FL): StatPearls Publishing LLC.

Pettit, N. N., Alonso, V., Wojcik, E., Anyanwu, E. C., Ebara, L., & Benoit, J. L. (2016). Possible serotonin syndrome with carbidopa-levodopa and linezolid. *Journal of Clinical Pharmacy and Therapeutics, 41*(1), 101–103. https://doi.org/10.1111/jcpt.12352.

Piroth, L., Cottenet, J., Mariet, A. S., Bonniaud, P., Blot, M., Tubert-Bitter, P., et al. (2021). Comparison of the characteristics, morbidity, and mortality of COVID-19 and seasonal influenza: A nationwide, population-based retrospective cohort study. *The Lancet Respiratory Medicine, 9*(3), 251–259. https://doi.org/10.1016/s2213-2600(20)30527-0.

Poewe, W., Seppi, K., Tanner, C. M., Halliday, G. M., Brundin, P., Volkmann, J., et al. (2017). Parkinson disease. *Nature Reviews Disease Primers, 3*(1), 17013. https://doi.org/10.1038/nrdp.2017.13.

Prediger, R. D., Matheus, F. C., Schwarzbold, M. L., Lima, M. M., & Vital, M. A. (2012). Anxiety in Parkinson's disease: A critical review of experimental and clinical studies. *Neuropharmacology, 62*(1), 115–124. https://doi.org/10.1016/j.neuropharm.2011.08.039.

Pun, B. T., Badenes, R., Heras La Calle, G., Orun, O. M., Chen, W., Raman, R., et al. (2021). Prevalence and risk factors for delirium in critically ill patients with COVID-19 (COVID-D): A multicentre cohort study. *The Lancet. Respiratory Medicine, 9*(3), 239–250. https://doi.org/10.1016/S2213-2600(20)30552-X.

Quincho-Lopez, A., Benites-Ibarra, C. A., Hilario-Gomez, M. M., Quijano-Escate, R., & Taype-Rondan, A. (2021). Self-medication practices to prevent or manage COVID-19: A systematic review. *PLoS One, 16*(11), e0259317. https://doi.org/10.1371/journal.pone.0259317.

Quinn, D. K., & Stern, T. A. (2009). Linezolid and serotonin syndrome. *Primary Care Companion to the Journal of Clinical Psychiatry, 11*(6), 353–356. https://doi.org/10.4088/PCC.09r00853.

Raeder, V., Boura, I., Leta, V., Jenner, P., Reichmann, H., Trenkwalder, C., et al. (2021). Rotigotine transdermal patch for motor and non-motor Parkinson's disease: A review of 12 years' clinical experience. *CNS Drugs, 35*(2), 215–231. https://doi.org/10.1007/s40263-020-00788-4.

Reddymasu, S. C., Soykan, I., & McCallum, R. W. (2007). Domperidone: Review of pharmacology and clinical applications in gastroenterology. *The American Journal of Gastroenterology, 102*(9), 2036–2045. https://doi.org/10.1111/j.1572-0241.2007.01255.x.

Reuter, S., Deuschl, G., Berg, D., Helmers, A., Falk, D., & Witt, K. (2018). Life-threatening DBS withdrawal syndrome in Parkinson's disease can be treated with early reimplantation. *Parkinsonism & Related Disorders, 56*, 88–92. https://doi.org/10.1016/j.parkreldis.2018.06.035.

Richter, D., Scherbaum, R., Bartig, D., Gold, R., Krogias, C., & Tönges, L. (2021). Analysis of nationwide multimodal complex treatment and drug pump therapy in Parkinson's disease in times of COVID-19 pandemic in Germany. *Parkinsonism & Related Disorders, 85*, 109–113. https://doi.org/10.1016/j.parkreldis.2021.03.006.

Roberts, D. P., & Lewis, S. J. G. (2018). Considerations for general anaesthesia in Parkinson's disease. *Journal of Clinical Neuroscience, 48*, 34–41. https://doi.org/10.1016/j.jocn.2017.10.062.

Rogers, J. P., Chesney, E., Oliver, D., Pollak, T. A., McGuire, P., Fusar-Poli, P., et al. (2020). Psychiatric and neuropsychiatric presentations associated with severe coronavirus infections: A systematic review and meta-analysis with comparison to the COVID-19 pandemic. *The Lancet. Psychiatry*, 7(7), 611–627. https://doi.org/10.1016/S2215-0366(20)30203-0.

Romano, S., Savva, G. M., Bedarf, J. R., Charles, I. G., Hildebrand, F., & Narbad, A. (2021). Meta-analysis of the Parkinson's disease gut microbiome suggests alterations linked to intestinal inflammation. *NPJ Parkinson's Disease*, 7(1), 27. https://doi.org/10.1038/s41531-021-00156-z.

Schapira, A. H. V., Chaudhuri, K. R., & Jenner, P. (2017). Non-motor features of Parkinson disease. *Nature Reviews. Neuroscience*, *18*(8), 509. https://doi.org/10.1038/nrn.2017.91.

Seppi, K., Ray Chaudhuri, K., Coelho, M., Fox, S. H., Katzenschlager, R., Perez Lloret, S., et al. (2019). Update on treatments for nonmotor symptoms of Parkinson's disease-an evidence-based medicine review. *Movement Disorders: Official Journal of the Movement Disorder Society*, *34*(2), 180–198. https://doi.org/10.1002/mds.27602.

Serrano-Dueñas, M. (2003). Neuroleptic malignant syndrome-like, or–dopaminergic malignant syndrome–due to levodopa therapy withdrawal. Clinical features in 11 patients. *Parkinsonism & Related Disorders*, *9*(3), 175–178. https://doi.org/10.1016/s1353-8020(02)00035-4.

Shaikh, S. I., & Verma, H. (2011). Parkinson's disease and anaesthesia. *Indian Journal of Anaesthesia*, *55*(3), 228–234. https://doi.org/10.4103/0019-5049.82658.

Sharma, V. D., Safarpour, D., Mehta, S. H., Vanegas-Arroyave, N., Weiss, D., Cooney, J. W., et al. (2021). Telemedicine and deep brain stimulation—Current practices and recommendations. *Parkinsonism & Related Disorders*, *89*, 199–205. https://doi.org/10.1016/j.parkreldis.2021.07.001.

Siddiqui, M. S., Jimenez-Shahed, J., Mari, Z., Walter, B. L., De Jesus, S., Panov, F., et al. (2021). North American survey on impact of the COVID-19 pandemic shutdown on DBS care. *Parkinsonism & Related Disorders*, *92*, 41–45. https://doi.org/10.1016/j.parkreldis.2021.10.011.

Simonet, C., Tolosa, E., Camara, A., & Valldeoriola, F. (2020). Emergencies and critical issues in Parkinson's disease. *Practical Neurology*, *20*(1), 15. https://doi.org/10.1136/practneurol-2018-002075.

Siskind, D., Honer, W. G., Clark, S., Correll, C. U., Hasan, A., Howes, O., et al. (2020). Consensus statement on the use of clozapine during the COVID-19 pandemic. *Journal of Psychiatry & Neuroscience: JPN*, *45*(3), 222–223. https://doi.org/10.1503/jpn.200061.

Smith, M. D., & Maani, C. V. (2021). Norepinephrine. In *StatPearls*. Treasure Island (FL): StatPearls Publishing LLC.

Suttrup, I., & Warnecke, T. (2016). Dysphagia in Parkinson's disease. *Dysphagia*, *31*(1), 24–32. https://doi.org/10.1007/s00455-015-9671-9.

Takubo, H., Harada, T., Hashimoto, T., Inaba, Y., Kanazawa, I., Kuno, S., et al. (2003). A collaborative study on the malignant syndrome in Parkinson's disease and related disorders. *Parkinsonism & Related Disorders*, *9*(Suppl. 1), S31–S41. https://doi.org/10.1016/s1353-8020(02)00122-0.

Tan, A. H., Mahadeva, S., Marras, C., Thalha, A. M., Kiew, C. K., Yeat, C. M., et al. (2015). Helicobacter pylori infection is associated with worse severity of Parkinson's disease. *Parkinsonism & Related Disorders*, *21*(3), 221–225. https://doi.org/10.1016/j.parkreldis.2014.12.009.

Tentillier, N., Etzerodt, A., Olesen, M. N., Rizalar, F. S., Jacobsen, J., Bender, D., et al. (2016). Anti-inflammatory modulation of microglia via CD163-targeted glucocorticoids protects dopaminergic neurons in the 6-OHDA Parkinson's disease model. *The Journal of Neuroscience*, *36*(36), 9375–9390. https://doi.org/10.1523/jneurosci.1636-16.2016.

Ternák, G., Kuti, D., & Kovács, K. J. (2020). Dysbiosis in Parkinson's disease might be triggered by certain antibiotics. *Medical Hypotheses*, *137*, 109564. https://doi.org/10.1016/j.mehy.2020.109564.

Ticinesi, A., Cerundolo, N., Parise, A., Nouvenne, A., Prati, B., Guerra, A., et al. (2020). Delirium in COVID-19: Epidemiology and clinical correlations in a large group of patients admitted to an academic hospital. *Aging Clinical and Experimental Research*, *32*(10), 2159–2166. https://doi.org/10.1007/s40520-020-01699-6.

Tisdale, J. E. (2016). Drug-induced QT interval prolongation and torsades de pointes: Role of the pharmacist in risk assessment, prevention and management. *Canadian Pharmacists Journal: CPJ = Revue des pharmaciens du Canada: RPC*, *149*(3), 139–152. https://doi.org/10.1177/1715163516641136.

Touret, F., & de Lamballerie, X. (2020). Of chloroquine and COVID-19. *Antiviral Research*, *177*, 104762. https://doi.org/10.1016/j.antiviral.2020.104762.

Touret, F., Gilles, M., Barral, K., Nougairède, A., van Helden, J., Decroly, E., et al. (2020). In vitro screening of a FDA approved chemical library reveals potential inhibitors of SARS-CoV-2 replication. *Scientific Reports*, *10*(1), 13093. https://doi.org/10.1038/s41598-020-70143-6.

Travagli, R. A., Browning, K. N., & Camilleri, M. (2020). Parkinson disease and the gut: New insights into pathogenesis and clinical relevance. *Nature Reviews. Gastroenterology & Hepatology*, *17*(11), 673–685. https://doi.org/10.1038/s41575-020-0339-z.

Tufekci, K. U., Meuwissen, R., Genc, S., & Genc, K. (2012). Inflammation in Parkinson's disease. *Advances in Protein Chemistry and Structural Biology*, *88*, 69–132. https://doi.org/10.1016/b978-0-12-398314-5.00004-0.

van Wamelen, D. J., Leta, V., Johnson, J., Ocampo, C. L., Podlewska, A. M., Rukavina, K., et al. (2020). Drooling in Parkinson's disease: Prevalence and progression from the non-motor international longitudinal study. *Dysphagia*, *35*(6), 955–961. https://doi.org/10.1007/s00455-020-10102-5.

Varatharaj, A., Thomas, N., Ellul, M. A., Davies, N. W. S., Pollak, T. A., Tenorio, E. L., et al. (2020). Neurological and neuropsychiatric complications of COVID-19 in 153 patients: A UK-wide surveillance study. *The Lancet. Psychiatry*, 7(10), 875–882. https://doi.org/10.1016/S2215-0366(20)30287-X.

Vardy, E. R., Teodorczuk, A., & Yarnall, A. J. (2015). Review of delirium in patients with Parkinson's disease. *Journal of Neurology*, *262*(11), 2401–2410. https://doi.org/10.1007/s00415-015-7760-1.

Vijayan, S., Singh, B., Ghosh, S., Stell, R., & Mastaglia, F. L. (2020). Brainstem ventilatory dysfunction: A plausible mechanism for dyspnea in Parkinson's disease? *Movement Disorders*, *35*(3), 379–388. https://doi.org/10.1002/mds.27932.

Volpi-Abadie, J., Kaye, A. M., & Kaye, A. D. (2013). Serotonin syndrome. *The Ochsner Journal*, *13*(4), 533–540. Retrieved from https://pubmed.ncbi.nlm.nih.gov/24358002. https://www.ncbi.nlm.nih.gov/pmc/articles/PMC3865832/.

Wada, K., Yamada, N., Sato, T., Suzuki, H., Miki, M., Lee, Y., et al. (2001). Corticosteroid-induced psychotic and mood disorders: Diagnosis defined by DSM-IV and clinical pictures. *Psychosomatics*, *42*(6), 461–466. https://doi.org/10.1176/appi.psy.42.6.461.

Wakefield, B. J. (2002). Risk for acute confusion on hospital admission. *Clinical Nursing Research*, *11*(2), 153–172. https://doi.org/10.1177/105477380201100205.

Warrington, T. P., & Bostwick, J. M. (2006). Psychiatric adverse effects of corticosteroids. *Mayo Clinic Proceedings*, *81*(10), 1361–1367. https://doi.org/10.4065/81.10.1361.

Weinreich, D. M., Sivapalasingam, S., Norton, T., Ali, S., Gao, H., Bhore, R., et al. (2021). REGN-COV2, a neutralizing antibody cocktail, in outpatients with Covid-19. *The New England Journal of Medicine*, *384*(3), 238–251. https://doi.org/10.1056/NEJMoa2035002.

Wilde, M. I., & Markham, A. (1996). Ondansetron. A review of its pharmacology and preliminary clinical findings in novel applications. *Drugs*, *52*(5), 773–794. https://doi.org/10.2165/00003495-199652050-00010.

Won, J. H., Byun, S. J., Oh, B. M., Park, S. J., & Seo, H. G. (2021). Risk and mortality of aspiration pneumonia in Parkinson's disease: A nationwide database study. *Scientific Reports*, *11*(1), 6597. https://doi.org/10.1038/s41598-021-86011-w.

Wright, J. J., Goodnight, P. D., & McEvoy, M. D. (2009). The utility of ketamine for the preoperative management of a patient with Parkinson's disease. *Anesthesia and Analgesia*, *108*(3), 980–982. https://doi.org/10.1213/ane.0b013e3181924025.

Wu, Z., Li, H., Liao, K., & Wang, Y. (2021). Association between dexamethasone and delirium in critically ill patients: A retrospective cohort study of a large clinical database. *The Journal of Surgical Research*, *263*, 89–101. https://doi.org/10.1016/j.jss.2021.01.027.

Yadav, N., Thakur, A. K., Shekhar, N., & Ayushi. (2021). Potential of antibiotics for the treatment and Management of Parkinson's disease: An overview. *Current Drug Research Reviews*, *13*(3), 166–171. https://doi.org/10.2174/2589977513666210315095133.

Yan, A., & Bryant, E. E. (2021). Quinolones. In *StatPearls*. Treasure Island (FL): StatPearls Publishing LLC.

Yu, X. X., & Fernandez, H. H. (2017). Dopamine agonist withdrawal syndrome: A comprehensive review. *Journal of the Neurological Sciences*, *374*, 53–55. https://doi.org/10.1016/j.jns.2016.12.070.

Zesiewicz, T. A., Hauser, R. A., Freeman, A., Sullivan, K. L., Miller, A. M., & Halim, T. (2009). Fentanyl-induced bradykinesia and rigidity after deep brain stimulation in a patient with Parkinson disease. *Clinical Neuropharmacology*, *32*(1), 48–50. https://doi.org/10.1097/WNF.0b013e31817e23e3.

Zhang, H., Penninger, J. M., Li, Y., Zhong, N., & Slutsky, A. S. (2020). Angiotensin-converting enzyme 2 (ACE2) as a SARS-CoV-2 receptor: Molecular mechanisms and potential therapeutic target. *Intensive Care Medicine*, *46*(4), 586–590. https://doi.org/10.1007/s00134-020-05985-9.

Zhang, Q., Schultz, J. L., Aldridge, G. M., Simmering, J. E., & Narayanan, N. S. (2020). Coronavirus disease 2019 case fatality and Parkinson's disease. *Movement Disorders*, *35*(11), 1914–1915. https://doi.org/10.1002/mds.28325.

Zhu, J., Zhong, Z., Ji, P., Li, H., Li, B., Pang, J., et al. (2020). Clinicopathological characteristics of 8697 patients with COVID-19 in China: A meta-analysis. *Family Medicine and Community Health*, *8*(2). https://doi.org/10.1136/fmch-2020-000406.

Zoremba, N., & Coburn, M. (2019). Acute confusional states in hospital. *Deutsches Ärzteblatt International*, *116*(7), 101–106. https://doi.org/10.3238/arztebl.2019.0101.

Zornberg, G. L., Bodkin, J. A., & Cohen, B. M. (1991). Severe adverse interaction between pethidine and selegiline. *Lancet*, *337*(8735), 246. https://doi.org/10.1016/0140-6736(91)92219-r.

CHAPTER EIGHT

Covid-19 and Parkinson's disease: Nursing care, vaccination and impact on advanced therapies

Anna Roszmann[a,*], Aleksandra M. Podlewska[b,c], Yue Hui Lau[b,c], Iro Boura[b,c,d], and Annette Hand[e]

[a]Department of Neuro-Psychiatric Nursing, Medical University of Gdańsk, Gdańsk, Poland
[b]Department of Neurosciences, Institute of Psychiatry, Psychology & Neuroscience, King's College London, London, United Kingdom
[c]Parkinson's Foundation Centre of Excellence, King's College Hospital NHS Foundation Trust, London, United Kingdom
[d]Medical School, University of Crete, Heraklion, Crete, Greece
[e]Newcastle upon Tyne Hospitals NHS Foundation Trust and Northumbria University, Newcastle upon Tyne, United Kingdom
*Corresponding author: e-mail address: anna.roszmann@gumed.edu.pl

Contents

Abstract

The Coronavirus Disease 2019 (Covid-19) pandemic has created many challenges for the Parkinson's Disease (PD) care service delivery, which has been established over the past decades. The need for rapid adjustments to the new conditions has highlighted

International Review of Neurobiology, Volume 165
ISSN 0074-7742
https://doi.org/10.1016/bs.irn.2022.04.005
Copyright © 2022 Elsevier Inc. All rights reserved.

the role of technology, which can act as an enabler both in patient-facing aspects of care, such as clinical consultations, as well as in professional development and training. The Parkinson's Disease Nurse Specialists (PNSs) play a vital role in the effective management of people with PD (PwP). Maintaining optimum functionality and availability of device aided therapies is essential in order to ensure patients' quality of life. PwP are particularly recommended to use vaccination as a basic protection from the virus. The long-term consequences of this pandemic on PwP are highly uncertain, and education, support and reassurance of patients and their families may help ease their burden.

1. Introduction

Over the past 2 years the pandemic caused by Severe Acute Respiratory Syndrome Coronavirus-2 (SARS-CoV-2) has spread globally, profoundly affecting morbidity and mortality of the general population (Abate, Checkol, & Mantefardo, 2021; Piroth et al., 2021). People with Parkinson's Disease (PwP, PD) are often older and have comorbidities, which might increase the risk for severe or critical Coronavirus Disease 2019 (Covid-19) (Cascella, Rajnik, Aleem, Dulebohn, & Di Napoli, 2021). This might be particularly true for those with advanced PD, as they usually experience a heterogeneous set of motor and non-motor symptoms and follow complicated therapeutic regimens.

Optimal care of PwP should involve a multidisciplinary team (MDT), with PD Nurse Specialists (PNSs) being key members. During the pandemic, PNSs have worked hard in order to assist PwP in self-isolation consequences and symptom management, reducing the risk of both hospital admission and extended hospital stays (Simpson & Doyle, 2020). The role of a PNS is particularly prominent in the care of advanced PwP, who are suitable for device-aided therapies. Moreover, the International Parkinson and Movement Disorder Society (MDS) has recommended in favor of the Covid-19 vaccinations, recognizing their paramount role in the protection of PwP, physicians and associated healthcare providers in this critical situation of the global pandemic (Bloem et al., 2021).

2. The role of the Parkinson's disease nurse specialist

Within the United Kingdom (UK) the first PNS came into post in 1989, following recommendations from a commissioned research project, investigating ways to provide up-to-date care and support to PwP (MacMahon, 1999). The PNS role has been developed with the aim of improving PD prognosis, through better education and support, and reducing the impact of PD on patients, but also their families and caregivers. A task force for the PD Society has suggested that each individual with PD will be assigned a healthcare professional, holding a leading role in coordinating the patients' care and treatment changes in collaboration with a multidisciplinary team (Brown et al., 2020). Today, the PNS role is still recognized as pivotal (Cook, McNamee, McFetridge, & Deeny, 2007) with around 450 PNSs working in primary, secondary, and tertiary care settings in the UK.

The PNS provides assistance with ongoing management and follow-up by providing medication review, clinical leadership, help with post-diagnostic counseling, guidance to other services, education about PD, support, and advice for PwP, their caregivers and other involving staff (Axelrod et al., 2010). Many PNSs run their own clinics, organize PD helplines, make home visits, refer to other experts and co-ordinate care packages according to patients' needs. PNSs are often the first point of contact for PwP, ensuring fast access to specialist care, whilst relieving some of the pressure from neurologists/ geriatricians with a special interest in PD, who are in short supply (Axelrod et al., 2010). The National Institute for Health and Care Excellence (NICE) guidance (Rogers, Davies, Pink, & Cooper, 2017) recommends that PwP should have regular access to:

- Clinical monitoring and medication adjustment.
- A continuing point of contact for support, including home visits when appropriate.
- Reliable sources of information about clinical and social matters of concern to PwP, their family members and their caregivers (as appropriate).

Within the UK, these recommendations are often achieved due to the services provided by the PNSs, who consitute, thus, vital team members of the PD specialist service.

2.1 Covid-19 and the impact on people with Parkinson's

In April 2020, the UK Department of Health and Social Care (DHSC) announced that all people should "Stay at home, protect the National Health Service (NHS), save lives." Further recommendations were given to people aged over 70, or those with an underlying health condition, to stay at home at all times and not to leave, even to buy food, collect medicine or exercise. Within the UK this advice lasted for months, and people have experienced a number of "lock downs" as the situation changed with the appearance of different variants of SARS-CoV-2. Particular guidance has been given to people at high risk of developing severe complications from Covid-19, who were described as "clinically extremely vulnerable." These individuals were contacted via the NHS and provided with extra support to ensure they were able to remain at home. PwP have been classified as "moderately vulnerable," which means they were only recommended to leave their home in order to buy food or medicine, or to exercise once daily.

Staying home and isolating from others has led in dramatic changes in daily routine and lifestyle for everyone. Activities previously taken for granted, such as going shopping, attending social gatherings, collecting medications, providing childcare or attending a day center, had to stop, imposing drastic changes in people's daily schedule. Flexible adaptations to new circumstances have been acknowledged as cognitive operations, which depend on normal dopaminergic functioning (Helmich & Bloem, 2020). An insightful paper by Helmich and Bloem (2020), based on a large body of literature, discusses how PwP experience cognitive and motor inflexibility, as a result of nigrostriatal dopamine depletion, which forms the pathophysiological substrate of PD (Helmich, Aarts, de Lange, Bloem, & Toni, 2009). Douma and de Kloet (2020) have also hypothesized that dopamine-dependent adaptation is a prerequisite for successful coping, which, when deficient, might lead to a sense of loss of control and increased psychological stress (Douma & de Kloet, 2020).

Up to now, numerous papers have been published, citing the indirect impact of Covid-19 on individuals, including stress, self-isolation, depression, and anxiety, along with the consequences of prolonged immobility due to the lockdown (Helmich & Bloem, 2020; Prasad et al., 2020). Helmich and Bloem (2020), have also identified that these increased levels of stress during the Covid-19 pandemic may have both short- and long-term negative sequelae for PwP. Indeed, high levels of psychological stress have been associated with temporary aggravation of a number of motor PD symptoms, such as tremor, freezing of gait, or dyskinesias (Macht et al., 2007; Prasad et al., 2020; Zach, Dirkx, Pasman, Bloem, & Helmich, 2017), along with a reduced efficacy of dopaminergic medication (Zach et al., 2017).

2.2 Covid-19 and the impact on Parkinson's disease services

Due to increasing numbers of Covid-19 patients being admitted in hospitals, the pressure on healthcare professionals has been enhanced, with many specialist nurses, including PNSs, being redeployed to different clinical areas in order to support the front line staff. As a result, many PwP have been left without their vital link to specialist knowledge, information, and support for prolonged periods of time. Regular PNSs reviews had to be canceled, helplines were closed and PwP lost their prior access to specialist service. These changes have likely contributed to the high levels of stress and anxiety already experienced by PwP.

The few PNSs, who were left to manage the specialist services, had to rapidly re-design and adjust to the new circumstances. National guidance for healthcare professionals was to avoid all face-to-face contact, unless there was a clear urgent need not to do so. This meant that many Outpatient Clinics, including Movement Disorders Outpatient Clinics, were canceled and the ability to provide standard PD care was severely compromised by the strain on healthcare systems brought about by this pandemic (Papa et al., 2020). For those services who were able to continue to asses or review PwP face-to-face, restrictions were often applied, including less new or regular patients assessed per clinic in order to allow for social distancing regulations and to ensure appropriate room cleaning between appointments.

Many PD Specialist services were only able to use telehealth and telephone-based services to continue to monitor and support PwP. Many PNSs were no longer able to conduct home visits or review those living in sheltered accommodation, residential or nursing care facilities. Many PNSs developed protocols for telephone consultations to ensure that a safe practice was followed and vital symptoms were not missed. Once PNSs were able to review patients, particularly those with complex needs or those approaching end of life, within their own home or in a community setting, appropriate personal protective equipment (PPE) guidelines were devised and followed. During this time, the use of wearable technologies was further encouraged and will be discussed later in this chapter.

Like in other specialties, further pressure has been imposed on PD specialist services, as some PNSs might be required to "self-isolate" due to their medical history or have to look after children or dependents during lockdown periods (Nune, Iyengar, Ahmed, & Sapkota, 2020). Under these circumstances, many PNSs have opted to work from home, trying to ensure continuity of care for PwP. Overall, redeployment of staff and self-isolation regulations have greatly reduced the availability of PNSs during the pandemic.

2.3 The role of the PNS in supporting people with Parkinson's disease during the pandemic

Since the onset of the pandemic, misinformation, conflicting advice, and rapidly changing instructions have contributed to increased stress and anxiety (Geldsetzer, 2020; Loomba, de Figueiredo, Piatek, de Graaf, & Larson, 2021). Misinformation has also led individuals across the globe to question the safety, effectiveness, and importance of the Covid-19 vaccine (de Figueiredo, Simas, Karafillakis, Paterson, & Larson, 2020). The PD Society of the UK (Parkinson's UK), a research and support charity within the UK, has worked closely with PNSs and other healthcare specialists to provide information, support and guidance to PwP, their families and friends. A regular roundup of news, containing the latest information and advice on what PwP need to know, as they continue to live with Covid-19, has also been published online. Furthermore, due to potential changes in people's access to PD specialist services, it was fully appreciated that contacting healthcare professionals may be different or difficult during the Covid-19 crisis. A number of resources, including information sheets, forums, videos and blogs, and webinars, have been developed, such as:

- How to get the support you need from healthcare professionals, including who to contact and how the communication with a healthcare professional might be modified due to the use of telephone or video consultations.
- Information on anti-parkinsonian medication, including guidance on how review appointments may change and what to do if an individual is unable to obtain their usual anti-PD medication.
- Advice on what to do if an individual is experiencing new or worsening PD symptom.
- Information and advice for PwP or their caregivers, related to working, furlough and redundancy due to Covid-19.
- Benefits to support those who may have found that their financial situation changed due to Covid-19.
- Information on the Covid-19 vaccine in order to answer any queries or concerns PwP might have.

Support, information, and advice have also been provided via social media sites and on-line support groups, as a way for people to stay in touch, reduce isolation and engage with others.

During the Covid-19 pandemic, many PwP have seen their health deteriorate and have experienced an aggravation of their parkinsonian symptoms, including tremor, stiffness, anxiety, hallucinations and others (Simpson & Doyle, 2020). Family members and friends of PwP have also taken on more caring responsibilities, such as assistance with personal care and mobility, with reported impacts on their own physical and mental health (Simpson & Doyle, 2020). Access of PwP to a multi-disciplinary team is essential to improve self-management, limit deterioration of any symptoms and lower the risk of exposure to Covid-19. Those PNSs who were able to continue in their usual clinical role, have assisted individuals with self-isolation and have provided symptoms management, reducing both the risk of hospital admission and extended hospital stays. More specifically, PNSs may:

- support PwP in overall clinical management of their condition and guidance, including provision of therapy, via remote consultation. Potential misunderstandings or delays in the doses or regimen of anti-parkinsonian drugs and other medication commonly used in PD, such as anti-depressants or anti-psychotics, can lead to adverse incidents and life-threatening situations.
- refer PwP to other medical care and social services, when needed, and facilitate their navigation across multiple referral pathways.
- recognize and advise on PD symptoms that could mimic Covid-19 manifestations, such as loss of smell, nausea, or hallucinations (Andrews, Cai, Rudd, & Sanger, 2021; Parra et al., 2020).
- provide advice and proper guidance to PwP and their caregivers in case of medical emergencies (e.g., who to contact, how to proceed).
- support other healthcare professionals, such as General Practitioners (GPs) and non-specialist secondary care teams, by providing expert consultation when needed.
- assist PwP in registering to and familiarizing with Parkinson's UK and other support organizations.

For all the above reasons, Parkinson's UK has called for PD specialist services to be retained as a matter of priority. Parkinson's UK, with support from the leaders of the Parkinson's Nurse Associations across the UK, has produced a national statement urging managers and commissioners that the role of the PNS is crucial during this time and that PNSs should not be redeployed, but maintain their original role during the pandemic.

3. Covid-19 vaccinations

3.1 General information on the available Covid-19 vaccines

As the prevalence of Covid-19 continues to surge globally, constituting a major global public health threat, and with limited therapeutic interventions available up to now, the development of Covid-19 vaccines has brought hope to humanity, especially in low- and middle-income countries (Choi, 2021), as an effective mean to prevent severe disease and control the pandemic. As of January 2022, at least nine Covid-19 vaccines have obtained Emergency Use Listing (EUL) by the World Health Organization (WHO), based on their safety and efficacy characteristics (World Health Organization: Coronavirus disease (COVID-19): Vaccines, 2022), while more than 15 vaccines are currently investigated or await approval (Phanhdone et al., 2021). Nevertheless, authorizing the use of specific Covid-19 vaccines on a local basis remains at the discretion of each country and complies to national regulations.

Currently, there are few main types of Covid-19 vaccines available globally, including mRNA-1273 (Moderna INC., USA), BNT162b2 mRNA (Pfizer, New York, USA), AZD1222 (Oxford/ AstraZeneca, UK), Janssen Ad26.CoV2·S (Johnson & Johnson, New Jersey, USA), and others (World Health Organization: Coronavirus disease (COVID-19): Vaccines, 2022). Both Pfizer and Moderna vaccines are mRNA-based vaccines and are estimated to have an efficacy rate of more than 94% (Hippisley-Cox et al., 2021). As of March 14, 2022, 4.46 billion people worldwide have completed the whole course of vaccination, with 65.1% of the global population having received at least one vaccine dose (Coronavirus (COVID-19) Vaccinations, 2022; Holder, 2022). Six types of Covid-19 vaccines, including BBIBP-CorV (Sinopharm, Shanghai, China), WIBP-CorV (Sinopharm, Shanghai, China), Ad5-nCoV (CanSinoBIO, Tianjin, China), CoronaVac (Sinovac Biotech, Beijing, China), have been approved in China (6 Vaccines Approved for Use in China, 2022).

It is now acknowledged that SARS-CoV-2 is constantly evolving and mutating, producing multiple novel variants with varying characteristics. Most of these mutations are expected to have an unremarkable clinical effect, however, a minority of them is anticipated to significantly differentiate from the original pattern of the virus interaction with the immune system of human hosts (Harvey et al., 2021). Recently, the newly recognized

Delta and Omicron variants have attracted considerable attention due to reported alterations in transmissibility, the duration of incubation period, along with the potential for severe clinical manifestations (Bai, Gu, Liu, & Zhou, 2021). Moreover, one of the biggest concerns of the scientific community lays on whether the emergence of these new variants might lower the Covid-19 vaccine efficiency, leading to insufficient immunization cover around the world (Karim & Karim, 2021).

According to a recent, large study analyzing surveillance data collected from the Center for Disease Control and Prevention (CDC) from 13 United States of America (US) jurisdictions, the rate of deaths among unvaccinated citizens was found to be about 11 times higher compared to those fully vaccinated since the Delta variant became the dominant strain in the affected population (Scobie, Johnson, & Suthar, 2021). Although results from various studies have suggested a constant vaccine effectiveness against severe Covid-19 from the Delta variant, a relative decline in vaccine efficacy was found compared to previous variants (Bruxvoort et al., 2021; Lopez Bernal et al., 2021). Another US study, analyzing data from 32,867 adults derived from nine states during the summer months of 2021, when the Delta variant was predominant in the country, revealed that vaccinated people were significantly less likely to be hospitalized due to Covid-10 or contract SARS-CoV-2 than those unvaccinated (Grannis et al., 2021). Interestingly, vaccines were 89% (85% to 92%) effective in preventing hospital admissions due to Covid-19 among adults aged under 75, but 76% (64% to 84%) effective in those older than 75. According to the researchers, these tendencies might reflect a declining immunity in people who were vaccinated early, as older vaccine recipients were, indeed, given priority in the vaccination campaigns around the world. Following rapid distribution and administration of the mRNA Covid-19 vaccines (Pfizer-BioNTech and Moderna) under an Emergency Use Authorization by the Food and Drug Administration (FDA), early observational studies among nursing home residents have demonstrated a vaccine effectiveness ranging from 53% to 92% against SARS-CoV-2 infection, a percentage which dropped significantly to 53.1% after the predominance of the Delta variant (Nanduri, Pilishvili, & Derado, 2021). The UK Health Security Agency revealed early estimates indicating that the Omicron variant has significantly reduced the effectiveness of vaccines against symptomatic infection, when compared with the previously dominant Delta infection; however, a booster dose was expected to lead to a moderate to high vaccine effectiveness ranging from 70% to 75% (UK Health Security

Agency: COVID-19 variants identified in the UK, 2022). A large study, using data from more than 1.1 million people aged over 60 (30 July to 31 August 2021), reported that a third dose of the Pfizer vaccine was found to substantially reduce rates of infection and of severe illness in this population, when compared to those who only had two doses (Mahase, 2021).

3.2 Covid-19 vaccines and people with Parkinson's disease

Increasing evidence suggests that PwP are particularly vulnerable to the sequelae of SARS-CoV-2 infection due to intrinsic features of PD, including, but not limited to, rigidity of respiratory muscles, abnormal posture and impaired cough reflex (Lau, Lau, & Ibrahim, 2021; van Wamelen et al., 2020), while they are likely to experience an aggravation of their motor and non-motor symptoms, especially in cases of advanced PD (Merello, Bhatia, & Obeso, 2021). Restrictions of mobility, often accompanying PD, limited access to healthcare resources, and reduced social interactions due to the imposed lockdown regulations might further deteriorate patients' health and quality of life.

Among PwP, pneumonia has been reported as the most common cause of death and inpatient admissions (Okunoye, Kojima, Marston, Walters, & Schrag, 2020), while PwP are considered at increased risk of vaccine-preventable infections of the respiratory system, including influenza and pneumococcal pneumonia (Leibson et al., 2006; Pilishvili & Bennett, 2015). Nevertheless, in a recent study of 143 homebound and ambulatory PwP, almost 10% of individuals reported having missed all influenza vaccinations within the last 5 years and almost one out of three had never been vaccinated against pneumococcus (Phanhdone et al., 2021). More specifically, approximately 35% and 19% of the participants reported hesitation or refusal, respectively to be vaccinated, while, surprisingly, 13% thought vaccination is contraindicated in PD. The study has also highlighted that a significant percentage of household members of PwP would opt out from getting vaccinated as well, thus, posing a threat to more vulnerable individuals (Nordström, Ballin, & Nordström, 2021).

Vaccination against SARS-CoV-2 should be a priority in the management of PwP during the pandemic (Lau et al., 2021). The MDS has released a Covid-19 vaccine statement to address aspects of Covid-19 vaccination among PwP (Bloem et al., 2021). More specifically, PwP are strongly encouraged to receive Covid-19 vaccination, unless there is a specific contraindication. The MDS highlights that the benefits and risks of PwP being vaccinated against Covid-19 are similar to those reported in the general, age-matched population, while the administration of Covid-19 vaccination is not expected to interfere with regular anti-parkinsonian medications and is not known to interact with any subjacent neurodegenerative processes observed in PD (Bloem et al., 2021). Of notice, it has been reported that even though vaccines seem safe for older adults, it is important to remain cautious when administering the vaccine to very frail and terminally ill elderly patients, as a small number of deaths has been reported shortly after vaccination (within 6 days) in markedly frail patients over 75 years old (Torjesen, 2021). Cosentino and colleagues have reported two cases of patients with a PD diagnosis, who developed some degree of temporary aggravation of their motor symptoms soon after the first dose of the Pfizer/bioNTech vaccine with symptoms subsiding spontaneously without any interventions or changes in the patients' regular regimen (Cosentino et al., 2022). More specifically, the first patient manifested a 2-day increase in rigidity and gait impairment, while the second patient exhibited an increased resting tremor, lasting approximately two weeks. Erro and colleagues had earlier reported two cases of PwP, who presented with severe dyskinesia after receiving the BNT162b2 mRNA vaccine (Erro, Buonomo, Barone, & Pellecchia, 2021). In both cases, dyskinesias were managed by reducing the total levodopa dose received by the patients, with the first patient exhibiting a good response to this modification. Considering the second patient, dyskinesias were also accompanied by fever, confusion, delusions and increased levels of D-dimers. Although severe symptoms did abate after 2 weeks, mild confusion and worse than baseline dyskinesias persisted. The reasons underlying these phenomena remain unclear, however, it has been postulated that a subjacent systemic inflammatory response, triggered by excessive anxiety or due to an interaction with regular anti-dopaminergic medication, might have contributed. These cases represent a very low incidence of adverse events following vaccination in PwP and, therefore, PwP should not be discouraged from receiving Covid-19 vaccines. Booster doses should also be offered to PwP according to international guidelines in order to strengthen their immunity against SARS-CoV-2.

Many countries have rolled out vaccination programs, as well as booster doses for their citizens, with a clear priority given to the elderly, especially those with reported comorbidities, including PD. Vaccinated PwP are still expected to strictly comply to the national health guidelines to reduce exposure to and transmission of SARS-CoV-2 (Beauchamp, Finkelstein, Bush, Evans, & Barnham, 2020).

Effective communication strategies between the treating neurologist, PwP, and their household members may limit misinformation and vaccine hesitancy in the community, encourage people to get vaccinated, increase immunization cover against SARS-CoV-2 and, thus, strengthen public health, by reducing rates of Covid-19 morbidity and mortality (Harrison & Wu, 2020). Importantly, the role of nursing care is integral to the success of the Covid-19 vaccines, as it significantly contributes to patients' and caregiviers' education or the promotion of outreached programs with regards to vaccination, as well as identifying any barriers to PwP receiving vaccination. For instance, nursing home and long-term care facility residents live in congregate settings and are often elderly and frail, thus, at increased risk of Covid-19 (Nanduri et al., 2021). Nursing home residents remain vulnerable despite vaccination, and, hence, Covid-19 prevention strategies, including infection control, frequent testing, and vaccination of nursing home staff members, residents, and visitors are critical. These interventions could be delivered by dedicated nurses through organizing focused campaigns, providing relevant reading or audio-visual material with regards to Covid-19 and PD, and advocating vaccination.

Difficulty traveling to specialized clinical settings (Outpatients Clinics, vaccination centers) has been identified as an obstacle to administering Covid-19 vaccinations to homebound PwP (Harrison & Wu, 2020). Furthermore, homebound women with advanced PD were more often found to live alone and encounter considerable difficulties in accessing healthcare facilities (Nwabuobi et al., 2019). In one recent meta-analysis, living alone was strongly associated with lower vaccine uptake (van der Heide, Meinders, Bloem, & Helmich, 2020). PwP, particularly those with advanced disease and/ or homebound have been reported to be at an increased risk of complications from Covid-19 (Brown et al., 2020; Del Prete et al., 2021; van der Heide et al., 2020). Trained nurses could overcome such difficulties by providing vaccination through home visits to this group of patients (Harrison & Wu, 2020).

4. Impact on advanced therapies

4.1 Management of advanced therapies in times of Covid-19

Device-aided therapies constitute effective therapeutic options for patients with advanced PD, whose symptoms are so severe that are no longer sufficiently controlled with optimum medical therapy. These treatments, including continuous subcutaneous apomorphine infusion (CSAI), levodopa/ carbidopa intestinal gel (LCIG) infusion, and deep brain stimulation (DBS), have been found to decrease the intensity and frequency of motor fluctuations, improve non-motor aspects of PD and ameliorate the patients' and their caregivers' quality of life (Dafsari et al., 2019; Santos-García, Añón, Fuster-Sanjurjo, & de la Fuente-Fernández, 2012), while some benefits are maintained in the long-term (Antonini et al., 2021; Limousin & Foltynie, 2019). Although device-aided therapies, especially DBS, appear to be cost-effective in the long-term (Smilowska et al., 2021), regular follow-up visits are still required, while the role of nursing care in the management of these patients, including home-visits, is essential (Antonini, Mirò, Castiglioni, & Pezzoli, 2008; De Rosa, Tessitore, Bilo, Peluso, & De Michele, 2016).

As mentioned above, the pandemic of Covid-19 has taken a toll on PwP using advanced therapies, not only because of the direct effects of the virus on these vulnerable patients (worse Covid-19 outcome due to older age, comorbidities and inherent characteristics of PD, including the fact that advanced stage PwP might be frail, bedridden, incontinent or cognitively impaired), but also due to limitations considering their access to healthcare services and specialized personnel (Fasano et al., 2020).

In order to overcome the Covid-19-related reductions in scheduled visits, many changes have been implemented considering care and management of PwP on device-aided therapies, with healthcare professionals having to re-design services and update medical procedures according to individual situations, while struggling to minimize the risk for infections. Keeping as a priority to provide high quality of care with a concurrent feeling of safety, along with psychological comfort, are of the utmost importance, especially nowadays. Depending on the type of advanced therapy, the product manufacturers have implemented novel possibilities for phone communication, personal assistance and device replacement, when needed. Consequently, and similarly to measures applied in periods predisposing to flu, especially during autumn and winter, direct contacts, not only in consultation rooms, but also with homebound patients and those in nursing homes, had to be reduced. Movement disorders specialists and PNSs have begun to take advantage of modern telemedicine applications to support the needs of PwP on advanced therapies, while carefully picking those specific cases who might need face-to-face assessments at patients' homes with respect to safety rules.

To ensure and maintain constant care of PwP on advanced therapies, it has been recommended to use video or telephone consultations, strengthen the caregiver's role and authorize patients to use electronic prescriptions with simultaneous home delivering of medication via the caregiver or courier. In situations requiring a home visit and a direct contact with the patient, special caution measures have been recommended:

1. A more detailed history, along with a distant assessment of the patient's physical and neurological condition can take place via phone, before the actual face-to-face visit. The collected information and given instructions might suffice for the patient's management at the moment or lead the healthcare professional to schedule a traditional visit.
2. Informing patient and their caregiver about the possibility of providing services through tele- or video-consultation, plus suggesting possible methods of conducting it, such as video-calls using available and patient-friendly platforms.
3. Setting up convenient times and settings (especially for patients and caregivers) to conduct the tele- or video-consultation.
4. Informing patients and their caregivers about:
   a. the need to report any symptoms of health deterioration by phone.
   b. refrain from visiting the Emergency Department and/ or Outpatient Clinics when not necessary.

4.2 Initiating device-aided therapies during the Covid-19 pandemic

The Covid-19 pandemic had a major impact both on the potential of introducing new patients to advanced therapies, as well as on ensuring continuation of treatment in patients who are already enrolled. More specifically, the pandemic has significantly reduced the number of new patients included in device-aided practices, mainly due to the widespread limitations on scheduled hospitalizations, which in most countries are necessary for the qualification and initiation of DBS, LCIG infusion or CSAI. The hospitalization period is expected to last an average of 4–6 days for LCIG infusion (including the naso-jejunal test phase, the percutaneous endoscopic gastrostomy (PEG) with jejunal extension tube (J-tube; PEG-J) insertion procedure, and the titration phase), 3–4 days for the CSAI (preparation for initiation, starting infusion, and titration phase),1 day after the implementation (neurosurgery ward) and 2–3 days for the DBS stimulation setting (Mikos et al., 2010), although timelines might differ among various clinical settings.

For infusion therapies, many difficulties have arisen in patients' and caregivers' face-to-face education considering the pump, as the pandemic restrictions have limited the opportunities for family and caregivers to participate in the initial education and training needed for the self-management and self-care support of patients. In the pre-pandemic era, such processes would normally take place during the patient's hospitalization. These processes constitute key elements, affecting infusion therapies long-lasting success, as they allow caregivers to expand their skills in real life conditions in order to cope with the needs of home care. Routine care would, thus, be much faster and better results would be produced with the reinforcement of caregivers' self-esteem and self-management abilities at home. Both infusion therapies require training in pump operation, skin care routine and many other important, practical issues. During the Covid-19 pandemic, the educational training is carried out mainly at the end of hospitalization, when the patient is about to be discharged.

4.3 Supporting patients and caregivers during Covid-19

4.3.1 Initiation phase for advanced therapies

Placement of device-aided therapies and dose titration in infusion therapies traditionally take place in a hospital. However, there are examples of countries where significant parts of the initiation phase, including medication titration, might take place at patients' homes using telemedicine services or during home visits by healthcare professionals (Willows et al., 2017). Such initiatives might be more comfortable for the patient and are expected to reduce any infection risks in the Covid-19 era. Nevertheless, implementing such measures on a worldwide basis might not be easy or even possible in many countries due to regulatory issues or shortage in personnel.

Movement disorder centers were forced to work out their own possible individual solutions for continuing enrollments and after-care management. While PwP are hospitalized in a ward for the device implementation, it is challenging to secure them from infections during the Covid-19 pandemic. Despite any precautionary measures taken by the hospital authorities and healthcare professionals, the infection risk remains high. The majority of adverse events occurring during the first few post-procedural weeks after implementing LCIG infusion can be, in most cases, easily managed by movement disorders specialists or PNSs by phone (Fasano et al., 2020; Wirdefeldt, Odin, & Nyholm, 2016).

4.3.2 Educational phase for advanced therapies

A possible way to cope with the difficulties encountered by patients and caregivers during the educational and preparation process of device-aided therapies might be a simulation—focused medical training, which would offer useful opportunities to improve learners' competence and confidence, maintaining patient safety. Specially prepared educational material (e.g., an abdomen phantom, dummy infusion pumps and PEG-J tubes), may successfully substitute for patients' and/or caregivers' training, as it is necessary for them to practice skin care routine, drug connection and flushing, possible ways of puncturing the abdomen skin and pump management. Patients could also be equipped with a list of recommendations for possible troublesome issues (concerning the pump, skin care hygiene, high pressure alarm). Finally, patients and their caregivers should be aware of any alarming or dysregulation symptoms and should be properly informed about the advised procedures that need to be followed in case of an emergency (e.g., contact details of treating physician or nurse).

4.3.3 Continuation phase

All device-aided therapies are recommended to be continued in the event of contracting a SARS-CoV-2 infection, as undisturbed continuation of treatment is expected to keep the patients' motor and non-motor PD symptoms under control and reduce OFF periods and fluctuations. Moreover, a non-oral administration of dopaminergic treatment might be more desirable and adequate for patients suffering from conditions impending oral medication, like dysphagia.

4.4 LCIG infusion therapy

Using tele- or video-consultation might be very convenient for LCIG infusion patients, as photos or video visualization of the pump, PEG and abdomen are not difficult to access. Changing pump settings can be done on an unlocked pump by the caregiver with proper guidance by phone, although this must be based on good cooperation with a caregiver or a patient without a subjacent dysregulation syndrome. Patients should be aware of the need to keep at home a stock of oral medications (with individual administration instructions) in case of pump or system failure. LCIG infusion treatment may need PEG-J tubes exchange due to tube blockage, loops or any other adverse event which may occur any time. If continuing the therapy through the stomach tube is possible, this might allow the patient some time to book an appointment in order to have the complication properly addressed. Sometimes, though, an acute situation may require immediate intervention.

It is also of notice that during the pandemic hospitals have extra obligatory safety procedures; typically patients and their accompanying caregivers need to be fully vaccinated and/ or a negative PCR or antigen test must be demonstrated on entrance. These extra procedures are time-consuming and might feel uncomfortable for the patients.

4.5 Continuous Subcutaneous Apomorphine Infusion therapy

CSAI seems to be the easiest among the device-aided therapies to be implemented. However, it still needs careful preparation and education in order to cope with the most frequently encountered difficulties and complications. Patients and caregivers should be aware of potential skin problems, nausea, somnolence, orthostatic hypotension, neuropsychiatric issues and the rarely occurring, but severe, drug—induced hemolytic anemia or eosinophilia. Such situations require regular check-up examinations of blood results and caregiver knowledge about the symptoms of anemia. Routine laboratory tests can be postponed after confirming patients' status or can be done at patient's home. Similarly, to LCIG infusion therapy, it is advised to continue CSAI if the patient is admitted in the Intensive Care Unit (ICU). Sudden withdrawal of CSAI might be very dangerous, as patients might experience acute lethargy, malignant akinesia, and symptoms of dopaminergic agonist withdrawal syndrome (prominent psychiatric features, dysautonomia, generalized pain) (Chaudhuri et al., 2015). Under these circumstances, caregivers should be properly informed and prepared in order to proceed with substitution treatment.

4.6 Deep brain stimulation

DBS, like other deviced-aided therapies, constitutes an established treatment option during the pandemic. A possible limitation might be related to artifacts in conducting an electrocardiogram or electroencephalogram, but this can be solved by turning DBS off during the time of the procedure. Patients should be educated on how to use their handheld controller for DBS and be aware of the situations that it might be useful.

Another issue in DBS therapy comes with battery replacement, as delayed or postponed outpatient visits may cause an abrupt cessation of DBS function. Limited access to ambulatory consultations and to scheduled surgeries might also restrict battery exchange procedures. Postponing follow-up visits can result in depletion of the implanted pulse generator (IPG), leading to the life—threatening situation of DBS—withdrawal syndrome, characterized by a severe akinetic state, mimicking the parkinsonism-pyrexia syndrome (Reuter et al., 2018). It is, thus, highly recommended to regularly perform tele- or video-consultations to check on the DBS status and be vigilant for any signs indicative of abnormal function.

4.7 Outlook for best care and caregiver support

Clinical experience supports the use of device-aided therapies for PwP who are severely affected by pneumonia, possibly needing ventilation, and for those in a palliative stage (Antonini, Leta, Teo, & Chaudhuri, 2020; Chaudhuri, 2020).

The Covid-19 pandemic has had a big impact on lifestyle, with many changes forced by isolation, restriction of social contacts, limitations in regular stationary exercise classes, physiotherapy, functional or rehabilitation therapy and other relevant sessions. PwP, including those using device-aided therapies, have been forced to stay at home, a situation that has brought more duties for the caregiver, burdening them even more. Consequently, caregivers might be in extra need for support and free time. For example, the time spent by the patient during the aforementioned sessions (exercise, physiotherapy etc.) could also be used by the caregiver to rest.

Telemedicine services can be of great use during the Covid-19 pandemic, supporting PwP in their routine and promoting a healthy lifestyle. Reorganization is needed, for example establishment of comprehensive rehabilitation programs suitable for tele- or video-consultations with PwP (Langer et al., 2021; Vellata et al., 2021). Especially for PwP on advanced therapies, a positive attitude and proper information technology (IT) infrastructure to conduct tele- or video-consultations is deemed necessary, although the application of these services might not always be feasible for this group of patients due to practical difficulties. In some situations, patients may not have computers, tablets, or phones, and sometimes may require assistance in using the relevant software. Families and caregivers are often able to operate the system on their own, but sometimes may need prior training in this field. Taking actions and looking for solutions is worthwhile, because studies on continuing care and therapy in a remote fashion, like PwP attending online dancing sessions, have had a beneficial effect on PwP, reducing their anxiety levels (Morris et al., 2021).

Our own experience from clinical practice indicates that during the Covid-19 pandemic, patients' needs have been focused on closer communication with assurance of maintaining an ongoing care. There was a need to educate patients for video consultations, but only minor problems have arisen during this process. Some patients might have even enjoyed video consultations and the feeling of being home, while receiving safe and professional consultation.

4.8 Conclusion

In conclusion, embracing the new norm with a practical and careful approach towards the care of PwP is of utmost importance to ensure that patients are not unnecessarily exposed to SARS-CoV-2 and that all aspects of PD management are optimized. The role of the PNS in organizing this multi-faceted task is invaluable. Given the widespread impact of the Covid-19 pandemic among PwP, it is crucial that timely vaccination must be provided to reduce morbidity and mortality, while PNSs hold an important role in thoroughly informing PwP and facilitating vaccinations. The device-aided therapies-related procedures in PD, especially when hospitalization is required, have become more complicated during the pandemic, highlighting even more the need for coordination by the PNS in order to make these processes as comfortable as possible for patients and caregivers, but also for the implicated healthcare professionals.

References

Abate, S. M., Checkol, Y. A., & Mantefardo, B. (2021). Global prevalence and determinants of mortality among patients with COVID-19: A systematic review and meta-analysis. *Annals of Medicine and Surgery*, *64*, 102204. https://doi.org/10.1016/j.amsu.2021.102204.

Andrews, P. L. R., Cai, W., Rudd, J. A., & Sanger, G. J. (2021). COVID-19, nausea, and vomiting. *Journal of Gastroenterology and Hepatology*, *36*(3), 646–656. https://doi.org/10.1111/jgh.15261.

Antonini, A., Leta, V., Teo, J., & Chaudhuri, K. R. (2020). Outcome of Parkinson's disease patients affected by COVID-19. *Movement Disorders*.

Antonini, A., Mirò, L., Castiglioni, C., & Pezzoli, G. (2008). The rationale for improved integration between home care and neurology hospital services in patients with advanced Parkinson's disease. *Neurological sciences : official journal of the Italian Neurological Society and of the Italian Society of Clinical Neurophysiology*, *29*(Suppl 5), S392–S396. https://doi.org/10.1007/s10072-008-1056-5.

Antonini, A., Odin, P., Pahwa, R., Aldred, J., Alobaidi, A., Jalundhwala, Y. J., et al. (2021). The long-term impact of levodopa/carbidopa intestinal gel on 'Off'-time in patients with advanced Parkinson's disease: A systematic review. *Advances in Therapy*, *38*(6), 2854–2890. https://doi.org/10.1007/s12325-021-01747-1.

Axelrod, L., Gage, H., Kaye, J., Bryan, K., Trend, P., & Wade, D. (2010). Workloads of Parkinson's specialist nurses: Implications for implementing national service guidelines in England. *Journal of Clinical Nursing, 19*(23–24), 3575–3580. https://doi.org/10.1111/j.1365-2702.2010.03279.x.

Bai, W., Gu, Y., Liu, H., & Zhou, L. (2021). Epidemiology features and effectiveness of vaccination and non-pharmaceutical interventions of Delta and lambda SARS-CoV-2 variants. *China CDC Wkly, 3*(46), 977–982. https://doi.org/10.46234/ccdcw2021.216.

Beauchamp, L. C., Finkelstein, D. I., Bush, A. I., Evans, A. H., & Barnham, K. J. (2020). Parkinsonism as a third wave of the COVID-19 pandemic? *Journal of Parkinson's Disease, 10*(4), 1343–1353. https://doi.org/10.3233/JPD-202211.

Bloem, B. R., Trenkwalder, C., Sanchez-Ferro, A., Kalia, L. V., Alcalay, R., Chiang, H. L., et al. (2021). COVID-19 vaccination for persons with Parkinson's disease: Light at the end of the tunnel? *Journal of Parkinson's Disease, 11*(1), 3–8. https://doi.org/10.3233/jpd-212573.

Brown, E. G., Chahine, L. M., Goldman, S. M., Korell, M., Mann, E., Kinel, D. R., et al. (2020). The effect of the COVID-19 pandemic on people with Parkinson's disease. *Journal of Parkinson's Disease, 10*(4), 1365–1377. https://doi.org/10.3233/jpd-202249.

Brown, S., Dalkin, S. M., Bate, A., Bradford, R., Allen, C., Brittain, K., et al. (2020). Exploring and understanding the scope and value of the Parkinson's nurse in the UK (the USP project): A realist economic evaluation protocol. *BMJ Open, 10*(10), e037224. https://doi.org/10.1136/bmjopen-2020-037224.

Bruxvoort, K. J., Sy, L. S., Qian, L., Ackerson, B. K., Luo, Y., Lee, G. S., et al. (2021). Effectiveness of mRNA-1273 against delta, mu, and other emerging variants of SARS-CoV-2: Test negative case-control study. *BMJ, 375*, e068848. https://doi.org/10.1136/bmj-2021-068848.

Cascella, M., Rajnik, M., Aleem, A., Dulebohn, S. C., & Di Napoli, R. (2021). Features, evaluation, and treatment of coronavirus (COVID-19). In *StatPearls*. Treasure Island, FL: StatPearls Publishing. Copyright © 2021, StatPearls Publishing LLC.

Chaudhuri, R. (2020). Covid-19 and Parkinson's disease. *Kinetic, 2*, 4–5.

Chaudhuri, K. R., Todorova, A., Nirenberg, M. J., Parry, M., Martin, A., Martinez-Martin, P., et al. (2015). A pilot prospective, multicenter observational study of dopamine agonist withdrawal syndrome in Parkinson's disease. *Movement Disorders Clinical Practice, 2*(2), 170–174. https://doi.org/10.1002/mdc3.12141.

Choi, E. M. (2021). COVID-19 vaccines for low- and middle-income countries. *Transactions of the Royal Society of Tropical Medicine and Hygiene, 115*(5), 447–456. https://doi.org/10.1093/trstmh/trab045.

Cook, N., McNamee, D., McFetridge, B., & Deeny, P. (2007). The role of a Parkinson's disease nurse specialist as perceived by professionals in Northern Ireland. *British Journal of Neuroscience Nursing, 3*(10), 472–479. https://doi.org/10.12968/bjnn.2007.3.10.27275.

Coronavirus (COVID-19) Vaccinations. (2022). Retrieved from https://ourworldindata.org/covid-vaccinations?country=OWID_WRL.

Cosentino, C., Torres, L., Vélez, M., Nuñez, Y., Sánchez, D., Armas, C., et al. (2022). SARS-CoV-2 vaccines and motor symptoms in Parkinson's disease. *Movement Disorders, 37*(1), 233. https://doi.org/10.1002/mds.28851.

Dafsari, H. S., Martinez-Martin, P., Rizos, A., Trost, M., Dos Santos Ghilardi, M. G., Reddy, P., et al. (2019). EuroInf 2: Subthalamic stimulation, apomorphine, and levodopa infusion in Parkinson's disease. *Movement Disorders, 34*(3), 353–365. https://doi.org/10.1002/mds.27626.

de Figueiredo, A., Simas, C., Karafillakis, E., Paterson, P., & Larson, H. J. (2020). Mapping global trends in vaccine confidence and investigating barriers to vaccine uptake: A large-scale retrospective temporal modelling study. *Lancet, 396*(10255), 898–908. https://doi.org/10.1016/s0140-6736(20)31558-0.

De Rosa, A., Tessitore, A., Bilo, L., Peluso, S., & De Michele, G. (2016). Infusion treatments and deep brain stimulation in Parkinson's disease: The role of nursing. *Geriatric Nursing, 37*(6), 434–439. https://doi.org/10.1016/j.gerinurse.2016.06.012.

Del Prete, E., Francesconi, A., Palermo, G., Mazzucchi, S., Frosini, D., Morganti, R., et al. (2021). Prevalence and impact of COVID-19 in Parkinson's disease: Evidence from a multi-center survey in Tuscany region. *Journal of Neurology, 268*(4), 1179–1187. https://doi.org/10.1007/s00415-020-10002-6.

Douma, E. H., & de Kloet, E. R. (2020). Stress-induced plasticity and functioning of ventral tegmental dopamine neurons. *Neuroscience and Biobehavioral Reviews, 108*, 48–77. https://doi.org/10.1016/j.neubiorev.2019.10.015.

Erro, R., Buonomo, A. R., Barone, P., & Pellecchia, M. T. (2021). Severe dyskinesia after administration of SARS-CoV2 mRNA vaccine in Parkinson's disease. *Movement Disorders, 36*(10), 2219. https://doi.org/10.1002/mds.28772.

Fasano, A., Antonini, A., Katzenschlager, R., Krack, P., Odin, P., Evans, A. H., et al. (2020). Management of Advanced Therapies in Parkinson's disease patients in times of humanitarian crisis: The COVID-19 experience. *Movement Disorders Clinical Practice*, 7(4), 361–372. https://doi.org/10.1002/mdc3.12965.

Geldsetzer, P. (2020). Knowledge and perceptions of COVID-19 among the general public in the United States and the United Kingdom: A cross-sectional online survey. *Annals of Internal Medicine, 173*(2), 157–160. https://doi.org/10.7326/M20-0912.

Grannis, S. J., Rowley, E. A., Ong, T. C., Stenehjem, E., Klein, N. P., DeSilva, M. B., et al. (2021). Interim estimates of COVID-19 vaccine effectiveness against COVID-19-associated emergency department or urgent care clinic encounters and hospitalizations among adults during SARS-CoV-2 B.1.617.2 (Delta) variant predominance—nine states, June-August 2021. *MMWR. Morbidity and Mortality Weekly Report, 70*(37), 1291–1293. https://doi.org/10.15585/mmwr.mm7037e2.

Harrison, E. A., & Wu, J. W. (2020). Vaccine confidence in the time of COVID-19. *European Journal of Epidemiology, 35*(4), 325–330. https://doi.org/10.1007/s10654-020-00634-3.

Harvey, W. T., Carabelli, A. M., Jackson, B., Gupta, R. K., Thomson, E. C., Harrison, E. M., et al. (2021). SARS-CoV-2 variants, spike mutations and immune escape. *Nature Reviews. Microbiology, 19*(7), 409–424. https://doi.org/10.1038/s41579-021-00573-0.

Helmich, R. C., Aarts, E., de Lange, F. P., Bloem, B. R., & Toni, I. (2009). Increased dependence of action selection on recent motor history in Parkinson's disease. *Journal of Neuroscience, 29*(19), 6105–6113.

Helmich, R. C., & Bloem, B. R. (2020). The impact of the COVID-19 pandemic on Parkinson's disease: Hidden sorrows and emerging opportunities. *Journal of Parkinson's Disease, 10*(2), 351–354.

Hippisley-Cox, J., Coupland, C. A. C., Mehta, N., Keogh, R. H., Diaz-Ordaz, K., Khunti, K., et al. (2021). Risk prediction of covid-19 related death and hospital admission in adults after covid-19 vaccination: National prospective cohort study. *BMJ, 374*, n2244. https://doi.org/10.1136/bmj.n2244.

Holder, J. (2022). *Tracking Coronavirus Vaccinations Around the World*. Retrieved from https://www.nytimes.com/interactive/2021/world/covid-vaccinations-tracker.html.

Karim, S. S. A., & Karim, Q. A. (2021). Omicron SARS-CoV-2 variant: A new chapter in the COVID-19 pandemic. *Lancet, 398*(10317), 2126–2128. https://doi.org/10.1016/s0140-6736(21)02758-6.

Langer, A., Gassner, L., Flotz, A., Hasenauer, S., Gruber, J., Wizany, L., et al. (2021). How COVID-19 will boost remote25 exercise-based treatment in Parkinson's disease: A narrative review. *Npj. Parkinson's Disease*, 7(1), 25–33. https://doi.org/10.1038/s41531-021-00160-3.

Lau, Y. H., Lau, K. M., & Ibrahim, N. M. (2021). Management of Parkinson's disease in the COVID-19 pandemic and future perspectives in the era of vaccination. *Journal of Movement Disorders, 14*(3), 177–183. https://doi.org/10.14802/jmd.21034.

Leibson, C. L., Maraganore, D. M., Bower, J. H., Ransom, J. E., O'Brien, P., & C., & Rocca, W. A. (2006). Comorbid conditions associated with Parkinson's disease: A population-based study. *Movement Disorders, 21*(4), 446–455. https://doi.org/10.1002/mds.20685.

Limousin, P., & Foltynie, T. (2019). Long-term outcomes of deep brain stimulation in Parkinson disease. *Nature Reviews. Neurology, 15*(4), 234–242. https://doi.org/10.1038/s41582-019-0145-9.

Loomba, S., de Figueiredo, A., Piatek, S. J., de Graaf, K., & Larson, H. J. (2021). Measuring the impact of COVID-19 vaccine misinformation on vaccination intent in the UK and USA. *Nature Human Behaviour, 5*(3), 337–348. https://doi.org/10.1038/s41562-021-01056-1.

Lopez Bernal, J., Andrews, N., Gower, C., Gallagher, E., Simmons, R., Thelwall, S., et al. (2021). Effectiveness of Covid-19 vaccines against the B.1.617.2 (Delta) variant. *The New England Journal of Medicine, 385*(7), 585–594. https://doi.org/10.1056/NEJMoa2108891.

Macht, M., Kaussner, Y., Möller, J. C., Stiasny-Kolster, K., Eggert, K. M., Krüger, H. P., et al. (2007). Predictors of freezing in Parkinson's disease: A survey of 6,620 patients. *Movement Disorders, 22*(7), 953–956.

MacMahon, D. G. (1999). Parkinson's disease nurse specialists: An important role in disease management. *Neurology, 52*(7 Suppl 3), S21–S25.

Mahase, E. (2021). Covid-19: Booster dose reduces infections and severe illness in over 60s, Israeli study reports. *BMJ, 374*, n2297. https://doi.org/10.1136/bmj.n2297.

Merello, M., Bhatia, K. P., & Obeso, J. A. (2021). SARS-CoV-2 and the risk of Parkinson's disease: Facts and fantasy. *Lancet Neurology, 20*(2), 94–95. https://doi.org/10.1016/s1474-4422(20)30442-7.

Mikos, A., Pavon, J., Bowers, D., Foote, K. D., Resnick, A. S., Fernandez, H. H., et al. (2010). Factors related to extended hospital stays following deep brain stimulation for Parkinson's disease. *Parkinsonism & Related Disorders, 16*(5), 324–328. https://doi.org/10.1016/j.parkreldis.2010.02.002.

Morris, M. E., Slade, S. C., Wittwer, J. E., Blackberry, I., Haines, S., Hackney, M. E., et al. (2021). Online dance therapy for people with Parkinson's disease: Feasibility and impact on consumer engagement. *Neurorehabilitation and Neural Repair, 35*(12), 1076–1087. https://doi.org/10.1177/15459683211046254.

Nanduri, S., Pilishvili, T., Derado, G., et al. (2021). *Effectiveness of Pfizer-BioNTech and moderna vaccines in preventing SARS-CoV-2 infection among nursing home residents before and during widespread circulation of the SARS-CoV-2 B.1.617.2 (delta) variant—National Healthcare Safety Network, March 1–August 1, 2021*. MMWR. Morbidity and mortality weekly report. Retrieved from https://www.cdc.gov/mmwr/volumes/70/wr/mm7034e3.htm#suggestedcitation.

Nordström, P., Ballin, M., & Nordström, A. (2021). Association between risk of COVID-19 infection in nonimmune individuals and COVID-19 immunity in their family members. *JAMA Internal Medicine, 181*(12), 1589–1595. https://doi.org/10.1001/jamainternmed.2021.5814.

Nune, A., Iyengar, K., Ahmed, A., & Sapkota, H. (2020). Challenges in delivering rheumatology care during COVID-19 pandemic. *Clinical Rheumatology, 39*(9), 2817–2821. https://doi.org/10.1007/s10067-020-05312-z.

Nwabuobi, L., Barbosa, W., Sweeney, M., Oyler, S., Meisel, T., Di Rocco, A., et al. (2019). Sex-related differences in homebound advanced Parkinson's disease patients. *Clinical Interventions in Aging, 14*, 1371–1377. https://doi.org/10.2147/cia.S203690.

Okunoye, O., Kojima, G., Marston, L., Walters, K., & Schrag, A. (2020). Factors associated with hospitalisation among people with Parkinson's disease - A systematic review and meta-analysis. *Parkinsonism & Related Disorders*, *71*, 66–72. https://doi.org/10.1016/j.parkreldis.2020.02.018.

Papa, S. M., Brundin, P., Fung, V. S., Kang, U. J., Burn, D. J., Colosimo, C., et al. (2020). Impact of the COVID-19 pandemic on Parkinson's disease and movement disorders. *Movement Disorders*.

Parra, A., Juanes, A., Losada, C. P., Álvarez-Sesmero, S., Santana, V. D., Martí, I., et al. (2020). Psychotic symptoms in COVID-19 patients. A retrospective descriptive study. *Psychiatry Research*, *291*, 113254. https://doi.org/10.1016/j.psychres.2020.113254.

Phanhdone, T., Drummond, P., Meisel, T., Friede, N., Di Rocco, A., Chodosh, J., et al. (2021). Barriers to vaccination among people with Parkinson's disease and implications for COVID-19. *Journal of Parkinson's Disease*, *11*(3), 1057–1065. https://doi.org/10.3233/JPD-202497.

Pilishvili, T., & Bennett, N. M. (2015). Pneumococcal disease prevention among adults: Strategies for the use of pneumococcal vaccines. *Vaccine*, *33*(Suppl 4), D60–D65. https://doi.org/10.1016/j.vaccine.2015.05.102.

Piroth, L., Cottenet, J., Mariet, A. S., Bonniaud, P., Blot, M., Tubert-Bitter, P., et al. (2021). Comparison of the characteristics, morbidity, and mortality of COVID-19 and seasonal influenza: A nationwide, population-based retrospective cohort study. *The Lancet Respiratory Medicine*, *9*(3), 251–259. https://doi.org/10.1016/s2213-2600(20)30527-0.

Prasad, S., Holla, V. V., Neeraja, K., Surisetti, B. K., Kamble, N., Yadav, R., et al. (2020). Parkinson's disease and COVID-19: Perceptions and implications in patients and caregivers. *Movement Disorders*.

Reuter, S., Deuschl, G., Berg, D., Helmers, A., Falk, D., & Witt, K. (2018). Life-threatening DBS withdrawal syndrome in Parkinson's disease can be treated with early reimplantation. *Parkinsonism & Related Disorders*, *56*, 88–92. https://doi.org/10.1016/j.parkreldis.2018.06.035.

Rogers, G., Davies, D., Pink, J., & Cooper, P. (2017). Parkinson's disease: Summary of updated NICE guidance. *BMJ*, *358*, j1951. https://doi.org/10.1136/bmj.j1951.

Santos-García, D., Añón, M. J., Fuster-Sanjurjo, L., & de la Fuente-Fernández, R. (2012). Duodenal levodopa/carbidopa infusion therapy in patients with advanced Parkinson's disease leads to improvement in caregivers' stress and burden. *European Journal of Neurology*, *19*(9), 1261–1265. https://doi.org/10.1111/j.1468-1331.2011.03630.x.

Scobie, H. M., Johnson, A. G., Suthar, A. B., et al. (2021). *Monitoring incidence of COVID-19 cases, hospitalizations, and deaths, by vaccination status—13 U.S. Jurisdictions*. Retrieved from https://www.cdc.gov/mmwr/volumes/70/wr/mm7037e1.htm#suggestedcitation.

Simpson, J., & Doyle, C. (2020). *The impact of coronavirus restrictions on people affected by Parkinson's: The findings from a survey by Parkinson's UK*. Lancaster: Lancaster University. https://eprints.lancs.ac.uk/id/eprint/146591/.

Smilowska, K., van Wamelen, D. J., Pietrzykowski, T., Calvano, A., Rodriguez-Blazquez, C., Martinez-Martin, P., et al. (2021). Cost-effectiveness of device-aided therapies in Parkinson's disease: A structured review. *Journal of Parkinson's Disease*, *11*(2), 475–489. https://doi.org/10.3233/jpd-202348.

Torjesen, I. (2021). Covid-19: Norway investigates 23 deaths in frail elderly patients after vaccination. *BMJ*, *372*, n149. https://doi.org/10.1136/bmj.n149.

UK Health Security Agency: COVID-19 variants identified in the UK. (2022). Retrieved from https://www.gov.uk/government/news/covid-19-variants-identified-in-the-uk.

Vaccines Approved for Use in China. (2022). Retrieved from https://covid19.trackvaccines.org/country/china/.

van der Heide, A., Meinders, M. J., Bloem, B. R., & Helmich, R. C. (2020). The impact of the COVID-19 pandemic on psychological distress, physical activity, and symptom severity in Parkinson's disease. *Journal of Parkinson's Disease, 10*(4), 1355–1364. https://doi.org/10.3233/jpd-202251.

van Wamelen, D. J., Leta, V., Johnson, J., Ocampo, C. L., Podlewska, A. M., Rukavina, K., et al. (2020). Drooling in Parkinson's disease: Prevalence and progression from the non-motor international longitudinal study. *Dysphagia, 35*(6), 955–961. https://doi.org/10.1007/s00455-020-10102-5.

Vellata, C., Belli, S., Balsamo, F., Giordano, A., Colombo, R., & Maggioni, G. (2021). Effectiveness of Telerehabilitation on motor impairments, non-motor symptoms and compliance in patients with Parkinson's disease: A systematic review. *Frontiers in Neurology, 12*, 627999. https://doi.org/10.3389/fneur.2021.627999.

Willows, T., Dizdar, N., Nyholm, D., Widner, H., Grenholm, P., Schmiauke, U., et al. (2017). Initiation of levodopa-carbidopa intestinal gel infusion using telemedicine (video communication system) facilitates efficient and well-accepted home titration in patients with advanced Parkinson's disease. *Journal of Parkinson's Disease*, 7(4), 719–728. https://doi.org/10.3233/jpd-161048.

Wirdefeldt, K., Odin, P., & Nyholm, D. (2016). Levodopa-carbidopa intestinal gel in patients with Parkinson's disease: A systematic review. *CNS Drugs, 30*(5), 381–404. https://doi.org/10.1007/s40263-016-0336-5.

World Health Organization: Coronavirus disease (COVID-19): Vaccines. (2022). Retrieved from https://www.who.int/news-room/questions-and-answers/item/coronavirus-disease-(covid-19)-vaccines?gclid=Cj0KCQiAxoiQBhCRARIsAPsvo-wta-wNEhh7pahc4m4UrnwrpJelPds-vTgMgf2PO3cY1yCem3Ju5gUaAhTCEALw_wcB&topicsurvey=v8kj13).

Zach, H., Dirkx, M. F., Pasman, J. W., Bloem, B. R., & Helmich, R. C. (2017). Cognitive stress reduces the effect of levodopa on Parkinson's resting tremor. *CNS Neuroscience & Therapeutics, 23*(3), 209–215.

CHAPTER NINE

Social isolation, loneliness and mental health sequelae of the Covid-19 pandemic in Parkinson's disease

Bradley McDaniels[a] **and Indu Subramanian**[b,c,*]
[a]Department of Rehabilitation and Health Services, University of North Texas, Denton, TX, United States
[b]David Geffen School of Medicine, UCLA, Department of Neurology, Los Angeles, CA, United States
[c]PADRECC, West Los Angeles, Veterans Administration, Los Angeles, CA, United States
*Corresponding author: e-mail address: isubramanian@mednet.ucla.edu

Contents

Abstract

People living with Parkinson Disease (PwP) have been at risk for the negative effects of loneliness even before the Coronavirus Disease 2019 (Covid-19) pandemic. Despite some similarities with previous outbreaks, the Covid-19 pandemic is significantly more wide-spread, long-lasting, and deadly, which likely means demonstrably more negative mental health issues. Although PwP are not any more likely to contract Covid-19 than those without, the indirect negative sequelae of isolation, loneliness, mental health issues, and worsening motor and non-motor features remains to be fully

International Review of Neurobiology, Volume 165
ISSN 0074-7742
https://doi.org/10.1016/bs.irn.2022.03.003
Copyright © 2022 Elsevier Inc.
All rights reserved.

realized. Loneliness is not an isolated problem; the preliminary evidence indicates that loneliness associated with the Covid-19 restrictions has dramatically increased in nearly all countries around the world.

1. Introduction

In March 2020, the World Health Organization (WHO) declared the Coronavirus disease 2019 (Covid-19), caused by the novel Severe Acute Respiratory Syndrome Coronavirus-2 (SARS-CoV-2), a global pandemic (World Health Organization, 2020), which has since wreaked havoc on health of people around the world and has resulted in an undeniable health crisis. Not only is the risk of contracting the virus concerning, but the consequences resulting from the mitigation efforts are alarming and poorly understood. The international literature concludes that among the potential mental health consequence imposed by social distancing, loneliness may be the most problematic (Killgore, Cloonan, Taylor, & Dailey, 2020; Luchetti et al., 2020). Loneliness is undoubtedly a concern for the general population, but for people with chronic illnesses, including Parkinson's disease (PD), it can compound health problems and be particularly detrimental. A growing body of research reports on the myriad complications associated with loneliness among people with PD (e.g., depression, fatigue, anxiety, apathy) (Antonini, Leta, Teo, & Chaudhuri, 2020; Brooks, Weston, & Greenberg, 2021; Prasad et al., 2020). Screening patients for loneliness and identifying appropriate interventions to mitigate its effects is critical as the pandemic persists. Importantly, loneliness is not confined to the Western World; it is rather recognized as a burgeoning health epidemic affecting the communities around the globe (Holt-Lunstad, 2017) and is considered among the greatest health challenges of this generation (World Health Organization, 2015).

2. Loneliness

People living with Parkinson Disease (PwP) have been at risk for the negative effects of loneliness even before the Covid-19 pandemic. Despite some similarities with previous outbreaks, the Covid-19 pandemic is significantly more wide-spread, long-lasting, and deadly, which likely means demonstrably more negative mental health issues (Mesa Vieira, Franco, Gómez Restrepo, & Abel, 2020). Although PwP are not any more likely to contract SARS-CoV-2 that those without PD, the indirect negative sequelae of isolation and loneliness, including mental health issues and worsening motor and non-motor features, remain to be fully realized (Prasad et al., 2020; Stoessl, Bhatia, & Merello, 2020). Loneliness is not an isolated problem; the preliminary evidence indicates that loneliness associated with the Covid-19 restrictions has dramatically increased in nearly all countries around the world (Folk, Okabe-Miyamoto, Dunn, & Lyubomirsky, 2020; Krendl & Perry, 2021; Macdonald & Hülür, 2021; McGinty, Presskreischer, Han, & Barry, 2020).

Although the risks associated with contracting SARS-CoV-2 have been well-documented, literature continues to evolve, describing the considerable psychosocial sequelae affecting both well-being and quality of life resulting from the unprecedented, strict social distancing measures (Antonini, 2020). International pandemics can evoke significant emotional disturbances, even among individuals who are seemingly at low risk (Brooks et al., 2020). As a result of the world-wide stay-at-home and social distancing mandates, social isolation and loneliness rose precipitously and the long-term effects remain unknown. Covid-19 is not the first time that quarantining (e.g., social distancing, isolation) has been employed to reduce the spread of an infectious disease—it was used successfully to contain severe acute respiratory syndrome (SARS), the first pandemic of the 20th century. As a result of the mitigation efforts, individuals who experienced social isolation during SARS reported significant increases in both posttraumatic stress disorder (PTSD) and depression with longer time in isolation resulting in increased symptomatology (Hawryluck et al., 2004; Reynolds et al., 2008). Additionally, likely due to the effective coping strategies, increased alcohol abuse/dependence has been reported for up to three years after being exposed to the isolation associated with the SARS mitigation efforts (Wu et al., 2008). Similarly, following the rapid spread of Middle East Respiratory Syndrome (MERS) and subsequent short periods of isolation (e.g., 6 months), people reported increased mental health issues (Jeong et al., 2016). The common denominator between the last three major viral pandemics is the use of social distancing measures, which resulted in isolation, loneliness, and increased mental health problems.

All people have a fundamental need for social connection and meaningful interpersonal relationships without which human flourishing is unlikely to be attained (Baumeister & Leary, 1995; Cacioppo, Capitanio, & Cacioppo, 2014). The instinctive drive for social connectedness is evident throughout the lifespan—"from cradle to grave" (Bowlby, 1979, p. 179). Moreover, social relationships are essential to promote positive health and functioning. Social isolation and loneliness are frequently used interchangeably; however, they are distinct concepts. Social isolation can be described as the *objective* feeling of the inadequacy of social connections, resulting in a decreased frequency of social contact (Coyle & Dugan, 2012). Social isolation is particularly problematic for the aging population due to diminishing financial resources, declining physical health, and death of friends and loved ones (Steptoe, Shankar, Demakakos, & Wardle, 2013). Loneliness, however, is a *subjective* emotional state associated with the lack of social connectedness (Leigh-Hunt et al., 2017). Therefore, one can be socially isolated without feeling lonely and can feel lonely despite being socially well-connected.

Although loneliness is largely discussed as a singular construct, researchers have delineated three dimensions or spheres of loneliness that identify the missing relationship. Intimate or emotional loneliness is the craving for a close, intimate relationship or emotional partner. Relational or social loneliness includes the absence of close friendships and a longing for social companionship. Collective loneliness refers to the lack of a structured network of people who all share a common interest and purpose—a community (Michela, Peplau, & Weeks, 1982; Murthy, 2020). Hence, a person can still be lonely if they lack a circle of friends or a community with a shared purpose outside their home even if they are happily married (Table 1).

Social isolation can undoubtedly exacerbate loneliness, but one's perception of loneliness is more closely associated with the *quality* of social interactions rather than the *quantity* (Hawkley et al., 2008). This may be due to the fact that loneliness is influenced by a number of factors, which are largely

Table 1 Social isolation vs. loneliness.

Social isolation	**Loneliness**
• *Objective* lack of social connections/social contact	• *Subjective* emotional state associated with lack of social connectedness
• No involvement in outside activities	• Potential consequence of social isolation
• Associated with loss of close relationships	• Quality of contact rather than quantity
	Three types:
	i. Intimate/Emotional
	ii. Relational/Social
	iii. Collective

unrelated to one's objective view, including heritability (Boomsma, Willemsen, Dolan, Hawkley, & Cacioppo, 2005), environmental factors (Bartels, Cacioppo, Hudziak, & Boomsma, 2008), cultural norms (Cacioppo & Patrick, 2008), individual social needs (Dykstra & Fokkema, 2007), presence of a chronic illness or disability (Hawkley et al., 2008; Mushtaq, 2014), poor coping strategies (Deckx, van den Akker, Buntinx, & van Driel, 2018), lower socioeconomic status (Wee et al., 2019), being female (Solmi et al., 2020), and the disparity between experienced versus desired relationships (Cacioppo & Hawkley, 2009). More likely as a result of stigma and systemic societal issues, racial, ethnic, and LGBTQ+ minorities are more prone to loneliness (Jeste, Lee, & Cacioppo, 2020). Loneliness is clearly a multidimensional construct, but, based on our evolutional heritage and available literature, the brain has been fashioned to seek meaningful connections with others.

3. Physiology of loneliness

Because early humans had a greater chance of survival if they existed in groups, evolutionary adaptation favored the preference for close human bonds by selecting genes that result in pleasure in company and feelings of unease when involuntarily alone (Solmi et al., 2020). Not only does social connectedness portend positive outcomes, but being deprived of them may lead to profound insecurity and loneliness. Most of the research on loneliness has been focused on surveys and animal studies to evaluate the effect of social deprivation on neuroendocrine activity. Research demonstrates an association between loneliness and stress-related inflammatory and neuroendocrine responses in adults (Steptoe et al., 2013). Loneliness can also affect cognitive processing and result in hypervigilance, which may lead to anxiety (Cacioppo et al., 2014). Additionally, loneliness is associated with increased hypothalamic-pituitary-adrenal (HPA) axis activity. For instance, loneliness is linked with greater cortisol awakening response in middle-aged adults (Steptoe et al., 2013), and previous day loneliness results in heightened cortisol awakening response in older adults (Adam, Hawkley, Kudielka, & Cacioppo, 2006). In PwP, cortisol-related pathological responses to stress has been shown to negatively affects both motor and non-motor symptoms of PD potentially even noticeable in the prodromal stage (van Wamelen, Wan, Chaudhuri, & Jenner, 2020).

A robust body of literature exists explicating the role of social support in buffering stress (Ditzen & Heinrichs, 2014; Holt-Lunstad, 2018; Nitschke et al., 2021; Sandi & Haller, 2015). During times of limited social support and isolation, people are increasingly susceptible to the damaging effects of stress. Loneliness, in addition to an inadequate social support system, is a predictor of a poor immune response to stress; however, the weakest immune response occurs in people who report being both lonely and lacking a reasonable social network (Heinrich & Gullone, 2006; Lim, Eres, & Vasan, 2020). Despite these reports, the impact of loneliness on people and society at large remains underrecognized and poorly understood.

4. Loneliness and the brain

Clinical evidence supports the notion that the human brain is wired to desire social connection and to have meaningful interactions with others. Specifically, neuroimaging studies report that the brain responds to the social pain of loneliness similar to how it responds to physical pain (Cacioppo & Patrick, 2008). Research has begun to attempt to understand the role of loneliness on brain health and preliminary evidence has identified connected regions of the brain midline and medial temporal lobes, including the hippocampus, which are hypothesized to provide a potential neural substrate, linking loneliness, aging, and brain health (Mwilambwe-Tshilobo et al., 2019; Spreng et al., 2020). Moreover, loneliness has been associated with higher amyloid burden - a neuropathological feature of Alzheimer's disease (Donovan et al., 2016). Loneliness not only results in a variety of psychosocial issues, but it may also be associated with pathophysiological changes in the brain.

5. General health effects of loneliness

Loneliness can result in significant negative health outcomes, including increased risk of depression, anxiety, cognitive decline, type 2 diabetes, poor sleep, arthritis (Dossey, 2020; Hawkley & Cacioppo, 2010; Holt-Lunstad, Smith, Baker, Harris, & Stephenson, 2015), and even cancer (Nausheen, Gidron, Peveler, & Moss-Morris, 2009) and suicide (Brooks et al., 2020). Concerns have been raised about the association between loneliness and potentially problematic alcohol consumption, as alcohol sales rose by 54% during the initial shelter-in-place orders (Grossman, Benjamin-Neelon, & Sonnenschein, 2020). Loneliness is associated with a 29% increase in the incidence of coronary heart disease and a 32% increase in the risk of stroke (Valtorta, Kanaan, Gilbody, Ronzi, & Hanratty, 2016). Accordingly, a recent meta-analysis revealed that lonely people have an estimated 26% increased risk of all-cause mortality compared to non-lonely individuals (Holt-Lunstad et al., 2015). An estimated 162,000 Americans die annually as a result of social isolation (Veazie, Gilbert, Winchell, Paynter, & Guise, 2019) and loneliness has been reported to be more detrimental to health than obesity, smoking, lack of exercise, and excessive alcohol consumption *combined* (Lynch, 2000).

6. Loneliness and mental and cognitive health

Not only is there a variety of negative physical sequelae of loneliness, but mental and cognitive health are also significantly affected, possibly in a more damaging way. Research has implicated loneliness in depressive symptomology (Cacioppo, Hawkley, & Thisted, 2010). In addition to depression, loneliness is also associated with increased social anxiety and paranoia (Lim, Rodebaugh, Zyphur, & Gleeson, 2016). Emotional regulation is a strategy to manage or cope with emotions and has been shown to be negatively affected by loneliness, which results in less expression and enjoyment of positive feelings (Hawkley, Thisted, & Cacioppo, 2009). Furthermore, loneliness is associated with greater cognitive impairment among community dwelling adults and, even after controlling for confounders (e.g., sociodemographic, health conditions, depression), loneliness is implicated in accelerated cognitive decline later in life (Donovan et al., 2017; Poey, Burr, & Roberts, 2017). People who reported higher loneliness were 64% more likely to develop dementia than those who were not lonely (Holwerda et al., 2012) and were twice as likely to develop Alzheimer's disease, even after controlling for social isolation, social support, and living alone (Zhou, Wang, & Fang, 2018).

7. Loneliness and the aging population

The body of literature assessing the myriad health impacts of loneliness is large and unequivocal. While loneliness is associated with a variety of negative outcomes for nearly everyone, for older adults loneliness may be particularly injurious (Beam & Kim, 2020; Czaja, Moxley, & Rogers, 2021; Heckhausen, Wrosch, & Schulz, 2013; Pinquart & Sorensen, 2001). In particular, the United States Centers for Disease Control and Prevention (CDC) report that older adults suffering from loneliness are at higher risk for dementia, depression, and anxiety (Centers for Disease Control and Prevention, 2021). Loneliness is reported to have a prevalence of 30–40% in older individuals (Courtin & Knapp, 2017). As people age, they experience an increase in sensory impairments (e.g., hearing, visual), which can preclude them from fully participating in social activities and interactions, putting them at a greater risk for loneliness (Eggar, Spencer, Anderson, & Hiller, 2002). This is of particular relevance for PwP, as the vast majority of them are diagnosed at age 60 or older.

8. Loneliness and chronic illness

The negative effects of loneliness on health are well-established for the general population, but loneliness may be detrimental for people with serious health conditions (Perissinotto, Holt-Lunstad, Periyakoil, & Covinsky, 2019). Research asserts that loneliness is the single most significant predictor of psychological distress among people with chronic illness and disability (CID). Additionally, not only was loneliness associated with one's own CID, spousal CID was also correlated with higher levels of loneliness and subsequent poor health outcomes (Ditzen & Heinrichs, 2014; Holt-Lunstad, 2018; Nitschke et al., 2021; Sandi & Haller, 2015). The International Classification of Functioning, Disability, and Health (ICF; World Health Organization [WHO], 2001) was established in an attempt to provide a comprehensive framework for holistically conceptualizing health and has been used to describe the effects of neurologic conditions on health outcomes (McDaniels & Bishop, 2021; McDaniels, Bishop, & Rumrill Jr., 2021). The ICF model centers on the interaction between psychological, biological, and social components and how they, collectively, affect functioning (Peterson, 2016) (Fig. 1). For example, when individuals have limited interaction with others and are lonely (environmental factors), decreased community participation may occur leading to increased anxiety and depression (health condition). For people with CID, their health condition alone (e.g., PD) may result in impairment on body functions and structures (e.g., gait impairment, tremor, cognitive issues), precluding full community participation; however, when problems like loneliness arise in additional areas of one's life, consequences compound and may lead to additional health consequences and diagnoses.

The data for the negative effects of loneliness on the general population is clear (Xiong et al., 2020), but how the lockdown measures have affected those with neurologic conditions is still developing. Emerging data, however, suggests that individuals with an accompanying chronic disease are more susceptible to the psychological effects of the Covid-19 social distancing mandates, which may implicate loneliness as one of the predictors of

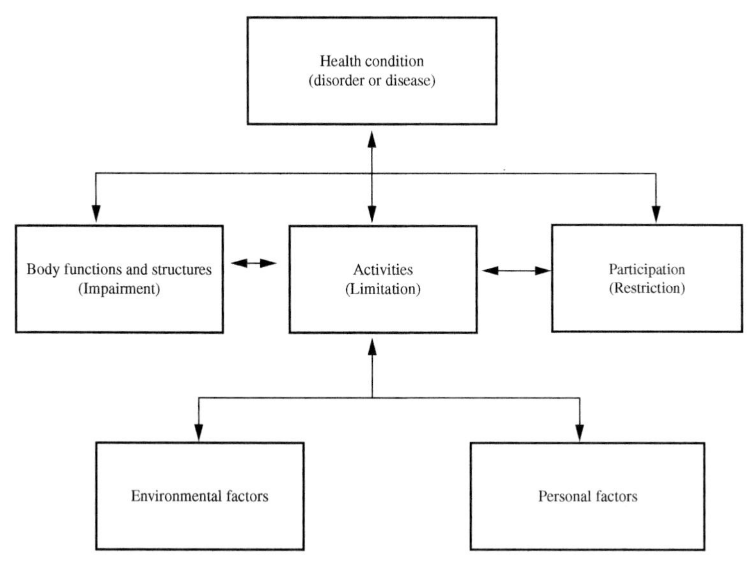

Fig. 1 Interactions between the components of the ICF model. *Adapted from World Health Organization. (2001).* The international classification of functioning, disability, and health. *World Health Organization.*

poorer psychological health outcomes (Özdin & Bayrak Özdin, 2020). Among people with multiple sclerosis (PwMS), data reports increased levels of depression and anxiety from pre-pandemic to intra-pandemic, with loneliness being considered one of the precipitating factors (Alschuler, Roberts, Herring, & Ehde, 2021; Naser Moghadasi, 2020; Shaygannejad, Afshari-Safavi, & Hatef, 2021; Stojanov et al., 2020). Specifically, 73.3% of PwMS with increased anxiety and depression during the Covid-19 outbreak reported increased loneliness (Garjani et al., 2021). Similar findings have been published for people with epilepsy (Tashakori-Miyanroudi et al., 2021), visual impairments (Heinze et al., 2021), rheumatic diseases (Kool & Geenen, 2012), and generally for all people with CID (Elran-Barak & Mozeikov, 2020; Horesh, Kapel Lev-Ari, & Hasson-Ohayon, 2020; Wong et al., 2020).

9. Loneliness and Parkinson's disease

The fact that PwP are largely in an older demographic group, where loneliness is more common, means that this population may be particularly at risk for the effects of loneliness. In general, feeling socially isolated and lonely is not uncommon for PwP, as there can be stigma associated with the diagnosis of PD, which may cause patients to socially withdraw (Maffoni, Giardini, Pierobon, Ferrazzoli, & Frazzitta, 2017; Subramanian et al., 2021). Other reasons for social withdrawal can include difficulty communicating due to hypophonia, dysarthria, and facial masking (Subramanian, Farahnik, & Mischley, 2020). The presence of tremor and/or dyskinesia can also result in feelings of embarrassment, making one more reticent to engage socially (Soleimani, Negarandeh, Bastani, & Greysen, 2014). Drooling, difficulty handling utensils, unpredictable bowel and bladder issues (e.g., incontinence), the use of a cane or walker, poor balance, freezing of gate, and risk of falling are all predictors of loneliness among PwP (Sjödahl Hammarlund, Westergren, Åström, Edberg, & Hagell, 2018).

Mental health issues, such as apathy, depression and anxiety, can contribute to lack of motivation to seek social connection and a diminished desire to actively engage when found in such situations (Perepezko et al., 2019). Accumulating evidence supports the relationship between depression and social isolation among PwP (Helmich & Bloem, 2020; Janiri et al., 2020; Kitani-Morii et al., 2021; Subramanian et al., 2020). Not surprisingly, the added social consequences of the mitigation efforts associated with the pandemic have further compounded the effects of loneliness on the overall well-being and quality of life of PwP (Antonini, 2020; Subramanian et al., 2020). Subramanian et al. (2020) also reported that among PwP, those who are lonely mention greater severity of the symptoms of bradykinesia, pain, memory, depression, anxiety, and fatigue among others.

10. Non-motor aspects of Parkinson's disease

Despite the historical focus on the motor features of PD, a variety of non-motor features are commonly encountered and are a leading cause of poor quality of life for PwP (Chaudhuri et al., 2015; Kalia & Lang, 2015). Non-motor symptoms affect nearly all (>90%) PwP and are often considered to be more burdensome than motor symptoms (Antonini et al., 2012; Martinez-Martin et al., 2011). Among the most problematic non-motor features are cognitive impairment, depression, sleep disorders, anxiety, and apathy (Weintraub & Mamikonyan, 2019). Although prevalence reports vary, dementia affects about 50% of PwP (Williams-Gray et al., 2013), while mild cognitive impairment (MCI) may be present in 25–50% (Aarsland et al., 2010). Cognitive decline is common, typically slow and insidious, and may occur across all stages of PD, increasing the risk of early dementia (Aarsland et al., 2021). Among mood disorders, depression, anxiety, and apathy are the most widely reported and occur in 35% (Schapira, Chaudhuri, & Jenner, 2017), 31% (Broen, Narayen, Kuijf, Dissanayaka, & Leentjens, 2016), and 60% of PwP (Schapira et al., 2017), respectively. Fatigue is another commonly reported non-motor feature of PD and reports indicate prevalence rates of around 50% (Siciliano et al., 2018). Sleep disorders are emerging among the most common and heterogeneous non-motor manifestations of PD, occurring in up to 90% of PwP (Chaudhuri et al., 2010). These disorders present significant challenges; however, when coupled with the effects of the mitigation efforts associated with a global pandemic and the resultant loneliness, their negative influence on quality of life is likely compounded.

11. Covid-19 and Parkinson's disease

Data on the effects of the Covid-19 pandemic on PwP is increasing, but appears to clearly portend a worsening of nearly all PD-related symptoms. Among the most common potentiators of neuropsychiatric symptoms manifestation during a pandemic are fear of contracting the virus (Jeong et al., 2016) and isolation or lack of social connections (Hawryluck et al., 2004), all of which have been experienced by PwP throughout the past two years. Considering the unusual external conditions that contribute to increased anxiety, depression, and stress, along with the daily stress of living with and managing a chronic neurologic condition, identifying specific contributing factors and potential interventions is critical (McDaniels, Novak, Braitsch, & Chitnis, 2021; Subramanian et al., 2021).

Fear of health emergencies can contribute to severe emotional distress and anxiety among low risk populations, but for PwP the ramifications become more problematic (Schapira et al., 2017). Stress and anxiety are consistently among the most frequently reported non-motor features of PD; however, considering the challenges associated with the pandemic, there was a significant rise in these mental health issues. PwP are acutely sensitive to increased stress and reduction in physical activity, which can increase both motor and non-motor features of PD (Helmich & Bloem, 2020). One proposed mechanism is that mood symptoms (e.g., apathy, depression, anxiety) are believed to be associated with dysfunction of the dopaminergic, serotonergic, and noradrenergic system along with the changes in serum levels of dopaminergic medications (Bomasang-Layno, Fadlon, Murray, & Himelhoch, 2015; Cooney & Stacy, 2016; Espay, LeWitt, & Kaufmann, 2014).

Exercise and physical therapy are considered among the most highly encouraged and beneficial interventions for PD management, as they have been associated with improved mood and cognition (Douma & de Kloet, 2020) and may even improve all-cause mortality (Yoon, Suh, Yang, Han, & Kim, 2021). Periods of inactivity are positively correlated with worsening of both motor and non-motor features of PD (e.g., psychological stress, insomnia, depression; Helmich & Bloem, 2020). PwP were unable to engage with their medical providers via in-person appointments to discuss problematic progression of motor and non-motor symptoms, the evolution of new issues, or receive essential programming adjustments for deep brain stimulation (DBS) devices. Additionally, PwP experienced a dramatic reduction in their ability to remain active in community support groups and to interact with others in the community.

Loneliness and the lack of social connection stemming from Covid-19 clearly increases the likelihood of developing or exacerbating neuropsychiatric symptoms (Antonini et al., 2020; Prasad et al., 2020). A better understanding of the neurobiological sequalae of increased stress and the compounding effects of loneliness and isolation is critical not only for PwP, but for all individuals with CID. The pandemic has undoubtedly wreaked havoc on the lives of all populations worldwide, resulting in increased levels of stress, but those with PD and other CID are particularly at risk, as they rely on a variety of in-person support mechanisms to facilitate physical functioning and general well-being. Remaining both physically and socially engaged is imperative for the preservation of health, functioning, and quality of life among older individuals and PwP (Brady et al., 2020; Hajek et al., 2017). Thus, identifying novel means of remaining engaged with exercise and to socially connect is essential to stave off the negative effects of loneliness. The mental health issues associated with social isolation and loneliness are far-reaching, significant, and consistent around the globe (Table 2).

12. Next steps in addressing loneliness among PwP

12.1 Accurate identification

As the evidence recognizing loneliness as a major risk factor for a variety of adverse psychological and physical health outcomes mounts, identifying - evidence-based interventions aimed at mitigating loneliness is critical. Acknowledging the wide-spread problem of loneliness and its negative

Table 2 Global Non-Motor Effects of Isolation in PwP.

Study (authors, year)	**Country**	**Sample size**	**Mental health outcomes**
Anghelescu, Bruno, Martino, and Roach (2021)	Canada	22	Anxiety, irritability, sleep
Balci, Aktar, Buran, Tas, and Donmez Colakoglu (2021)	Turkey	45	Anxiety, cognition, depression, sleep
Baschi et al. (2020)	Italy	65	Anxiety, apathy, cognition, depression, fatigue
Brown et al. (2020)	USA (80%)	5429	Anxiety, apathy, cognition, depression, fatigue, sleep
Del Prete et al. (2021)	Italy	740	Anxiety, mood, sleep
de Rus Jacquet et al. (2021)	Canada	417	Anxiety, cognition, depression
Feeney et al. (2021)	USA	1342	Anxiety, cognition
Guo et al. (2020)	China	113	Depression, fatigue, sleep
Janiri et al. (2020)	Italy	134	Apathy, cognition, irritability, sleep
Kapel, Serdoner, Fabiani, and Velnar (2021)	Slovenia	42	Anxiety/stress, apathy
Kitani-Morii et al. (2021)	Japan	39	Anxiety, depression, sleep
Kumar et al. (2020)	India	832	Anxiety, cognition, depression, fatigue, sleep
Oppo et al. (2020)	Italy	32	Cognitive issues, fatigue, mood, sleep
Palermo et al. (2020)	Italy	28	Anxiety, cognition, sleep
Prasad et al. (2020)	India	100	Anxiety, depression, sleep, fatigue
Salari et al. (2020)	Iran	137	Anxiety
Santos-García et al. (2020)	Spain	570	Psychological stress, sleep
Schirinzi et al. (2020)	Italy	162	Anxiety, other neuropsychiatric symptoms

Continued

Table 2 Global Non-Motor Effects of Isolation in PwP.—cont'd

Study (authors, year)	Country	Sample size	Mental health outcomes
Shalash et al. (2020)	Egypt	38	Anxiety/stress, depression
Simpson, Lekwuwa, and Crawford (2013)	England, Scotland, Wales	1741	Anxiety, cognition, sleep, fatigue
Song et al. (2020)	South Korea	100	Depression, fatigue, sleep, stress
Suzuki et al. (2021)	Japan	100	Cognition, depression, sleep, stress
Thomsen, Wallerstedt, Winge, and Bergquist (2021)	Sweden/Netherlands	67	Apathy, anxiety, depression, sleep
Zipprich, Teschner, Witte, Schönenberg, and Prell (2020)	Germany	99	Anxiety

sequelae may be a silver-lining of the Covid-19 pandemic. Unfortunately, the medical community has been slow to realize the adverse health impact of loneliness and, hence, has not provided patients with vital interventions needed to address it (Perissinotto et al., 2019). Healthcare practitioners and social scientists have been recently driven to further identify ways to proactively alleviate worldwide loneliness.

The first step toward is raising awareness and accurate screening for loneliness. A variety of international initiatives to draw awareness to the problem have begun in an attempt to increase the visibility of this challenging issue (cf., Cacioppo, Grippo, London, Goossens, & Cacioppo, 2015). In 2014, the Institute of Medicine (IOM) convened a team charged with the inclusion of a psychosocial "vital sign" in electronic health records (EHRs; Matthews, Adler, Forrest, & Stead, 2016). As such, the IOM recognized "Social Connections and Social Isolation" as an essential domain of inclusion and the evidence supporting this addition was equivalent to that of race, education, physical activity, tobacco use, and neighborhood characteristics. Loneliness can be adequately assessed using the UCLA three-item loneliness scale (Domènech-Abella et al., 2017; Musich, Wang, Hawkins, & Yeh, 2015) or the IOM recommended Berkman-Syme Social Network Index (Institute of Medicine, 2014) (Table 3).

Table 3 Tips for clinicians.

1. *Awareness*—Become more aware of the clinical importance of loneliness
2. *Screening*—Ask patients about their social health (e.g., the three spheres of loneliness)
   a. Ask, "Do you sometimes feel left out?"
   b. Ask, "Do you sometimes feel isolated from others?"
3. *Intervening*—Tailor the appropriate intervention for this specific patient taking into account their cultural context
   a. Social Prescribing (e.g., virtual community groups, virtual exercise classes, and outdoor recreation activities)
   b. Psychological Interventions (e.g., Mindfulness, CBT, ACT)
   c. Wellness Strategies (e.g., diet, sleep, exercise, mind/body approaches)

12.2 Interventions

Although the research describing the risk factors and health outcomes associated with loneliness is robust, limited research is available to inform mitigation efforts. Novel interventions geared toward attenuating loneliness have been proposed for this growing global issue, which itself has been termed a pandemic, but the work is far from complete (Gardiner, Geldenhuys, & Gott, 2018). The following is an overview of the interventions that have demonstrated cursory effectiveness for targeting loneliness.

12.2.1 Psychological interventions

Considering that loneliness is highly subjective and perceptual, it could be hypothesized that psychological interventions may constitute an effective tool. Improving social contact alone does not necessarily assuage the maladaptive emotional response leading to loneliness (Käll et al., 2020). Additionally, counseling-based interventions are effective for reducing anxiety and depression, involving mental processes which can overlap with the cognitive changes associated with loneliness, and it is believed that positively affecting a person's mental processes may result in decreased loneliness (Mann et al., 2017).

Cognitive Behavioral Therapy (CBT). Among the most evaluated psychological interventions for loneliness is cognitive behavioral therapy (CBT; Hickin et al., 2021). The perceptual and cognitive biases that result in hypervigilance to negative social information, a common precursor to loneliness, is targeted by CBT interventions (J. T. Cacioppo & Hawkley, 2009; S. Cacioppo et al., 2015). Specifically, the commonly used CBT technique of cognitive restructuring appears to be useful in helping people reframe perceptions of loneliness and self-efficacy with the aim of decreasing loneliness (Käll, Backlund, Shafran, & Andersson, 2020). In addition to in-person individual and group CBT, emerging evidence supports the feasibility of a newly developed CBT intervention for loneliness delivered via telehealth with guidance from a trained therapist (Käll, Backlund, et al., 2020).

Mindfulness. Mindfulness can be conceptualized as a process leading to a mental state of non-judgmental awareness of present moment experiences, including sensations, thoughts, bodily states, consciousness, and the environment, while fostering openness, curiosity, and acceptance (Käll, Backlund, et al., 2020). Mindfulness-based interventions are derived from Buddhist contemplative practices, emphasizing present-focused attention, and are nonreactive (Kabat-Zinn, 2013). The most well-known mindfulness-based intervention garnering empirical support is mindfulness-based stress reduction (MBSR, Kabat-Zinn, 1982). MBSR is an eight-week program that consists of weekly group-based classes with a trained instructor, daily audio-guided practice at home, and one day-long mindfulness retreat (Kabat-Zinn, 2013).

Evidence is available supporting the effectiveness of mindfulness/mediation, specifically utilizing a standardized eight-week MBSR program, for reducing the perception of loneliness in older adults (Creswell, 2017; Jazaieri, Goldin, Werner, Ziv, & Gross, 2012; Teoh, Letchumanan, & Lee, 2021; Veronese et al., 2021). In general, research appears to support that becoming more aware and present-focused may positively affect loneliness and social interactions (Lindsay, Young, Brown, Smyth, & Creswell, 2019). Mindfulness meditation has been associated with reduced depressive symptomology (Reangsing, Rittiwong, & Schneider, 2020), increased social cognition (Campos et al., 2019), and improved self-efficacy (Pandya, 2019), which may result in loneliness reduction. Additionally, preliminary evidence supports the effectiveness of mindfulness apps, thus, providing a convenient and inexpensive way for patients to engage in mindfulness in the comfort of their own homes (Boettcher et al., 2014; Figueroa & Aguilera, 2020; Lim, Condon, & DeSteno, 2015).

Acceptance and Commitment Therapy (ACT). Acceptance and commitment therapy (Hayes, Strosahl, & Wislon, 1999) is a psychological therapy that helps patients to evaluate their relationships with their thoughts and physical sensations through acceptance, mindfulness, and value-based action (Hayes, 2004). The overarching goal of ACT is to promote psychological flexibility (i.e., the ability to be mindful of experiences in the present moment) in an accepting and judgment-free manner, while maintaining a consistent behavior to one's values (Levin, Pistorello, Seeley, & Hayes, 2014). Limited but evolving evidence exists, delineating the negative association between psychological flexibility, ACT, and loneliness (Boman, Lundman, Nygren, Årestedt, & Santamäki Fischer, 2017; Frinking et al., 2020; Gardiner et al., 2018; Hamama-Raz & Hamama, 2015; Ziaee, Nejat, Amarghan, & Fariborzi, 2021), but the role of ACT in mitigating loneliness appears to be promising.

12.2.2 *Social prescribing*

Social prescribing is a holistic approach, consistent with the health and wellness precepts of the ICF model, empowering clinicians to refer patients for social support in their communities (Roland, Everington, & Marshall, 2020). Specialized clinicians have been long acquainted with social prescribing, as an intervention for addressing multifactorial physical and psychological health and social issues, with numerous publications reported over the years (Popay, Kowarzik, Mallinson, Mackian, & Barker, 2007). Specifically, there is increasing evidence that social prescribing improves patients reported well-being and reduces loneliness (Bickerdike, Booth, Wilson, Farley, & Wright, 2017; Foster et al., 2021; Kilgarriff-Foster & O'Cathain, 2015).

For PwP, there is a variety of ways to promote social engagement (e.g., boxing, dancing, yoga, meditation groups, music classes) (Popay et al., 2007), although the social distancing mandates associated with Covid-19 have led to cancellations of most in-person support and community groups, resulting in increased loneliness. However, through social prescribing, clinicians can direct patients to virtual exercise classes, virtual reality exercise games, and outdoor recreation activities of walking, hiking, biking, and running, thus, promoting connection with others in the community and encouraging PwP to continue exercising (Mirelman et al., 2016; Razani, Radhakrishna, & Chan, 2020; Schenkman et al., 2018; van der Kolk et al., 2019). Virtual support groups are growing at a rapid pace and provide PwP with tools for improving self-management of their disease and social connection (Subramanian et al., 2020; Visser, Bleijenbergh, Benschop, Van Riel, & Bloem, 2016).

Several social prescribing initiatives have been rolled out in the community and provide a potential avenue for improving loneliness. Specifically, the "Togetherness Program" at CareMore initiates home visits, regularly scheduled phone calls, and aims to bring patients in contact with appropriate social programs that already exist in the community (Masi, Chen, Hawkley, & Cacioppo, 2011; Murthy, 2020). In 2020, the Veteran's Administration began the "Compassionate Contact Corps Program" in response to the pandemic and has redeployed volunteers to make phone calls or virtual video visits to check on lonely veterans, while both the volunteers and those being called appear to benefit from this process (Winter & Gitlin, 2007). In the UK, the "Ways to Wellness" program (Newcastle Gateshed Clinical Commissioning Group, 2021) provides a model of social prescribing where a trained "Link Worker" is connected with a patient to identify health and wellness goals. Subsequently, the link worker connects the patient to community or volunteer groups, focusing on activities such as exercise, art-based therapy, gardening, and cooking (Moffatt, Steer, Lawson, Penn, & O'Brien, 2017). Considering the potential for social prescribing to translate into improved outcomes, proactive screening of loneliness by clinicians would be prudent (Subramanian et al., 2020).

12.3 Wellness strategies

Although a variety of interdisciplinary holistic health strategies have been implemented for PwP, they have almost exclusively focused on motor features (Subramanian et al., 2021). Helping PwP make lifestyle choices and creating structure and schedules to fill their daily lives during the pandemic has been posited as potentially helpful for preventing disruption of sleep pattern and exacerbation of mental health issues. Strategies such as exercise, health diet, sleep schedules, daily social activities, and mind-body approaches have been proposed as helpful interventions, even in the management of long-covid symptoms (Helmich & Bloem, 2020; Roth, Chan, & Jonas, 2021). Patients could be significantly benefited by some guidance in self-management in order to increase their sense of agency, when many other external factors related either to the pandemic or their disease are out of their control. Subramanian et al. (2021) have proposed several promising strategies designed to increase overall psychological wellness for PwP that include education, empowerment through teachable lifestyle choices, realigned health care team model, proactive outreach, social prescribing, and ultimately increased self-agency.

13. Conclusion

Due to the rapid spread of SARS-CoV-2 infections, the world has experienced an unprecedented global health crisis with tremendous morbidity and mortality. Covid-19 has dire implications in the psychosocial realm for all people around the world (Lai, Shih, Ko, Tang, & Hsueh, 2020), but for PwP the consequences may be particularly menacing. Although the risks of developing Covid-19 does not appear to be greater for PwP compared to the general population (Papa et al., 2020; Stoessl et al., 2020), the indirect impact of loneliness and social isolation have yet to be fully appreciated. Clinicians who treat PwP may not be aware of the negative health outcomes of loneliness, especially on non-motor symptoms, mental health, and quality of life (Subramanian et al., 2020). The added impact of the shelter-in-place and social distancing orders associated with the Covid-19 pandemic has further compounded the pandemic of loneliness. The interruption in social connection has led to increased reports of loneliness and associated increases in depression, anxiety, fatigue, cognitive decline, and apathy (Antonini et al., 2020; Brooks et al., 2021; Prasad et al., 2020). We propose a call to action in order to increase awareness of how to pro-actively screen for loneliness and to devise creative solutions for mitigating this problem and preventing the devastating consequences of loneliness in PwP.

References

Aarsland, D., Batzu, L., Halliday, G. M., Geurtsen, G. J., Ballard, C., Ray Chaudhuri, K., et al. (2021). Parkinson disease-associated cognitive impairment. *Nature Reviews. Disease Primers*, 7(1), 47. https://doi.org/10.1038/s41572-021-00280-3.

Aarsland, D., Bronnick, K., Williams-Gray, C., Weintraub, D., Marder, K., Kulisevsky, J., et al. (2010). Mild cognitive impairment in Parkinson disease: A multicenter pooled analysis. *Neurology*, *75*(12), 1062–1069. https://doi.org/10.1212/WNL.0b013e3181f39d0e.

Adam, E. K., Hawkley, L. C., Kudielka, B. M., & Cacioppo, J. T. (2006). Day-to-day dynamics of experience-cortisol associations in a population-based sample of older adults. *Proceedings of the National Academy of Sciences*, *103*(45), 17058–17063. https://doi.org/10.1073/pnas.0605053103.

Alschuler, K. N., Roberts, M. K., Herring, T. E., & Ehde, D. M. (2021). Distress and risk perception in people living with multiple sclerosis during the early phase of the COVID-19 pandemic. *Multiple Sclerosis and Related Disorders*, *47*, 102618. https://doi.org/10.1016/j.msard.2020.102618.

Anghelescu, B. A.-M., Bruno, V., Martino, D., & Roach, P. (2021). Effects of the COVID-19 pandemic on Parkinson's disease: A single-centered qualitative study. *Canadian Journal of Neurological Sciences/Journal Canadien Des Sciences Neurologiques*, *1–13*. https://doi.org/10.1017/cjn.2021.70.

Antonini, A. (2020). Health care for chronic neurological patients after COVID-19. *The Lancet Neurology*, *19*(7), 562–563. https://doi.org/10.1016/S1474-4422(20)30157-5.

Antonini, A., Barone, P., Marconi, R., Morgante, L., Zappulla, S., Pontieri, F. E., et al. (2012). The progression of non-motor symptoms in Parkinson's disease and their contribution to motor disability and quality of life. *Journal of Neurology*, *259*(12), 2621–2631. https://doi.org/10.1007/s00415-012-6557-8.

Antonini, A., Leta, V., Teo, J., & Chaudhuri, K. R. (2020). Outcome of Parkinson's disease patients affected by COVID-19. *Movement Disorders*, *35*(6), 905–908. https://doi.org/10.1002/mds.28104.

Balci, B., Aktar, B., Buran, S., Tas, M., & Donmez Colakoglu, B. (2021). Impact of the COVID-19 pandemic on physical activity, anxiety, and depression in patients with

Parkinson's disease. *International Journal of Rehabilitation Research, 44*(2), 173–176. https://doi.org/10.1097/MRR.0000000000000460.

Bartels, M., Cacioppo, J. T., Hudziak, J. J., & Boomsma, D. I. (2008). Genetic and environmental contributions to stability in loneliness throughout childhood. *American Journal of Medical Genetics Part B: Neuropsychiatric Genetics, 147B*(3), 385–391. https://doi.org/10.1002/ajmg.b.30608.

Baschi, R., Luca, A., Nicoletti, A., Caccamo, M., Cicero, C. E., D'Agate, C., et al. (2020). Changes in motor, cognitive, and behavioral symptoms in Parkinson's disease and mild cognitive impairment during the COVID-19 lockdown. *Frontiers in Psychiatry, 11*, 590134. https://doi.org/10.3389/fpsyt.2020.590134.

Baumeister, R. F., & Leary, M. R. (1995). The need to belong: Desire for interpersonal attachments as a fundamental human motivation. *Psychological Bulletin, 117*(3), 497–529.

Beam, C. R., & Kim, A. J. (2020). Psychological sequelae of social isolation and loneliness might be a larger problem in young adults than older adults. *Psychological Trauma Theory Research Practice and Policy, 12*(S1), S58–S60. https://doi.org/10.1037/tra0000774.

Bickerdike, L., Booth, A., Wilson, P. M., Farley, K., & Wright, K. (2017). Social prescribing: Less rhetoric and more reality. A systematic review of the evidence. *BMJ Open*, 7(4), e013384. https://doi.org/10.1136/bmjopen-2016-013384.

Boettcher, J., Åström, V., Påhlsson, D., Schenström, O., Andersson, G., & Carlbring, P. (2014). Internet-based mindfulness treatment for anxiety disorders: A randomized controlled trial. *Behavior Therapy, 45*(2), 241–253. https://doi.org/10.1016/j.beth.2013.11.003.

Boman, E., Lundman, B., Nygren, B., Årestedt, K., & Santamäki Fischer, R. (2017). Inner strength and its relationship to health threats in ageing-A cross-sectional study among community-dwelling older women. *Journal of Advanced Nursing, 73*(11), 2720–2729. https://doi.org/10.1111/jan.13341.

Bomasang-Layno, E., Fadlon, I., Murray, A. N., & Himelhoch, S. (2015). Antidepressive treatments for Parkinson's disease: A systematic review and meta-analysis. *Parkinsonism & Related Disorders, 21*(8), 833–842. https://doi.org/10.1016/j.parkreldis.2015.04.018.

Boomsma, D. I., Willemsen, G., Dolan, C. V., Hawkley, L. C., & Cacioppo, J. T. (2005). Genetic and environmental contributions to loneliness in adults: The Netherlands twin register study. *Behavior Genetics, 35*(6), 745–752. https://doi.org/10.1007/s10519-005-6040-8.

Bowlby, J. (1979). *The making and breaking of affectional bonds*. Tavistock.

Brady, S., D'Ambrosio, L. A., Felts, A., Rula, E. Y., Kell, K. P., & Coughlin, J. F. (2020). Reducing isolation and loneliness through membership in a fitness program for older adults: implications for health. *Journal of Applied Gerontology, 39*(3), 301–310. https://doi.org/10.1177/0733464818807820.

Broen, M. P. G., Narayen, N. E., Kuijf, M. L., Dissanayaka, N. N. W., & Leentjens, A. F. G. (2016). Prevalence of anxiety in Parkinson's disease: A systematic review and meta-analysis. *Movement Disorders, 31*(8), 1125–1133. https://doi.org/10.1002/mds.26643.

Brooks, S. K., Webster, R. K., Smith, L. E., Woodland, L., Wessely, S., Greenberg, N., et al. (2020). The psychological impact of quarantine and how to reduce it: Rapid review of the evidence. *The Lancet, 395*(10227), 912–920. https://doi.org/10.1016/S0140-6736(20)30460-8.

Brooks, S. K., Weston, D., & Greenberg, N. (2021). Social and psychological impact of the COVID-19 pandemic on people with Parkinson's disease: A scoping review. *Public Health, 199*, 77–86. https://doi.org/10.1016/j.puhe.2021.08.014.

Brown, E. G., Chahine, L. M., Goldman, S. M., Korell, M., Mann, E., Kinel, D. R., et al. (2020). The effect of the COVID-19 pandemic on people with Parkinson's disease [Preprint]. *Neurology*. https://doi.org/10.1101/2020.07.14.20153023.

Cacioppo, J. T., & Hawkley, L. C. (2009). Loneliness. In *Handbook of individual differences in social behavior* (pp. 227–240). Guilford Press.
Cacioppo, J. T., Hawkley, L. C., & Thisted, R. A. (2010). Perceived social isolation makes me sad: 5-year cross-lagged analyses of loneliness and depressive symptomatology in the Chicago Health, Aging, and Social Relations Study. *Psychology and Aging*, *25*(2), 453–463. https://doi.org/10.1037/a0017216.
Cacioppo, J. T., & Patrick, W. (2008). *Loneliness: Human nature and the need for social connection*. W.W. Norton.
Cacioppo, S., Capitanio, J. P., & Cacioppo, J. T. (2014). Toward a neurology of loneliness. *Psychological Bulletin*, *140*(6), 1464–1504. https://doi.org/10.1037/a0037618.
Cacioppo, S., Grippo, A. J., London, S., Goossens, L., & Cacioppo, J. T. (2015). Loneliness: Clinical import and interventions. *Perspectives on Psychological Science*, *10*(2), 238–249. https://doi.org/10.1177/1745691615570616.
Campos, D., Modrego-Alarcón, M., López-del-Hoyo, Y., González-Panzano, M., Van Gordon, W., Shonin, E., et al. (2019). Exploring the role of meditation and dispositional mindfulness on social cognition domains: A controlled study. *Frontiers in Psychology*, *10*, 809. https://doi.org/10.3389/fpsyg.2019.00809.
Centers for Disease Control and Prevention. (2021). *Alzheimer's disease and healthy aging: Loneliness and social isolation linked to serious health conditions*. Centers for Disease Control and Prevention. https://www.cdc.gov/aging/publications/features/lonely-older-adults.html.
Chaudhuri, K. R., Prieto-Jurcynska, C., Naidu, Y., Mitra, T., Frades-Payo, B., Tluk, S., et al. (2010). The nondeclaration of nonmotor symptoms of Parkinson's disease to health care professionals: An international study using the nonmotor symptoms questionnaire. *Movement Disorders*, *25*(6), 704–709. https://doi.org/10.1002/mds.22868.
Chaudhuri, K. R., Sauerbier, A., Rojo, J. M., Sethi, K., Schapira, A. H. V., Brown, R. G., et al. (2015). The burden of non-motor symptoms in Parkinson's disease using a self-completed non-motor questionnaire: A simple grading system. *Parkinsonism & Related Disorders*, *21*(3), 287–291. https://doi.org/10.1016/j.parkreldis.2014.12.031.
Cooney, J. W., & Stacy, M. (2016). Neuropsychiatric issues in Parkinson's disease. *Current Neurology and Neuroscience Reports*, *16*(5), 49. https://doi.org/10.1007/s11910-016-0647-4.
Courtin, E., & Knapp, M. (2017). Social isolation, loneliness and health in old age: A scoping review. *Health & Social Care in the Community*, *25*(3), 799–812. https://doi.org/10.1111/hsc.12311.
Coyle, C. E., & Dugan, E. (2012). Social isolation, loneliness and health among older adults. *Journal of Aging and Health*, *24*(8), 1346–1363. https://doi.org/10.1177/0898264312460275.
Creswell, J. D. (2017). Mindfulness interventions. *Annual Review of Psychology*, *68*(1), 491–516. https://doi.org/10.1146/annurev-psych-042716-051139.
Czaja, S. J., Moxley, J. H., & Rogers, W. A. (2021). Social support, isolation, loneliness, and health among older adults in the PRISM randomized controlled trial. *Frontiers in Psychology*, *12*, 728658. https://doi.org/10.3389/fpsyg.2021.728658.
de Rus Jacquet, A., Bogard, S., Normandeau, C. P., Degroot, C., Postuma, R. B., Dupré, N., et al. (2021). Clinical perception and management of Parkinson's disease during the COVID-19 pandemic: A Canadian experience. *Parkinsonism & Related Disorders*, *91*, 66–76. https://doi.org/10.1016/j.parkreldis.2021.08.018.
Deckx, L., van den Akker, M., Buntinx, F., & van Driel, M. (2018). A systematic literature review on the association between loneliness and coping strategies. *Psychology, Health & Medicine*, *23*(8), 899–916. https://doi.org/10.1080/13548506.2018.1446096.
Del Prete, E., Francesconi, A., Palermo, G., Mazzucchi, S., Frosini, D., Morganti, R., et al. (2021). Prevalence and impact of COVID-19 in Parkinson's disease: Evidence from a

multi-center survey in Tuscany region. *Journal of Neurology, 268*(4), 1179–1187. https://doi.org/10.1007/s00415-020-10002-6.

Ditzen, B., & Heinrichs, M. (2014). Psychobiology of social support: The social dimension of stress buffering. *Restorative Neurology and Neuroscience, 32*(1), 149–162. https://doi.org/10.3233/RNN-139008.

Domènech-Abella, J., Lara, E., Rubio-Valera, M., Olaya, B., Moneta, M. V., Rico-Uribe, L. A., et al. (2017). Loneliness and depression in the elderly: The role of social network. *Social Psychiatry and Psychiatric Epidemiology, 52*(4), 381–390. https://doi.org/10.1007/s00127-017-1339-3.

Donovan, N. J., Okereke, O. I., Vannini, P., Amariglio, R. E., Rentz, D. M., Marshall, G. A., et al. (2016). Association of higher cortical amyloid burden with loneliness in cognitively normal older adults. *JAMA Psychiatry, 73*(12), 1230. https://doi.org/10.1001/jamapsychiatry.2016.2657.

Donovan, N. J., Wu, Q., Rentz, D. M., Sperling, R. A., Marshall, G. A., & Glymour, M. M. (2017). Loneliness, depression and cognitive function in older U.S. adults: Loneliness, depression and cognition. *International Journal of Geriatric Psychiatry, 32*(5), 564–573. https://doi.org/10.1002/gps.4495.

Dossey, L. (2020). Loneliness and health. *EXPLORE, 16*(2), 75–78. https://doi.org/10.1016/j.explore.2019.12.005.

Douma, E. H., & de Kloet, E. R. (2020). Stress-induced plasticity and functioning of ventral tegmental dopamine neurons. *Neuroscience & Biobehavioral Reviews, 108*, 48–77. https://doi.org/10.1016/j.neubiorev.2019.10.015.

Dykstra, P. A., & Fokkema, T. (2007). Social and emotional loneliness among divorced and married men and women: Comparing the deficit and cognitive perspectives. *Basic and Applied Social Psychology, 29*(1), 1–12. https://doi.org/10.1080/01973530701330843.

Eggar, R., Spencer, A., Anderson, D., & Hiller, L. (2002). Views of elderly patients on cardiopulmonary resuscitation before and after treatment for depression. *International Journal of Geriatric Psychiatry, 17*(2), 170–174.

Elran-Barak, R., & Mozeikov, M. (2020). One month into the reinforcement of social distancing due to the COVID-19 outbreak: Subjective health, health behaviors, and loneliness among people with chronic medical conditions. *International Journal of Environmental Research and Public Health, 17*(15), 5403. https://doi.org/10.3390/ijerph17155403.

Espay, A. J., LeWitt, P. A., & Kaufmann, H. (2014). Norepinephrine deficiency in Parkinson's disease: The case for noradrenergic enhancement: Norepinephrine deficiency in PD. *Movement Disorders, 29*(14), 1710–1719. https://doi.org/10.1002/mds.26048.

Feeney, M. P., Xu, Y., Surface, M., Shah, H., Vanegas-Arroyave, N., Chan, A. K., et al. (2021). The impact of COVID-19 and social distancing on people with Parkinson's disease: A survey study. *Npj Parkinson's Disease*, 7(1), 10. https://doi.org/10.1038/s41531-020-00153-8.

Figueroa, C. A., & Aguilera, A. (2020). The need for a mental health technology revolution in the COVID-19 pandemic. *Frontiers in Psychiatry, 11*, 523. https://doi.org/10.3389/fpsyt.2020.00523.

Folk, D., Okabe-Miyamoto, K., Dunn, E., & Lyubomirsky, S. (2020). Did social connection decline during the first wave of COVID-19? The role of extraversion. *Collabra: Psychology, 6*(1), 37. https://doi.org/10.1525/collabra.365.

Foster, A., Thompson, J., Holding, E., Ariss, S., Mukuria, C., Jacques, R., et al. (2021). Impact of social prescribing to address loneliness: A mixed methods evaluation of a national social prescribing programme. *Health & Social Care in the Community, 29*(5), 1439–1449. https://doi.org/10.1111/hsc.13200.

Frinking, E., Jans-Beken, L., Janssens, M., Peeters, S., Lataster, J., Jacobs, N., et al. (2020). Gratitude and loneliness in adults over 40 years: Examining the role of psychological flexibility and engaged living. *Aging & Mental Health, 24*(12), 2117–2124. https://doi.org/10.1080/13607863.2019.1673309.

Gardiner, C., Geldenhuys, G., & Gott, M. (2018). Interventions to reduce social isolation and loneliness among older people: An integrative review. *Health & Social Care in the Community, 26*(2), 147–157. https://doi.org/10.1111/hsc.12367.

Garjani, A., Hunter, R., Law, G. R., Middleton, R. M., Tuite-Dalton, K. A., Dobson, R., et al. (2021). Mental health of people with multiple sclerosis during the COVID-19 outbreak: A prospective cohort and cross-sectional case–control study of the UK MS Register. *Multiple Sclerosis Journal*. https://doi.org/10.1177/13524585211020435. 1352 45852110204.

Grossman, E. R., Benjamin-Neelon, S. E., & Sonnenschein, S. (2020). Alcohol consumption during the COVID-19 pandemic: A cross-sectional survey of US adults. *International Journal of Environmental Research and Public Health, 17*(24), 9189. https://doi.org/10.3390/ijerph17249189.

Guo, D., Han, B., Lu, Y., Lv, C., Fang, X., Zhang, Z., et al. (2020). Influence of the COVID-19 pandemic on quality of life of patients with Parkinson's disease. *Parkinson's Disease, 2020*, 1–6. https://doi.org/10.1155/2020/1216568.

Hajek, A., Brettschneider, C., Mallon, T., Ernst, A., Mamone, S., Wiese, B., et al. (2017). The impact of social engagement on health-related quality of life and depressive symptoms in old age—Evidence from a multicenter prospective cohort study in Germany. *Health and Quality of Life Outcomes, 15*(1), 140. https://doi.org/10.1186/s12955-017-0715-8.

Hamama-Raz, Y., & Hamama, L. (2015). Quality of life among parents of children with epilepsy: A preliminary research study. *Epilepsy & Behavior, 45*, 271–276. https://doi.org/10.1016/j.yebeh.2014.12.003.

Hawkley, L. C., & Cacioppo, J. T. (2010). Loneliness matters: A theoretical and empirical review of consequences and mechanisms. *Annals of Behavioral Medicine, 40*(2), 218–227. https://doi.org/10.1007/s12160-010-9210-8.

Hawkley, L. C., Hughes, M. E., Waite, L. J., Masi, C. M., Thisted, R. A., & Cacioppo, J. T. (2008). From social structural factors to perceptions of relationship quality and loneliness: The Chicago health, aging, and social relations study. *The Journals of Gerontology Series B: Psychological Sciences and Social Sciences, 63*(6), S375–S384. https://doi.org/10.1093/geronb/63.6.S375.

Hawkley, L. C., Thisted, R. A., & Cacioppo, J. T. (2009). Loneliness predicts reduced physical activity: Cross-sectional & longitudinal analyses. *Health Psychology, 28*(3), 354–363. https://doi.org/10.1037/a0014400.

Hawryluck, L., Gold, W. L., Robinson, S., Pogorski, S., Galea, S., & Styra, R. (2004). SARS control and psychological effects of quarantine, Toronto, Canada. *Emerging Infectious Diseases, 10*(7), 7.

Hayes, S. C. (2004). Acceptance and commitment therapy, relational frame theory, and the third wave of behavioral and cognitive therapies. *Behavior Therapy, 35*, 639–665.

Hayes, S. C., Strosahl, K. D., & Wislon, K. G. (1999). *Acceptance and commitment therapy: An experiential approach to behavior change*. Guilford Press.

Heckhausen, J., Wrosch, C., & Schulz, R. (2013). A lines-of-defense model for managing health threats: A review. *Gerontology, 59*(5), 438–447. https://doi.org/10.1159/000351269.

Heinrich, L. M., & Gullone, E. (2006). The clinical significance of loneliness: A literature review. *Clinical Psychology Review, 26*(6), 695–718. https://doi.org/10.1016/j.cpr.2006.04.002.

Heinze, N., Hussain, S. F., Castle, C. L., Godier-McBard, L. R., Kempapidis, T., & Gomes, R. S. M. (2021). The long-term impact of the COVID-19 pandemic on loneliness in people living with disability and visual impairment. *Frontiers in Public Health, 9*, 738304. https://doi.org/10.3389/fpubh.2021.738304.

Helmich, R. C., & Bloem, B. R. (2020). The impact of the COVID-19 pandemic on Parkinson's disease: Hidden sorrows and emerging opportunities. *Journal of Parkinson's Disease*, *10*(2), 351–354. https://doi.org/10.3233/JPD-202038.

Hickin, N., Käll, A., Shafran, R., Sutcliffe, S., Manzotti, G., & Langan, D. (2021). The effectiveness of psychological interventions for loneliness: A systematic review and meta-analysis. *Clinical Psychology Review*, *88*, 102066. https://doi.org/10.1016/j.cpr.2021.102066.

Holt-Lunstad, J. (2017). The potential public health relevance of social isolation and loneliness: Prevalence, epidemiology, and risk factors. *Public Policy & Aging Report*, *27*(4), 127–130. https://doi.org/10.1093/ppar/prx030.

Holt-Lunstad, J. (2018). Why social relationships are important for physical health: A systems approach to understanding and modifying risk and protection. *Annual Review of Psychology*, *69*(1), 437–458. https://doi.org/10.1146/annurev-psych-122216-011902.

Holt-Lunstad, J., Smith, T. B., Baker, M., Harris, T., & Stephenson, D. (2015). Loneliness and social isolation as risk factors for mortality: A meta-analytic review. *Perspectives on Psychological Science*, *10*(2), 227–237. https://doi.org/10.1177/1745691614568352.

Holwerda, T. J., Beekman, A. T. F., Deeg, D. J. H., Stek, M. L., van Tilburg, T. G., Visser, P. J., et al. (2012). Increased risk of mortality associated with social isolation in older men: Only when feeling lonely? Results from the Amsterdam Study of the Elderly (AMSTEL). *Psychological Medicine*, *42*(4), 843–853. https://doi.org/10.1017/S0033291711001772.

Horesh, D., Kapel Lev-Ari, R., & Hasson-Ohayon, I. (2020). Risk factors for psychological distress during the COVID-19 pandemic in Israel: Loneliness, age, gender, and health status play an important role. *British Journal of Health Psychology*, *25*(4), 925–933. https://doi.org/10.1111/bjhp.12455.

Institute of Medicine. (2014). *Capturing social and behavioral domains in electronic health records: Phase 1* (p. 18709). National Academies Press. https://doi.org/10.17226/18709.

Janiri, D., Petracca, M., Moccia, L., Tricoli, L., Piano, C., Bove, F., et al. (2020). COVID-19 pandemic and psychiatric symptoms: The impact on Parkinson's disease in the elderly. *Frontiers in Psychiatry*, *11*, 581144. https://doi.org/10.3389/fpsyt.2020.581144.

Jazaieri, H., Goldin, P. R., Werner, K., Ziv, M., & Gross, J. J. (2012). A randomized trial of mbsr versus aerobic exercise for social anxiety disorder: MBSR v. AE in SAD. *Journal of Clinical Psychology*, *68*(7), 715–731. https://doi.org/10.1002/jclp.21863.

Jeong, H., Yim, H. W., Song, Y.-J., Ki, M., Min, J.-A., Cho, J., et al. (2016). Mental health status of people isolated due to Middle East Respiratory Syndrome. *Epidemiology and Health*, *38*, e2016048. https://doi.org/10.4178/epih.e2016048.

Jeste, D. V., Lee, E. E., & Cacioppo, S. (2020). Battling the modern behavioral epidemic of loneliness: Suggestions for research and interventions. *JAMA Psychiatry*, 77(6), 553. https://doi.org/10.1001/jamapsychiatry.2020.0027.

Kabat-Zinn, J. (1982). An outpatient program of behavioral medicine for chronic pain patients based on the practice of mindful meditation: Theoretical considerations and preliminary results. *General Hospital Psychaitry*, *4*, 33–47.

Kabat-Zinn, J. (2013). *Full catastrophe living: Using the wisdom of your body and mind to face stress, pain, and illness* (Revised ed.). Random House.

Kalia, L. V., & Lang, A. E. (2015). Parkinson's disease. *The Lancet*, *386*(9996), 896–912. https://doi.org/10.1016/S0140-6736(14)61393-3.

Käll, A., Backlund, U., Shafran, R., & Andersson, G. (2020). Lonesome no more? A two-year follow-up of internet-administered cognitive behavioral therapy for loneliness. *Internet Interventions*, *19*, 100301. https://doi.org/10.1016/j.invent.2019.100301.

Käll, A., Shafran, R., Lindegaard, T., Bennett, S., Cooper, Z., Coughtrey, A., et al. (2020). A common elements approach to the development of a modular cognitive behavioral theory for chronic loneliness. *Journal of Consulting and Clinical Psychology*, *88*(3), 269–282. https://doi.org/10.1037/ccp0000454.

Kapel, A., Serdoner, D., Fabiani, E., & Velnar, T. (2021). Impact of physiotherapy absence in COVID-19 pandemic on neurological state of patients with Parkinson disease. *Topics in Geriatric Rehabilitation*, *37*(1), 50–55. https://doi.org/10.1097/TGR.0000000000000304.

Kilgarriff-Foster, A., & O'Cathain, A. (2015). Exploring the components and impact of social prescribing. *Journal of Public Mental Health*, *14*(3), 127–134. https://doi.org/10.1108/JPMH-06-2014-0027.

Killgore, W. D. S., Cloonan, S. A., Taylor, E. C., & Dailey, N. S. (2020). Loneliness: A signature mental health concern in the era of COVID-19. *Psychiatry Research*, *290*, 113117. https://doi.org/10.1016/j.psychres.2020.113117.

Kitani-Morii, F., Kasai, T., Horiguchi, G., Teramukai, S., Ohmichi, T., Shinomoto, M., et al. (2021). Risk factors for neuropsychiatric symptoms in patients with Parkinson's disease during COVID-19 pandemic in Japan. *PLoS One*, *16*(1), e0245864. https://doi.org/10.1371/journal.pone.0245864.

Kool, M. B., & Geenen, R. (2012). Loneliness in patients with rheumatic diseases: The significance of invalidation and lack of social support. *The Journal of Psychology*, *146*(1–2), 229–241. https://doi.org/10.1080/00223980.2011.606434.

Krendl, A. C., & Perry, B. L. (2021). The impact of sheltering in place during the COVID-19 pandemic on older adults' social and mental well-being. *The Journals of Gerontology: Series B*, *76*(2), e53–e58. https://doi.org/10.1093/geronb/gbaa110.

Kumar, N., Gupta, R., Kumar, H., Mehta, S., Rajan, R., Kumar, D., et al. (2020). Impact of home confinement during COVID-19 pandemic on Parkinson's disease. *Parkinsonism & Related Disorders*, *80*, 32–34. https://doi.org/10.1016/j.parkreldis.2020.09.003.

Lai, C.-C., Shih, T.-P., Ko, W.-C., Tang, H.-J., & Hsueh, P.-R. (2020). Severe acute respiratory syndrome coronavirus 2 (SARS-CoV-2) and coronavirus disease-2019 (COVID-19): The epidemic and the challenges. *International Journal of Antimicrobial Agents*, *55*(3), 105924. https://doi.org/10.1016/j.ijantimicag.2020.105924.

Leigh-Hunt, N., Bagguley, D., Bash, K., Turner, V., Turnbull, S., Valtorta, N., et al. (2017). An overview of systematic reviews on the public health consequences of social isolation and loneliness. *Public Health*, *152*, 157–171. https://doi.org/10.1016/j.puhe.2017.07.035.

Levin, M. E., Pistorello, J., Seeley, J. R., & Hayes, S. C. (2014). Feasibility of a prototype web-based acceptance and commitment therapy prevention program for college students. *Journal of American College Health*, *62*(1), 20–30. https://doi.org/10.1080/07448481.2013.843533.

Lim, D., Condon, P., & DeSteno, D. (2015). Mindfulness and compassion: An examination of mechanism and scalability. *PLoS One*, *10*(2), e0118221. https://doi.org/10.1371/journal.pone.0118221.

Lim, M. H., Eres, R., & Vasan, S. (2020). Understanding loneliness in the twenty-first century: An update on correlates, risk factors, and potential solutions. *Social Psychiatry and Psychiatric Epidemiology*, *55*(7), 793–810. https://doi.org/10.1007/s00127-020-01889-7.

Lim, M. H., Rodebaugh, T. L., Zyphur, M. J., & Gleeson, J. F. M. (2016). Loneliness over time: The crucial role of social anxiety. *Journal of Abnormal Psychology*, *125*(5), 620–630. https://doi.org/10.1037/abn0000162.

Lindsay, E. K., Young, S., Brown, K. W., Smyth, J. M., & Creswell, J. D. (2019). Mindfulness training reduces loneliness and increases social contact in a randomized controlled trial. *Proceedings of the National Academy of Sciences*, *116*(9), 3488–3493. https://doi.org/10.1073/pnas.1813588116.

Luchetti, M., Lee, J. H., Aschwanden, D., Sesker, A., Strickhouser, J. E., Terracciano, A., et al. (2020). The trajectory of loneliness in response to COVID-19. *The American Psychologist*. https://doi.org/10.1037/amp0000690.

Lynch, J. J. (2000). *A cry unheard: New insights into the medical consequences of loneliness*. Bancroft Press.

Macdonald, B., & Hülür, G. (2021). Well-being and loneliness in Swiss older adults during the COVID-19 pandemic: The role of social relationships. *The Gerontologist, 61*(2), 240–250. https://doi.org/10.1093/geront/gnaa194.

Maffoni, M., Giardini, A., Pierobon, A., Ferrazzoli, D., & Frazzitta, G. (2017). Stigma experienced by Parkinson's disease patients: A descriptive review of qualitative studies. *Parkinson's Disease, 2017*, 1–7. https://doi.org/10.1155/2017/7203259.

Mann, F., Bone, J. K., Lloyd-Evans, B., Frerichs, J., Pinfold, V., Ma, R., et al. (2017). A life less lonely: The state of the art in interventions to reduce loneliness in people with mental health problems. *Social Psychiatry and Psychiatric Epidemiology*, *52*(6), 627–638. https://doi.org/10.1007/s00127-017-1392-y.

Martinez-Martin, P., Rodriguez-Blazquez, C., Kurtis, M. M., Chaudhuri, K. R., & on Behalf of the NMSS Validation Group. (2011). The impact of non-motor symptoms on health-related quality of life of patients with Parkinson's disease: Nms and HRQ O L in Parkinson's Disease. *Movement Disorders*, *26*(3), 399–406. https://doi.org/10.1002/mds.23462.

Masi, C. M., Chen, H.-Y., Hawkley, L. C., & Cacioppo, J. T. (2011). A meta-analysis of interventions to reduce loneliness. *Personality and Social Psychology Review*, *15*(3), 219–266. https://doi.org/10.1177/1088868310377394.

Matthews, K. A., Adler, N. E., Forrest, C. B., & Stead, W. W. (2016). Collecting psychosocial "vital signs" in electronic health records: Why now? What are they? What's new for psychology? *American Psychologist*, *71*(6), 497–504. https://doi.org/10.1037/a0040317.

McDaniels, B., & Bishop, M. (2021). Participation and psychological capital in adults with Parkinson's disease: Mediation analysis based on the international classification of functioning, disability, and health. *Rehabilitation Research, Policy, and Education*, *35*(3), 144–157. https://doi.org/10.1891/RE-21-03.

McDaniels, B., Bishop, M., & Rumrill, P. D., Jr. (2021). Quality of life in neurological disorders. In D. A. Harley, & C. Flaherty (Eds.), *Disability studies for human services: An interdisciplinary and intersectionality approach* (pp. 303–324). Springer.

McDaniels, B., Novak, D., Braitsch, M., & Chitnis, S. (2021). Management of Parkinson's disease during the COVID-19 pandemic. *Challenges and Opportunities*, *87*(1), 10.

McGinty, E. E., Presskreischer, R., Han, H., & Barry, C. L. (2020). Psychological distress and loneliness reported by US adults in 2018 and April 2020. *JAMA*, *324*(1), 93. https://doi.org/10.1001/jama.2020.9740.

Mesa Vieira, C., Franco, O. H., Gómez Restrepo, C., & Abel, T. (2020). COVID-19: The forgotten priorities of the pandemic. *Maturitas*, *136*, 38–41. https://doi.org/10.1016/j.maturitas.2020.04.004.

Michela, J. L., Peplau, L. A., & Weeks, D. G. (1982). Perceived dimensions of attributions for loneliness. *Journal of Personality and Social Psychology*, *43*(5), 929–936. https://doi.org/10.1037/0022-3514.43.5.929.

Mirelman, A., Rochester, L., Maidan, I., Del Din, S., Alcock, L., Nieuwhof, F., et al. (2016). Addition of a non-immersive virtual reality component to treadmill training to reduce fall risk in older adults (V-TIME): A randomised controlled trial. *The Lancet*, *388*(10050), 1170–1182. https://doi.org/10.1016/S0140-6736(16)31325-3.

Moffatt, S., Steer, M., Lawson, S., Penn, L., & O'Brien, N. (2017). Link worker social prescribing to improve health and well-being for people with long-term conditions: Qualitative study of service user perceptions. *BMJ Open*, 7(7), e015203. https://doi.org/10.1136/bmjopen-2016-015203.

Murthy, V. H. (2020). *Together: The healing power of human connection in a sometimes lonely world* (1st ed.). Harper Wave.

Mushtaq, R. (2014). Relationship between loneliness, psychiatric disorders and physical health? A review on the psychological aspects of loneliness. *Journal of Clinical and Diagnostic Research*. https://doi.org/10.7860/JCDR/2014/10077.4828.

Musich, S., Wang, S. S., Hawkins, K., & Yeh, C. S. (2015). The impact of loneliness on quality of life and patient satisfaction among older, sicker adults. *Gerontology and Geriatric Medicine*, *1*, 233372141558211. https://doi.org/10.1177/2333721415582119.

Mwilambwe-Tshilobo, L., Ge, T., Chong, M., Ferguson, M. A., Misic, B., Burrow, A. L., et al. (2019). Loneliness and meaning in life are reflected in the intrinsic network architecture of the brain. *Social Cognitive and Affective Neuroscience*, *14*(4), 423–433. https://doi.org/10.1093/scan/nsz021.

Naser Moghadasi, A. (2020). One aspect of Coronavirus disease (COVID-19) outbreak in Iran: High anxiety among MS patients. *Multiple Sclerosis and Related Disorders*, *41*, 102138. https://doi.org/10.1016/j.msard.2020.102138.

Nausheen, B., Gidron, Y., Peveler, R., & Moss-Morris, R. (2009). Social support and cancer progression: A systematic review. *Journal of Psychosomatic Research*, *67*(5), 403–415. https://doi.org/10.1016/j.jpsychores.2008.12.012.

Newcastle Gateshed Clinical Commissioning Group. (2021). *Ways to wellness*. https://waystowellness.org.uk/about/what-is-ways-to-wellness/.

Nitschke, J. P., Forbes, P. A. G., Ali, N., Cutler, J., Apps, M. A. J., Lockwood, P. L., et al. (2021). Resilience during uncertainty? Greater social connectedness during COVID-19 lockdown is associated with reduced distress and fatigue. *British Journal of Health Psychology*, *26*(2), 553–569. https://doi.org/10.1111/bjhp.12485.

Oppo, V., Serra, G., Fenu, G., Murgia, D., Ricciardi, L., Melis, M., et al. (2020). Parkinson's disease symptoms have a distinct impact on caregivers' and patients' stress: A study assessing the consequences of the COVID-19 lockdown. *Movement Disorders Clinical Practice*, 7(7), 865–867. https://doi.org/10.1002/mdc3.13030.

Özdin, S., & Bayrak Özdin, Ş. (2020). Levels and predictors of anxiety, depression and health anxiety during COVID-19 pandemic in Turkish society: The importance of gender. *International Journal of Social Psychiatry*, *66*(5), 504–511. https://doi.org/10.1177/0020764020927051.

Palermo, G., Tommasini, L., Baldacci, F., Del Prete, E., Siciliano, G., & Ceravolo, R. (2020). Impact of coronavirus disease 2019 pandemic on cognition in Parkinson's disease. *Movement Disorders*, *35*(10), 1717–1718. https://doi.org/10.1002/mds.28254.

Pandya, S. P. (2019). Meditation program enhances self-efficacy and resilience of home-based caregivers of older adults with Alzheimer's: A five-year follow-up study in two south Asian cities. *Journal of Gerontological Social Work*, *62*(16), 663–681.

Papa, S. M., Brundin, P., Fung, V. S. C., Kang, U. J., Burn, D. J., Colosimo, C., et al. (2020). Impact of the COVID -19 pandemic on Parkinson's disease and movement disorders. *Movement Disorders*, *35*(5), 711–715. https://doi.org/10.1002/mds.28067.

Perepezko, K., Hinkle, J. T., Shepard, M. D., Fischer, N., Broen, M. P. G., Leentjens, A. F. G., et al. (2019). Social role functioning in Parkinson's disease: A mixed-methods systematic review. *International Journal of Geriatric Psychiatry*, *34*(8), 1128–1138. https://doi.org/10.1002/gps.5137.

Perissinotto, C., Holt-Lunstad, J., Periyakoil, V. S., & Covinsky, K. (2019). A practical approach to assessing and mitigating loneliness and isolation in older adults: Loneliness and isolation in older adults. *Journal of the American Geriatrics Society*, *67*(4), 657–662. https://doi.org/10.1111/jgs.15746.

Peterson, D. B. (2016). The International Classification of Functioning, Disability, and Health: Applications for professional counselors. In *The professional counselor's desk reference* (2nd ed., pp. 329–336). Springer.

Pinquart, M., & Sorensen, S. (2001). Influences on loneliness in older adults: A meta-analysis. *Basic and Applied Social Psychology*, *23*(4), 245–266.

Poey, J. L., Burr, J. A., & Roberts, J. S. (2017). Social connectedness, perceived isolation, and dementia: Does the social environment moderate the relationship between genetic risk and cognitive well-being? *The Gerontologist*, *57*(6), 1031–1040. https://doi.org/10.1093/geront/gnw154.

Popay, J., Kowarzik, U., Mallinson, S., Mackian, S., & Barker, J. (2007). Social problems, primary care and pathways to help and support: Addressing health inequalities at the individual level. Part I: The GP perspective. *Journal of Epidemiology & Community Health*, *61*(11), 966–971. https://doi.org/10.1136/jech.2007.061937.

Prasad, S., Holla, V. V., Neeraja, K., Surisetti, B. K., Kamble, N., Yadav, R., et al. (2020). Parkinson's disease and COVID -19: Perceptions and implications in patients and caregivers. *Movement Disorders*, *35*(6), 912–914. https://doi.org/10.1002/mds.28088.

Razani, N., Radhakrishna, R., & Chan, C. (2020). Public lands are essential to public health during a pandemic. *Pediatrics*, *146*(2). https://doi.org/10.1542/peds.2020-1271, e20201271.

Reangsing, C., Rittiwong, T., & Schneider, J. K. (2020). Effects of mindfulness meditation interventions on depression in older adults: A meta-analysis. *Aging & Mental Health*, *15*, 1–10.

Reynolds, D. L., Garay, J. R., Deamond, S. L., Moran, M. K., Gold, W., & Styra, R. (2008). Understanding, compliance and psychological impact of the SARS quarantine experience. *Epidemiology and Infection*, *136*(7), 997–1007. https://doi.org/10.1017/S0950268807009156.

Roland, M., Everington, S., & Marshall, M. (2020). Social prescribing—Transforming the relationship between physicians and their patients. *New England Journal of Medicine*, *383*(2), 97–99. https://doi.org/10.1056/NEJMp1917060.

Roth, A., Chan, P. S., & Jonas, W. (2021). Addressing the long COVID crisis: Integrative health and long COVID. *Global Advances in Health and Medicine*, *10*, 216495612110565. https://doi.org/10.1177/21649561211056597.

Salari, M., Zali, A., Ashrafi, F., Etemadifar, M., Sharma, S., Hajizadeh, N., et al. (2020). Incidence of anxiety in Parkinson's disease during the Coronavirus Disease (COVID-19) pandemic. *Movement Disorders*, mds.28116. https://doi.org/10.1002/mds.28116.

Sandi, C., & Haller, J. (2015). Stress and the social brain: Behavioural effects and neurobiological mechanisms. *Nature Reviews Neuroscience*, *16*(5), 290–304. https://doi.org/10.1038/nrn3918.

Santos-García, D., Oreiro, M., Pérez, P., Fanjul, G., Paz González, J. M., Feal Painceiras, M. J., et al. (2020). Impact of Coronavirus Disease 2019 pandemic on Parkinson's disease: A cross-sectional survey of 568 Spanish patients. *Movement Disorders*, *35*(10), 1712–1716. https://doi.org/10.1002/mds.28261.

Schapira, A. H. V., Chaudhuri, K. R., & Jenner, P. (2017). Non-motor features of Parkinson disease. *Nature Reviews Neuroscience*, *18*(7), 435–450. https://doi.org/10.1038/nrn.2017.62.

Schenkman, M., Moore, C. G., Kohrt, W. M., Hall, D. A., Delitto, A., Comella, C. L., et al. (2018). Effect of high-intensity treadmill exercise on motor symptoms in patients with de novo Parkinson's disease: A phase 2 randomized clinical trial. *JAMA Neurology*, *75*(2), 219. https://doi.org/10.1001/jamaneurol.2017.3517.

Schirinzi, T., Cerroni, R., Di Lazzaro, G., Liguori, C., Scalise, S., Bovenzi, R., et al. (2020). Self-reported needs of patients with Parkinson's disease during COVID-19 emergency in Italy. *Neurological Sciences*, *41*(6), 1373–1375. https://doi.org/10.1007/s10072-020-04442-1.

Shalash, A., Roushdy, T., Essam, M., Fathy, M., Dawood, N. L., Abushady, E. M., et al. (2020). Mental health, physical activity, and quality of life in Parkinson's disease during COVID -19 pandemic. *Movement Disorders*, mds.28134. https://doi.org/10.1002/mds.28134.

Shaygannejad, V., Afshari-Safavi, A., & Hatef, B. (2021). Assessment of mental health, knowledge, and attitude of patients with multiple sclerosis and neuromyelitis optica spectrum disorder in response to 2019 novel coronavirus. *Neurological Sciences*, *42*(7), 2891–2901. https://doi.org/10.1007/s10072-020-04905-5.

Siciliano, M., Trojano, L., Santangelo, G., De Micco, R., Tedeschi, G., & Tessitore, A. (2018). Fatigue in Parkinson's disease: A systematic review and meta-analysis. *Movement Disorders*, *33*(11), 1712–1723. https://doi.org/10.1002/mds.27461.

Simpson, J., Lekwuwa, G., & Crawford, T. (2013). Illness beliefs and psychological outcome in people with Parkinson's disease. *Chronic Illness*, *9*(2), 165–176. https://doi.org/10.1177/1742395313478219.

Sjödahl Hammarlund, C., Westergren, A., Åström, I., Edberg, A.-K., & Hagell, P. (2018). The impact of living with Parkinson's disease: Balancing within a web of needs and demands. *Parkinson's Disease*, *2018*, 1–8. https://doi.org/10.1155/2018/4598651.

Soleimani, M. A., Negarandeh, R., Bastani, F., & Greysen, R. (2014). Disrupted social connectedness in people with Parkinson's disease. *British Journal of Community Nursing*, *19*(3), 136–141. https://doi.org/10.12968/bjcn.2014.19.3.136.

Solmi, M., Veronese, N., Galvano, D., Favaro, A., Ostinelli, E. G., Noventa, V., et al. (2020). Factors associated with loneliness: An umbrella review of observational studies. *Journal of Affective Disorders*, *271*, 131–138. https://doi.org/10.1016/j.jad.2020.03.075.

Song, J., Ahn, J. H., Choi, I., Mun, J. K., Cho, J. W., & Youn, J. (2020). The changes of exercise pattern and clinical symptoms in patients with Parkinson's disease in the era of COVID-19 pandemic. *Parkinsonism & Related Disorders*, *80*, 148–151. https://doi.org/10.1016/j.parkreldis.2020.09.034.

Spreng, R. N., Dimas, E., Mwilambwe-Tshilobo, L., Dagher, A., Koellinger, P., Nave, G., et al. (2020). The default network of the human brain is associated with perceived social isolation. *Nature Communications*, *11*(1), 6393. https://doi.org/10.1038/s41467-020-20039-w.

Steptoe, A., Shankar, A., Demakakos, P., & Wardle, J. (2013). Social isolation, loneliness, and all-cause mortality in older men and women. *Proceedings of the National Academy of Sciences*, *110*(15), 5797–5801. https://doi.org/10.1073/pnas.1219686110.

Stoessl, A. J., Bhatia, K. P., & Merello, M. (2020). Movement disorders in the world of COVID-19. *Movement Disorders*, *35*(5), 709–710. https://doi.org/10.1002/mds.28069.

Stojanov, A., Malobabic, M., Milosevic, V., Stojanov, J., Vojinovic, S., Stanojevic, G., et al. (2020). Psychological status of patients with relapsing-remitting multiple sclerosis during coronavirus disease-2019 outbreak. *Multiple Sclerosis and Related Disorders*, *45*, 102407. https://doi.org/10.1016/j.msard.2020.102407.

Subramanian, I., Farahnik, J., & Mischley, L. K. (2020). Synergy of pandemics-social isolation is associated with worsened Parkinson severity and quality of life. *Npj Parkinson's Disease*, *6*(1), 28. https://doi.org/10.1038/s41531-020-00128-9.

Subramanian, I., Hinkle, J. T., Chaudhuri, K. R., Mari, Z., Fernandez, H., & Pontone, G. M. (2021). Mind the gap: Inequalities in mental health care and lack of social support in Parkinson disease. *Parkinsonism & Related Disorders*. https://doi.org/10.1016/j.parkreldis.2021.11.015. S1353802021004193.

Suzuki, K., Numao, A., Komagamine, T., Haruyama, Y., Kawasaki, A., Funakoshi, K., et al. (2021). Impact of the COVID-19 pandemic on the quality of life of patients with Parkinson's disease and their caregivers: A single-center survey in Tochigi prefecture. *Journal of Parkinson's Disease*, *11*(3), 1047–1056. https://doi.org/10.3233/JPD-212560.

Tashakori-Miyanroudi, M., Souresrafil, A., Hashemi, P., Jafar Ehsanzadeh, S., Farrahizadeh, M., & Behroozi, Z. (2021). Prevalence of depression, anxiety, and psychological distress in patients with epilepsy during COVID-19: A systematic review. *Epilepsy & Behavior*, *125*, 108410. https://doi.org/10.1016/j.yebeh.2021.108410.

Teoh, S. L., Letchumanan, V., & Lee, L.-H. (2021). Can mindfulness help to alleviate loneliness? A systematic review and meta-analysis. *Frontiers in Psychology*, *12*, 633319. https://doi.org/10.3389/fpsyg.2021.633319.

Thomsen, T. H., Wallerstedt, S. M., Winge, K., & Bergquist, F. (2021). Life with Parkinson's disease during the COVID-19 pandemic: The pressure is "OFF". *Journal of Parkinson's Disease*, *11*(2), 491–495.

Valtorta, N. K., Kanaan, M., Gilbody, S., Ronzi, S., & Hanratty, B. (2016). Loneliness and social isolation as risk factors for coronary heart disease and stroke: Systematic review and meta-analysis of longitudinal observational studies. *Heart*, *102*(13), 1009–1016. https://doi.org/10.1136/heartjnl-2015-308790.

van der Kolk, N. M., de Vries, N. M., Kessels, R. P. C., Joosten, H., Zwinderman, A. H., Post, B., et al. (2019). Effectiveness of home-based and remotely supervised aerobic exercise in Parkinson's disease: A double-blind, randomised controlled trial. *The Lancet Neurology*, *18*(11), 998–1008. https://doi.org/10.1016/S1474-4422(19)30285-6.

van Wamelen, D. J., Wan, Y.-M., Chaudhuri, K. R., & Jenner, P. (2020). Chapter six—Stress and cortisol in Parkinson's disease. *International Review of Neurobiology*, *152*, 131–156. https://doi.org/10.1016/bs.irn.2020.01.005.

Veazie, S., Gilbert, J., Winchell, K., Paynter, R., & Guise, J.-M. (2019). *Addressing social isolation to improve the health of older adults: A rapid review*. Agency for Healthcare Research and Quality (AHRQ). https://doi.org/10.23970/AHRQEPC-RAPIDISOLATION.

Veronese, N., Galvano, D., D'Antiga, F., Vecchiato, C., Furegon, E., Allocco, R., et al. (2021). Interventions for reducing loneliness: An umbrella review of intervention studies. *Health & Social Care in the Community*, *29*(5). https://doi.org/10.1111/hsc.13248.

Visser, L. M., Bleijenbergh, I. L., Benschop, Y. W. M., Van Riel, A. C. R., & Bloem, B. R. (2016). Do online communities change power processes in healthcare? Using case studies to examine the use of online health communities by patients with Parkinson's disease: Table 1. *BMJ Open*, *6*(11), e012110. https://doi.org/10.1136/bmjopen-2016-012110.

Wee, L., Tsang, T., Yi, H., Toh, S., Lee, G., Yee, J., et al. (2019). Loneliness amongst low-socioeconomic status elderly Singaporeans and its association with perceptions of the neighbourhood environment. *International Journal of Environmental Research and Public Health*, *16*(6), 967. https://doi.org/10.3390/ijerph16060967.

Weintraub, D., & Mamikonyan, E. (2019). The Neuropsychiatry of Parkinson's disease: A perfect storm. *The American Journal of Geriatric Psychiatry*, *27*(9), 998–1018. https://doi.org/10.1016/j.jagp.2019.03.002.

Williams-Gray, C. H., Mason, S. L., Evans, J. R., Foltynie, T., Brayne, C., Robbins, T. W., et al. (2013). The CamPaIGN study of Parkinson's disease: 10-year outlook in an incident population-based cohort. *Journal of Neurology, Neurosurgery & Psychiatry*, *84*(11), 1258–1264. https://doi.org/10.1136/jnnp-2013-305277.

Winter, L., & Gitlin, L. N. (2007). Evaluation of a telephone-based support group intervention for female caregivers of community-dwelling individuals with dementia. *American Journal of Alzheimer's Disease and Other Dementias*, *21*(6), 391–397. https://doi.org/10.1177/1533317506291371.

Wong, S. Y. S., Zhang, D., Sit, R. W. S., Yip, B. H. K., Chung, R. Y., Wong, C. K. M., et al. (2020). Impact of COVID-19 on loneliness, mental health, and health service utilisation: A prospective cohort study of older adults with multimorbidity in primary care. *British Journal of General Practice*, *70*(700), e817–e824. https://doi.org/10.3399/bjgp20X713021.

World Health Organization. (2001). *International classification of functioning, disability, and health*. World Health Organization.

World Health Organization. (2015). *World report on ageing and health*. World Health Organization.

World Health Organization, G.T. (2020). *WHO director-general's opening remarks at the media briefing on COVID-19. 2020*. World Health Organization. https://www.who.int/director-general/speeches/detail/who-director-general-s-opening-remarks-at-the-media-briefing-on-covid-19- - -11-march-2020.

Wu, P., Liu, X., Fang, Y., Fan, B., Fuller, C. J., Guan, Z., et al. (2008). Alcohol abuse/dependence symptoms among hospital employees exposed to a SARS outbreak. *Alcohol and Alcoholism*, *43*(6), 706–712. https://doi.org/10.1093/alcalc/agn073.

Xiong, J., Lipsitz, O., Nasri, F., Lui, L. M. W., Gill, H., Phan, L., et al. (2020). Impact of COVID-19 pandemic on mental health in the general population: A systematic review. *Journal of Affective Disorders*, *277*, 55–64. https://doi.org/10.1016/j.jad.2020.08.001.

Yoon, S. Y., Suh, J. H., Yang, S. N., Han, K., & Kim, Y. W. (2021). Association of physical activity, including amount and maintenance, with all-cause mortality in Parkinson disease. *JAMA Neurology*. https://doi.org/10.1001/jamaneurol.2021.3926.

Zhou, Z., Wang, P., & Fang, Y. (2018). Loneliness and the risk of dementia among older Chinese adults: Gender differences. *Aging & Mental Health*, *22*(4), 519–525. https://doi.org/10.1080/13607863.2016.1277976.

Ziaee, A., Nejat, H., Amarghan, H. A., & Fariborzi, E. (2021). Existential therapy versus acceptance and commitment therapy for feelings of loneliness and irrational beliefs in male prisoners. *European Journal of Translational Myology*. https://doi.org/10.4081/ejtm.2022.10271.

Zipprich, H. M., Teschner, U., Witte, O. W., Schönenberg, A., & Prell, T. (2020). Knowledge, attitudes, practices, and burden during the COVID-19 pandemic in people with Parkinson's disease in Germany. *Journal of Clinical Medicine*, *9*(6), 1643. https://doi.org/10.3390/jcm9061643.

CHAPTER TEN

Parkinson's disease and Covid-19: Is there an impact of ethnicity and the need for palliative care

Katarina Rukavina[a,b,*], Victor McConvey[c], Kallol Ray Chaudhuri[a,b], and Janis Miyasaki[d]

[a]Institute of Psychiatry, Psychology & Neuroscience at King's College London and King's College Hospital NHS Foundation Trust, London, United Kingdom
[b]Parkinson's Foundation Centre of Excellence, King's College Hospital NHS Foundation Trust, London, United Kingdom
[c]Fight Parkinson's, Surrey Hills, VIC, Australia
[d]Parkinson and Movement Disorders Program and the Complex Neurologic Symptoms Clinic (Neuropalliative Care), University of Alberta, Edmonton, AB, Canada
*Corresponding author: e-mail address: katarina.rukavina@kcl.ac.uk

Contents

Abstract

Under the traditional models of care for People with Parkinson's Disease (PD, PwP), many of their needs remain unmet and a substantial burden of motor and non-motor symptoms they experience may not be tackled sufficiently. An introduction

International Review of Neurobiology, Volume 165
ISSN 0074-7742
https://doi.org/10.1016/bs.irn.2022.03.004
Copyright © 2022 Elsevier Inc.
All rights reserved.

of palliative care (PC) interventions early in the course of PD offers profound benefits: it may improve quality of life of patients, their families and caregivers through the prevention and relief of medical symptoms, while, at the same time, emphasizing their emotional needs and spiritual wellbeing, establishing goals of care, and engaging in the advance care planning (ACP).

The ongoing Coronavirus Disease 2019 (Covid-19) pandemic poses an unprecedented set of challenges for PwP and has in many ways (both directly and indirectly) magnified their suffering, thus rapidly raising the demand for PC interventions. Covid-19, as well as the repercussions of prolonged mobility restrictions and limited health-care access might exacerbate the severity of PD motor symptoms and interact negatively with a range of non-motor symptoms, with a detrimental effect on quality of life. Greater motor disability, higher amount of levodopa-induced motor fluctuations with an increased daily off-time, fatigue, anxiety, depression, sleep disturbances, pain and worsening of cognitive complaints might dominate the clinical presentation in PwP during the Covid-19 pandemic, alongside raising psychological and spiritual concerns and anticipatory grief.

Here, we aim to provide a foundation for pragmatic and clinically orientated PC approach to improve quality of life and relieve suffering of PwP in the context of the current, ongoing Covid-19 pandemic.

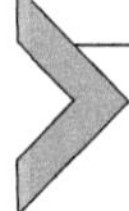

1. Palliative care: General principles and the importance for people with Parkinson's disease

Palliative care (PC) seeks to improve quality of life of patients, their families and caregivers through the prevention and relief of medical symptoms, while, at the same time, emphasizing their emotional needs and spiritual wellbeing, establishing goals of care, and engaging in the advance care planning (ACP) (Kluger & Quill, 2021; Miyasaki & Robinson, 2020). The PC interventions are holistic and "patient-centered," rather than "disease-centered," and exerted through a multidisciplinary approach, while actively integrating patients' values and prioritizing the symptoms according to the patient's preferences (Quill & Abernethy, 2013) (Chaudhuri et al., 2021). Although, historically, PC equaled a hospice tradition (defined as PC for individuals with a 6-month life expectancy), delivered only at the end of life, its role has increasingly been expanding and over the past decade the PC interventions are increasingly being delivered at earlier stages of the disease, alongside disease directed treatments in an outpatient setting (Kluger et al., 2020; Quill & Abernethy, 2013).

People with Parkinson's Disease (PD, PwP) may experience a substantial burden of motor and non-motor symptoms, both major determinants of the health-related quality of life (QoL), that may not be sufficiently tackled by a traditional model of care (a chronic illness model focused on motor symptoms and delivered through a patient–physician dyad) (Boersma et al., 2016; Chaudhuri et al., 2021; Rukavina et al., 2021). Gaps have also been indentified specially related to mental health issues unmasked during the Covid-19 pandemic (Subramanian et al., 2021).

PC is applicable early in the course of PD; its early introduction has been proven a powerful tool with profound benefits on QoL (including cognitive, social, emotional and spiritual aspects), which may enhance the overall satisfaction of PwP with care, reduce their hospital admission rates and offer support to caregiver (Di Luca et al., 2020). Furthermore, in PwP and patients with other movement disorders, PC may help to address non-traditional sources of suffering, including, for example, loss of identity, existential distress, and spiritual suffering (Miyasaki et al., 2021) (Fig. 1).

Yet, even in high-income countries, there is a practice gap between the evidence for PC effectiveness and its actual use in PD; despite significant increases in the past years, rates of PC referrals for PwP remain relatively low (less than 5% in certain settings) and PwP or those with atypical parkinsonism receive less end-of-life PC compared to patients with other neurological illnesses (Akbar et al., 2021; Dhamija, Saluja, & Miyasaki, 2021).

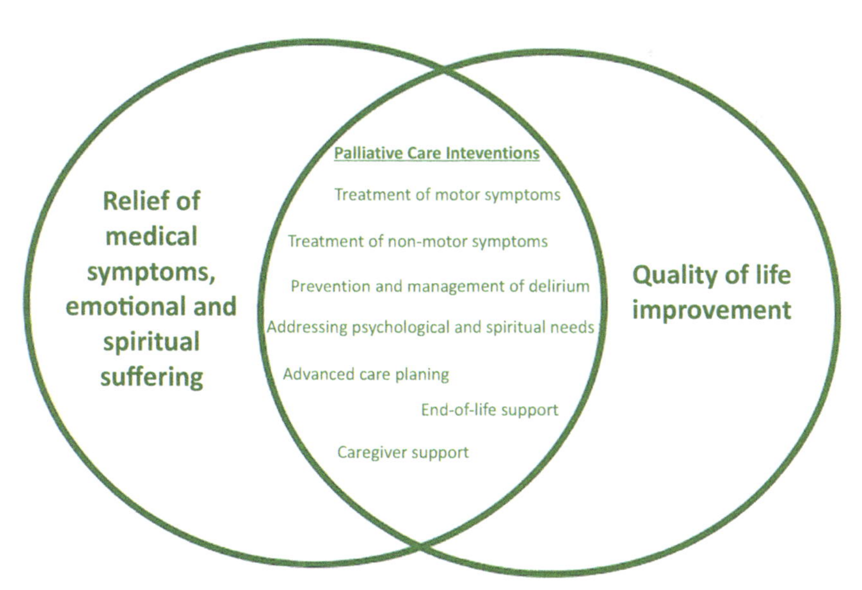

Fig. 1 Main principles of palliative care for People with Parkinson's during the COVID-19 pandemic. Proposed treatment algorithm prevention and management of delirium in People with Parkinson's disease. DA—dopamine agonists, MAOB-I—monoamine oxidase B inhibitors, COMT-I—catechol-o-methyltransferase inhibitors.

Importantly, the PC approach is not limited to palliative medicine specialists; it is a complementary and multidisciplinary approach, which can be provided while patients are receiving their usual, disease-specific care. There are various models of delivery of neuropalliative care for PwP and patients with other movement disorders, including *generalist PC* (PC approach adopted by clinicians without certification in palliative care), *specialist neuropalliative care* (provided by clinicians with a certification in both neurology and palliative care), and *collaborative care* (neurologist along-side a PC physician/clinician) (Oliver et al., 2020). In fact, all clinicians are providing certain degree of PC when seeking to enhance quality of life and to optimize their patients' functioning (Miyasaki & Robinson, 2020; Quill & Abernethy, 2013). Indeed, in a recent survey of the International Parkinson and Movement Disorders Society (MDS) membership, the majority of the participants agreed that neurologists should be involved in the care of PwP throughout the course of their illness and that healthcare professionals working with patients with movement disorders should adopt generalist PC approaches in their daily practice, while also expressing their will to learn more about PC (Miyasaki et al., 2021). At the same time, substantial systemic and individual barriers to PC for PwP were reported, including PC workforce shortages, traditional oncology focus of PC, patients' beliefs, PC physician attitudes and the lack of the role clarity, as well as clinicians' own discomfort with PC. In low income countries, limited financial support and infrastructure may pose additional challenges (Miyasaki et al., 2021).

The ongoing Covid-19 (the acute respiratory illness caused by severe acute respiratory syndrome coronavirus 2; SARS-CoV-2) pandemic poses an unprecedented set of challenges for PwP and has in many ways (both directly and indirectly; through physical illness and death, fears and anxieties, and financial and social instability as well as a major effect on mental health as alluded to previously) magnified their suffering, thus rapidly raising the demand for PC interventions (Subramanian & Vaughan, 2020).

Here, we propose a pragmatic, clinically orientated PC approach to improve quality of life and relieve suffering of PwP in the context of the current, ongoing Covid-19 pandemic with all its distinctive challenges, that can be applied by all health-care providers looking after PwP.

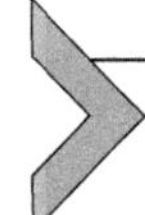

2. Parkinson's disease and Covid-19: Morbidity and mortality

Taken together, currently available evidence does not indicate that PD per se increases the susceptibility to Covid-19 or predisposes to higher mortality for those affected as the data remain inconclusive with mortality rates ranging from 5.2% to 100% (Fearon & Fasano, 2021). However, in general, Covid-19 affects elderly patients with chronic conditions to a greater extent. Thus, elderly PwP, particularly those at the advanced stages of the illness (with an impaired cough reflex and respiratory muscle involvement, axial manifestations, and abnormal posture), fall into a high-risk, vulnerable group, possibly susceptible to Covid-19 (Antonini, Leta, Teo, & Chaudhuri, 2020; Fasano et al., 2020). A significant burden of co-morbidities (particularly obesity, cardiovascular disease, and chronic obstructive pulmonary disease), polypharmacy and nursing home placements may further heighten the risk of adverse outcomes in PwP (Bhidayasiri, Virameteekul, Kim, Pal, & Chung, 2020; Fasano, Cereda, et al., 2020; Fearon & Fasano, 2021; McLean, Hindle, Guthrie, & Mercer, 2017). Table 1 summarizes some of the studies reporting mortality rates of Covid-19 in PwP.

In addition, the downstream consequences of the Covid-19 pandemic on PwP, beyond those directly attributable to the virus itself, are yet to be rigorously reported in large populations.

Table 1 Mortality rate due to COVID-19 in individuals with Parkinson's disease.

Study	***N***	**Outcome of the COVID-19 infection**
Antonini et al. (2020)	10	MR 40% in older PwP with longer DD; 50% in those on advanced therapies
Fasano et al. (2020)	117	MR 19.7%, linked with longer DD, higher age, and dementia
Fasano et al. (2020)		MR in patients with mild to moderate PD (mean H&Y 2.2 ± 0.8) does not differ significantly from general population (5.7% vs. 7.6%, $P = 0.77$)
Artusi et al. (2020)	8	MR 75%
Sainz-Amo et al. (2020)	39	MR 21%
Hainque and Grabli (2020)	2	MR 100%
Del Prete et al. (2020)	7	MR 14%
Parihar, Ferastraoaru, Galanopoulou, Geyer, and Kaufinan (2021)	53	MR 35.8%

MR—mortality rate, *PwP*—People with Parkinson's, *DD*—disease duration, *HY*—Hoehn & Yahr.

3. Increased vulnerability of Covid-19 in individuals from ethnic minority groups: Implications for Parkinson's disease

Ethnicity may be described as a complex entity, resulting from an interplay of distinctive genetic and epigenetic features, certain behavioral patterns, social set up and cultural identity (Lee, 2009). Differences in comorbidities (e.g., higher predisposition to the development of hypertension, coronary heart disease, stroke and type 2 diabetes) and in immune profiles among people of different ethnicities might drive differences in outcomes of an infection, as mirrored in previous pandemics (Otu, Ahinkorah, Ameyaw, Seidu, & Yaya, 2020). Furthermore, ethnicity might be a risk factor for other underlying conditions that might affect health, including socio-economic status, access to health care, and exposure to the virus related to occupation (e.g., through the engagement in frontline roles and as critical infrastructure workers, or other roles where implementing safe physical distancing measures may not be possible). In addition, individuals from ethnic minorities background are more likely to live in larger cities, usually within densely populated areas, and tend to reside in overcrowded multi-generational households, both facilitating the spread of infectious diseases, including Covid-19 (Otu et al., 2020). In the UK, the differential effect and higher morbidity caused by Covid-19 in Black and Asian origin subjects have been well publicized.

When affected by Covid-19, the mortality rate among people from ethnic minority background is disproportionally high: in the United Kingdom, this was established in July 2020, as in English hospitals the mortality rate due to Covid-19 was 3.5 times higher among people of Black African descent, and 2.7 and 1.7 times higher among those of Pakistani and Black Caribbean descent, respectively, compared to the mortality rates of White British people (Otu et al., 2020). Among the individuals living in the United States of America (USA), those from American Indian or Alaska Native groups were found to have 3.3 times higher risk of hospitalization and 2.2 higher risk of death due to Covid-19 compared to White, Non-Hispanic persons. For Black or African American individuals living in the USA, the risk of hospitalization was 2.6 times, and the risk of the death due to Covid-19 was 1.9 times greater than in White, Non-Hispanic individuals. These differences were evident after adjusting for age (Centers for Disease Control and Prevention, 2019).

Of note, PwP from ethnic minority backgrounds (in populations living in London) have recently been reported to exhibit greater motor disability, greater burden of non-motor symptoms (NMS), and a higher degree of cardiovascular comorbidities; the above features might further exacerbate the differences in risk for Covid-19 adverse outcomes between White and non-White PwP (Sauerbier et al., 2021). Indeed, a statistically significant higher mortality rate among Black/African American PwP with Covid-19 has been reported (Parihar et al., 2021). Moreover, the ethnic disparities are evident even beyond the direct impact of the virus per se. Although the majority of PwP, even when not infected by SARS-CoV-2, reported substantial disruptions in their medical care, belonging to a Non-White group was independently associated with difficulties obtaining medications—another concerning finding highlighting the barriers to healthcare access, that might have possibly been exacerbated during the pandemic (Brown et al., 2020; Rukavina et al., 2022).

Thus, amid the ongoing Covid-19 pandemic, PwP from ethnic minorities need to be identified and their needs properly addressed. Of note, PC interventions for PwP of an ethnic minority background need to be tailored in a personalized manner to tackle the distinctive cultural, religious, and spiritual needs of the communities involved (Chaudhuri et al., 2021). Enhancing the cross-cultural communication skills and training ethnically diverse health professionals in PC is urgently needed to meet this need (Kataoka-Yahiro, McFarlane, Koijane, & Li, 2017).

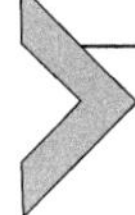

4. An increased need for palliative care pathways for people with Parkinson's disease during the Covid-19 pandemic

Under the traditional models of care for PwP, many of their needs remain unmet, and, in some countries, the benefits of introducing PC interventions early in the PD course have increasingly been acknowledged over the past decade (Boersma et al., 2016). Since the emergence of the Covid-19 pandemic, the emphasis on the PC needs of the individuals living with PD has gradually risen into the spotlight (Chaudhuri et al., 2021).

As the ongoing Covid-19 pandemic continues to intensify the suffering of individuals with PD, their families and caregivers worldwide (attributable to the SARS-CoV-2 infection directly, as well as to the collateral consequences of the pandemic), health-care providers are withstanding unprecedented challenges in a socially distant environment, as they aim to improve their patients' and caregivers' quality of life and achieve relief of physical, emotional, and spiritual suffering in alignment with the main PC principles (Miyasaki & Robinson, 2020; Subramanian & Vaughan, 2020).

In the following paragraphs, we aim to provide a foundation for pragmatic and clinically orientated, personalized PC approach strategies relevant to the setting of the ongoing pandemic globally.

5. Specific palliative care needs in people with Parkinson's disease during the Covid-19 pandemic

5.1 Management of motor symptoms

In general, infections might exacerbate the severity of motor symptoms of PD (through a variety of mechanisms, including systemic inflammation and altered pharmacodynamics of anti-parkinsonian drugs), often prompting an adjustment in dopaminergic medication (Fearon & Fasano, 2021). Covid-19 is not an exception. A substantial worsening of motor performance, greater motor disability and higher amount of levodopa-induced motor fluctuations with an increased daily off-time were reported in PwP affected by Covid-19; both those with mild-to-moderate illness managed in an outpatient setting and those who required a hospital admission (Cilia et al., 2020; de Marcaida et al., 2020). In some PwP, worsening of motor performance might be the presenting complaint of the SARS-CoV-2 infection (de Marcaida et al., 2020).

Thus, in PwP affected by Covid-19, management strategies should aim to prevent motor worsening and relieve an aggravated burden of motor symptoms (Chaudhuri et al., 2021). Keeping the strict timing of medications and continuation of levodopa for as long as possible is essential (Subramanian & Vaughan, 2020). Worryingly, in the context of emergency admissions of PwP to the general medical or elderly care wards, omitted or delayed administration of medication is common, potentially leading to immobility, complications and longer lengths of stay. In addition, PwP admitted to a non-neurological ward are more likely to receive contraindicated, anti-dopaminergic medication. Early neurologist consultation should be sought in such situations, and self-medication encouraged, whenever possible (Skelly et al., 2014). Adjustments in the standard regimen of dopaminergic medication, precisely tailored to match patients' needs, might be considered, where appropriate (Chaudhuri et al., 2021).

Of note, even in PwP not infected by SARS-CoV-2, the repercussions of prolonged mobility restrictions and limited health-care access might have implications on the mobility, such as balance worsening and greater risk of falls. Patients' education and raising awareness on the importance of physical activity by encouraging exercise programs delivered remotely is of paramount importance in order to maintain balance and preserve mobility (Luis-Martinez et al., 2021). Of note, while telehealth has proven to be a very helpful tool on such occasions, several barriers have been reported among PwP, potentially delaying delivery of care. It is thus essential to ensure that telehealth interventions are widely available to all patients from different clinical and sociodemographic backgrounds (van den Bergh, Bloem, Meinders, & Evers, 2021).

5.2 Management of non-motor symptoms

In PwP, SARS-CoV-2 infection, as well as pandemic-related social restrictions, may negatively interact with a range of NMS, increasing their overall burden and posing a detrimental effect on the quality of life (Chaudhuri et al., 2021).

For example, **fatigue**, an overwhelming sense of extreme and persistent tiredness and exhaustion (mental, physical or both), which interferes with normal function, is a common NMS of PD and an important determinant of quality of life, affecting the daily life activities in over 50% PwP, while its prevalence increases with PD progression (Lazcano-Ocampo et al., 2020). At the same time, fatigue is both among most commonly reported symptoms of an acute Covid-19 illness and an integral feature of the Post-Covid-19 syndrome, lingering for over 6 months in the aftermath of the Covid-19 infection in some patients (Davis et al., 2021; Leta et al., 2021).

Disruptions in normal life introduced by the Covid-19 pandemic generated considerable psychological stress in community-dwelling PwP, a population with limited abilities of flexible adaptation to rapid and drastic changes in daily routines (Dommershuijsen et al., 2021). Levels of **anxiety** (one of the most common neuropsychiatric features of PD) in PwP and their caregivers during the Covid-19 pandemic were significantly higher than in the general population; this appears to be associated with uncertainties regarding obtaining their medications and a perceived higher risk of contracting SARS-CoV-2 because of their underlying chronic medical condition (Salari et al., 2020). Similar observations have been made for **depression** (Shalash et al., 2020).

Heightened levels of anxiety and depression, together with other causes (including, but not limited to exacerbation of motor symptoms and disruptions of the routine health-care), might have contributed to an increased prevalence of **sleep disturbances** in PwP (significantly higher than in the general population) during the Covid-19 pandemic (Xia et al., 2020).

Social distancing and isolation have given rise to more **cognitive complaints** in a substantial proportion of PwP, or even generated new-onset cognitive disturbances in some of them (40% and 26%, respectively, according to the caregivers' reports). PwP with mild cognitive impairment might be particularly vulnerable (Baschi et al., 2020). Thus, "wellness strategies" to counteract a range of mental health issues in PD, ranging from social prescription, re-alignment of health care teams in the communities to teachable lifestyle choices and education has been proposed by Subramanian et al. (2021).

Along with the respiratory system, SARS-CoV-2 may affect the nervous system and skeletal muscles, causing neuropathic or mylagic **pain**, through direct or indirect mechanisms (Wang et al., 2020). In addition, it is possible that lack of exercise and restricted access to physical therapy might aggravate pain attributable to muscle stiffness, tremor or worsened dystonia (e.g., torticollis) in some PwP (Rukavina et al., 2019; Shalash et al., 2020). Indeed, in a small, single-centre study, 64% of PwP (across all Hoehn and Yahr stages) disclosed worsening of pain amid the ongoing Covid-19 emergency during their telehealth consultations. In 32% of patients, pain, mainly musculoskeletal, emerged during the pandemic for the first time (Chaudhuri et al., 2021).

Those NMS may dominate the clinical presentation in PwP during the Covid-19 pandemic and their accurate recognition constitutes the key step in order to select the most appropriate management strategy (Chaudhuri et al., 2021).

Among numerous accessible assessment tools (scales and questionnaires) that may capture and quantify distinctive NMS, the 30-item NMS Questionnaire (NMSQuest) is a simple and time-efficient, patient-completed outcome measure, which may be applied remotely to flag up the presence of certain NMS, and allows for a staging of the overall NMS burden (Chaudhuri et al., 2006; Rukavina et al., 2021).

A detailed discussion of the optimal management strategies to tackle distinctive NMS would be out of the scope of this book chapter - the Movement Disorder Society (MDS) Evidence Based Medicine Task Force recommendation and the American Academy of Neurology Guideline on Quality Measures for PD can be used as a guide to improve clinical care (Chou et al., 2021; Seppi et al., 2019).

5.3 Prevention and management of delirium

In PwP, an acute SARS-CoV-2 infection and hospital admission may trigger delirium - an acute onset, fluctuating neurocognitive disorder characterized by an altered level of consciousness, disorientation, along with disturbance in attention, awareness and other cognitive features (Ebersbach et al., 2019; Kennedy et al., 2020). In some PwP, delirium might be the primary or sole symptom of an infection (Kennedy et al., 2020). Delirium is significantly associated with an increased risk of poor hospital outcomes, including admission to an intensive care unit, discharge to a rehabilitation facility, an increased frequency of falls and death, generating high levels of distress for patients, their families and healthcare practitioners. It can thus serve as an important marker, flagging up vulnerable patients who might particularly benefit from PC interventions (Boland et al., 2019; Kennedy et al., 2020).

Systematic assessment and early recognition of delirium are essential, but, in a significant proportion of PwP, it may go undetected (particularly hypoactive delirium) due to phenomenological overlap between delirium and chronic neuropsychiatric features of PD (e.g., apathy, hallucinations, delusion, or dementia) or side effects of dopaminergic medication (Boland et al., 2019; Ebersbach et al., 2019).

The management of delirium is multifaceted and delivered through a multidisciplinary team. Following the assessment, any underlying causes (e.g., metabolic disturbances) should be identified and treated where appropriate and non-pharmacological interventions (including, but not limited to, optimal sound and/or lighting levels with bright light during day-time, dimming at night-time, an avoidance of background TV or other noise, gentle re-orientation, provision of vision and hearing aids, maintenance of hydration and nutrition, support of sleep–wake cycle, and early mobilization) should be applied. The presence of a family member should be encouraged; in the context of Covid-19 regular telephone calls should be offered and photographs and other items from the patient's home can be used to create a comforting and familiar environment (Boland et al., 2019; Ebersbach et al., 2019) (Fig. 2).

In terms of pharmacological treatment, continuing the exact personalized medication regimen is crucial (Gerlach, Winogrodzka, & Weber, 2011). In addition, medication should be carefully reviewed and, where appropriate, an anticholinergic load should be reduced, perhaps using the Anticholinergic cognitive burden scale as a guidance (Rukavina et al., 2021). The agents with the least anti-parkinsonian efficacy, which might

Non-pharmacological strategies:

Contact with loved ones
Optimal sound and/or lighting levels
(bright light during day-time, dimming at night-time)
Maintenance of sleep-wake cycle
Avoidance of background noise
Gentle reorientation
Provision of vision and hearing aids
Maintenance of hydration and nutrition
Early mobilization

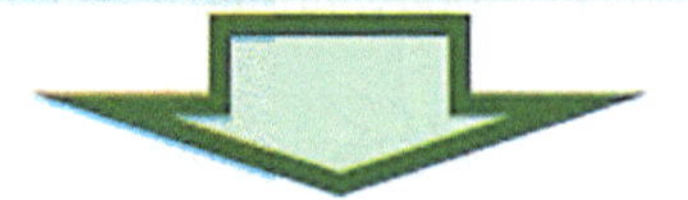

Pharmacological interventions:

Correction of underlaying causes
Countinuing the exact personal medication regime
Review of medication
Discontinuation of potentially contributing agents
(1. Anticholinergics, 2. Amantadine, 3. Selegiline, 4. DA and other MAOB-i, COMT-i)
Avoidance of anti-dopaminergic neuroleptics

Fig. 2 Proposed treatment algorithm for prevention and management of delirium in People with Parkinson's disease. DA—dopamine agonists, MAOB-I—monoamine oxidase B inhibitors, COMT-I—catechol-o-methyltransferase inhibitors.

potentially contribute to delirium should be discontinued (Proposed order: 1. Anticholinergics, 2. Amantadine, 3. Selegiline, 4. Dopamine agonists and other monoaminoxidase B inhibitors, catechol-o-methyltransferase inhibitors) (Ebersbach et al., 2019; Vardy, Teodorczuk, & Yarnall, 2015). Importantly, caution is required, as an abrupt discontinuation of amantadine, anticholinergics, or dopamine agonists may lead to withdrawal syndromes. Moreover, rapid cessation of dopaminergic medication can precipitate an akinetic crisis, mimicking a neuroleptic malignant-like syndrome or parkinsonism hyperpyrexia syndrome, those being life-threatening conditions prompting an urgent restoration of dopamine balance and supportive measures (e.g., hydration, antipyretics, non-oral use of DRT) (Chaudhuri et al., 2021). Of note, treatment algorithms commonly used in general or geriatric health care settings include anti-dopaminergic neuroleptics which should be avoided in PwP (Ebersbach et al., 2019) (Fig. 2).

5.4 Psychological and spiritual concerns, anticipatory grief

The ongoing Covid-19 pandemic continues to pose unprecedented barriers that force healthcare teams to deprioritize psychological and spiritual aspects of patient care. Across the globe, critically ill patients affected by Covid-19 are dying in isolation, without comfort by their loved ones and with both parties losing their rights to properly honor their end-of-life rites and rituals (Galbadage, Peterson, Wang, Wang, & Gunasekera, 2020).

While access to PC remains limited, an implementation of a palliative telehealth system providing spiritual (spiritual dialogue, joint prayer and virtual religious rites) and psychological support (active listening, psychosocial risk-factor assessment and emotional support), alongside compassionate family communication, may be a feasible and useful resource with a potential to protect patients' dignity, improve the "quality of death," and promote their relatives' mental health (e.g., reducing the incidence of mood disorders, post-traumatic stress disorders or complicated grief) (Palma et al., 2021).

5.5 Outcome measures in palliative care for PwP

PwP might experience repercussions of their condition going well beyond the symptoms clinicians and researchers typically consider (including, but not limited to a sense of loss, changing roles, relationship changes, altered concept of self, social embarrassment, grief, and uncertainties attributable to cognitive decline, the loss of independence, and deterioration of personal appearance) (Boersma et al., 2016). Many of these aspects might have been aggravated amid the ongoing pandemic (Chaudhuri et al., 2021). Traditionally used, disease-centered outcome measures, might not adequately capture some of those areas, prompting the use of patient-centered PC outcome measures (Chaudhuri et al., 2021).

The Palliative Outcome Scale (POS), although originally developed for individuals with advanced cancer, is now widely used for general PC purposes. Brief (with a completion time under 10 min) and easy to administer, POS consists of two questionnaires: patient-completed (10 items) and staff-completed (12 items), addressing the physical, psychological, and spiritual domains of life of both patients and their families, communication and information. It allows patients to directly declare their main concerns and list their symptoms and unmet needs (Hearn & Higginson, 1999).

Based on the POS, the IPOS Neuro-S8, an 8-item patient-completed PC outcome scale of symptom burden (encompassing both the severity of the symptoms and the patient's perception of their impact) has been developed specifically for individuals with the most common forms of progressive long-term neurological conditions. In PwP, IPOS Neuro-8 has shown moderate correlation with disease-specific assessment tools, like NMSQuest and Parkinson's Disease Questionnaire 8 (PDQ-8), but has not yet been through formal psychometric evaluation (Chaudhuri et al., 2006) (Jenkinson, Fitzpatrick, Peto, Greenhall, & Hyman, 1997) (Gao et al., 2016). This scale might be a helpful screening measure in non-palliative care settings, aiding in the identification of patients with PC needs and triggering an appropriate referral process (Gao et al., 2016).

5.6 Advance care planning

The Covid-19 pandemic has raised the awareness of the importance of discussion and documentation of care preferences in order to ensure that appropriate care, in alignment with patients' values, is delivered (Subramanian & Vaughan, 2020). PwP who receive integrated PC have higher rates and quality of advance directives completion (Kluger et al., 2020). ACP is an integral feature of PC; an ongoing communication process of discussing and refining the needs, wishes, and patients' preferences regarding disease-specific and end-of-life issues. It includes considerations about disease- and symptom-specific treatment, resuscitation and other life-prolonging modalities, treatment restrictions, end-of-life wishes and appointment of surrogate decision-makers (Walter, Seeber, Willems, & de Visser, 2018). ACP is most effective when initiated early, alongside curative therapies, allowing patients, their next-of-kin and caregivers to proactively address the challenges together, and is associated with better quality of life, reduced rates of stress, anxiety and depression and less hospital admissions (Kurpershoek et al., 2021; Walter et al., 2018).

However, although PwP prefer their health-care professionals to raise the subject of ACP, their desire to discuss advance care preferences ranges from early following the diagnosis to later stages of PD, and it might be challenging for clinicians to ascertain the level of readiness for such discussions

(Kurpershoek et al., 2021; Miyasaki & Robinson, 2020). For example, neurologists mostly initiate the ACP at the terminal stages of PD (Walter et al., 2018). Some clinicians, thus, may find it easier to discuss these topics framed within the pandemic (Subramanian & Vaughan, 2020). Courses and workshop delivered by experienced PC staff may help to equip clinicians with the knowledge, confidence and skills required to manage such difficult conversations (Brighton et al., 2018).

5.7 End-of-life support

Overall, mortality rate of Covid-19 in elderly and frail PwP at late stages of PD appears to be high, underscoring the need to prepare for events that could arise in PwP affected by Covid-19, and calling for end-of-life support strategies based on PC approaches to be readily available (Chaudhuri et al., 2021) (Miyasaki & Robinson, 2020). Of note, in many countries, the need to prioritize access to limited resources to those with greater chances of survival and anticipated shorter recovery led to the high rate of not-for-resuscitation and ceiling of care decisions in elderly, frail and multi-morbid patients, often including PwP (Straw et al., 2021). However, the decision on ceiling of care should not be based solely on the diagnosis of PD; instead, a personalized PC package, tailored in accordance with an individual's ACP, should be offered (Chaudhuri et al., 2021; Katz et al., 2018).

5.8 Caregiver support

Family caregivers provide substantial care for PwP, particularly those at later stages of the disease, often at great personal cost (Koljack et al., 2021; Schmotz, Richinger, & Lorenzl, 2017). Caregivers provide substantial support in the management of PD, directly and indirectly, and PwP with an identified caregiver are more likely to report a higher quality of life (Prizer et al., 2020).

A worsening of PD-related symptoms during the Covid-19 pandemic, as well as the pandemic-related restrictions and social isolation that increased the reliance of PwP on family members and other informal caregivers, have put a tremendous strain on PD caregivers (Suzuki et al., 2021). On the other hand, caregivers are the most important link for PwP; their potential infection by Covid-19 and the subsequent inability to fulfill their caregiving roles presents a crisis for PwP, which may even lead to a hospital admission

(Miyasaki & Robinson, 2020). PwP and their care partners in ambulatory clinics should thus be encouraged to prepare a contingency plan for such situations.

Increased caregiver burden (physical and emotional exhaustion resulting from a perceived obligation to provide care for their loved one with PD) may lead to increased caregiver morbidity, which in turn impacts the quality of life of PwP (Hudson & Aranda, 2014). Caregivers are affected by their role physically, emotionally, and financially; yet their psychosocial needs are frequently considered either secondary to those of the patient or overlooked (Hudson & Aranda, 2014). Evaluating caregiver burden in a comprehensive manner is necessary and PC approaches hold the potential to improve caregivers' quality of life through adequate preparation, information and support (Boersma et al., 2017; Hudson & Aranda, 2014; Macchi et al., 2020).

6. Conclusions

Under the traditional models of care for PwP many of their needs remain unmet. PwP experience a substantial burden of motor and non-motor symptoms, both major determinants of the health-related QoL. The ongoing Covid-19 pandemic has in many ways (both directly and indirectly) magnified their suffering, while, at the same time, it has forced healthcare teams to deprioritize psychological and spiritual aspects of patient care. In these times of unprecedented challenges, early implementation of PC interventions holds a potential to improve quality of life of patients, their families and caregivers through the prevention and relief of medical symptoms, while emphasizing the emotional needs and spiritual wellbeing, establishing goals of care and engaging in the ACP.

References

Akbar, U., McQueen, R. B., Bemski, J., Carter, J., Goy, E. R., Kutner, J., et al. (2021). Prognostic predictors relevant to end-of-life palliative care in Parkinson's disease and related disorders: A systematic review. *Journal of Neurology, Neurosurgery, and Psychiatry*. https://doi.org/10.1136/jnnp-2020-323939.

Antonini, A., Leta, V., Teo, J., & Chaudhuri, K. R. (2020). Outcome of Parkinson's disease patients affected by COVID-19. *Movement Disorders*, *35*(6), 905–908. https://doi.org/10.1002/mds.28104.

Artusi, C. A., Romagnolo, A., Imbalzano, G., Marchet, A., Zibetti, M., Rizzone, M. G., et al. (2020). COVID-19 in Parkinson's disease: Report on prevalence and outcome. *Parkinsonism & Related Disorders*, *80*, 7–9. https://doi.org/10.1016/j.parkreldis.2020.09.008.

Baschi, R., Luca, A., Nicoletti, A., Caccamo, M., Cicero, C. E., D'Agate, C., et al. (2020). Changes in motor, cognitive, and behavioral symptoms in Parkinson's disease and mild cognitive impairment during the COVID-19 lockdown. *Frontiers in Psychiatry*, *11*, 590134. https://doi.org/10.3389/fpsyt.2020.590134.

Bhidayasiri, R., Virameteekul, S., Kim, J. M., Pal, P. K., & Chung, S. J. (2020). COVID-19: An early review of its global impact and considerations for Parkinson's disease patient care. *Journal of Movement Disorders*, *13*(2), 105–114. https://doi.org/10.14802/jmd.20042.

Boersma, I., Jones, J., Carter, J., Bekelman, D., Miyasaki, J., Kutner, J., et al. (2016). Parkinson disease patients' perspectives on palliative care needs: What are they telling us? *Neurology Clinical Practice*, *6*(3), 209–219. https://doi.org/10.1212/CPJ.0000000000000233.

Boersma, I., Jones, J., Coughlan, C., Carter, J., Bekelman, D., Miyasaki, J., et al. (2017). Palliative care and Parkinson's disease: Caregiver perspectives. *Journal of Palliative Medicine*, *20*(9), 930–938. https://doi.org/10.1089/jpm.2016.0325.

Boland, J. W., Kabir, M., Bush, S. H., Spiller, J. A., Johnson, M. J., Agar, M., et al. (2019). Delirium management by palliative medicine specialists: A survey from the association for palliative medicine of Great Britain and Ireland. *BMJ Supportive & Palliative Care*. https://doi.org/10.1136/bmjspcare-2018-001586.

Brighton, L. J., Selman, L. E., Gough, N., Nadicksbernd, J. J., Bristowe, K., Millington-Sanders, C., et al. (2018). 'Difficult conversations': Evaluation of multi-professional training. *BMJ Supportive & Palliative Care*, *8*(1), 45–48. https://doi.org/10.1136/bmjspcare-2017-001447.

Brown, E. G., Chahine, L. M., Goldman, S. M., Korell, M., Mann, E., Kinel, D. R., et al. (2020). The effect of the COVID-19 pandemic on people with Parkinson's disease. *Journal of Parkinson's Disease*, *10*(4), 1365–1377. https://doi.org/10.3233/JPD-202249.

Centers for Disease Control and Prevention. (2019). Retrieved from https://www.cdc.gov/coronavirus/2019-ncov/Covid-data/investigations-discovery/hospitalization-death-by-race-ethnicity.html.

Chaudhuri, K. R., Martinez-Martin, P., Schapira, A. H., Stocchi, F., Sethi, K., Odin, P., et al. (2006). International multicenter pilot study of the first comprehensive self-completed nonmotor symptoms questionnaire for Parkinson's disease: The NMSQuest study. *Movement Disorders*, *21*(7), 916–923. https://doi.org/10.1002/mds.20844.

Chaudhuri, K. R., Rukavina, K., McConvey, V., Antonini, A., Lorenzl, S., Bhidayasiri, R., et al. (2021). The impact of COVID-19 on palliative care for people with Parkinson's and response to future pandemics. *Expert Review of Neurotherapeutics*, *21*(6), 615–623. https://doi.org/10.1080/14737175.2021.1923480.

Chou, K. L., Martello, J., Atem, J., Elrod, M., Foster, E. R., Freshwater, K., et al. (2021). Quality improvement in neurology: 2020 Parkinson disease quality measurement set update. *Neurology*, *97*(5), 239–245. https://doi.org/10.1212/WNL.0000000000012198.

Cilia, R., Bonvegna, S., Straccia, G., Andreasi, N. G., Elia, A. E., Romito, L. M., et al. (2020). Effects of COVID-19 on Parkinson's disease clinical features: A community-based case-control study. *Movement Disorders*, *35*(8), 1287–1292. https://doi.org/10.1002/mds.28170.

Davis, H. E., Assaf, G. S., McCorkell, L., Wei, H., Low, R. J., Re'em, Y., et al. (2021). Characterizing long COVID in an international cohort: 7 months of symptoms and their impact. *EClinicalMedicine*, *38*, 101019. https://doi.org/10.1016/j.eclinm.2021.101019.

de Marcaida, J. A., Lahrmann, J., Machado, D., Bluth, L., Dagostine, M., Moro-de Casillas, M., et al. (2020). Clinical characteristics of coronavirus disease 2019 (COVID-19) among patients at a Movement Disorders Center. *Geriatrics (Basel)*, *5*(3). https://doi.org/10.3390/geriatrics5030054.

Del Prete, E., Francesconi, A., Palermo, G., Mazzucchi, S., Frosini, D., Morganti, R., et al. (2020). Prevalence and impact of COVID-19 in Parkinson's disease: Evidence from a multi-center survey in Tuscany region. *Journal of Neurology*. https://doi.org/10.1007/s00415-020-10002-6.

Dhamija, R. K., Saluja, A., & Miyasaki, J. (2021). Advancing neuropalliative care. *Lancet Neurology*, *20*(11), 885–886. https://doi.org/10.1016/S1474-4422(21)00333-1.

Di Luca, D. G., Feldman, M., Jimsheleishvili, S., Margolesky, J., Cordeiro, J. G., Diaz, A., et al. (2020). Trends of inpatient palliative care use among hospitalized patients with Parkinson's disease. *Parkinsonism & Related Disorders*, 77, 13–17. https://doi.org/10.1016/j.parkreldis.2020.06.011.

Dommershuijsen, L. J., Van der Heide, A., Van den Berg, E. M., Labrecque, J. A., Ikram, M. K., Ikram, M. A., et al. (2021). Mental health in people with Parkinson's disease during the COVID-19 pandemic: Potential for targeted interventions? *NPJ Parkinsons Disorder*, 7(1), 95. https://doi.org/10.1038/s41531-021-00238-y.

Ebersbach, G., Ip, C. W., Klebe, S., Koschel, J., Lorenzl, S., Schrader, C., et al. (2019). Management of delirium in Parkinson's disease. *Journal of Neural Transmission (Vienna)*, *126*(7), 905–912. https://doi.org/10.1007/s00702-019-01980-7.

Fasano, A., Cereda, E., Barichella, M., Cassani, E., Ferri, V., Zecchinelli, A. L., et al. (2020). COVID-19 in Parkinson's disease patients living in Lombardy, Italy. *Movement Disorders*, *35*(7), 1089–1093. https://doi.org/10.1002/mds.28176.

Fasano, A., Elia, A. E., Dallocchio, C., Canesi, M., Alimonti, D., Sorbera, C., et al. (2020). Predictors of COVID-19 outcome in Parkinson's disease. *Parkinsonism & Related Disorders*, *78*, 134–137. https://doi.org/10.1016/j.parkreldis.2020.08.012.

Fearon, C., & Fasano, A. (2021). Parkinson's disease and the COVID-19 pandemic. *Journal of Parkinson's Disease*, *11*(2), 431–444. https://doi.org/10.3233/JPD-202320.

Galbadage, T., Peterson, B. M., Wang, D. C., Wang, J. S., & Gunasekera, R. S. (2020). Biopsychosocial and spiritual implications of patients with COVID-19 dying in isolation. *Frontiers in Psychology*, *11*, 588623. https://doi.org/10.3389/fpsyg.2020.588623.

Gao, W., Crosby, V., Wilcock, A., Burman, R., Silber, E., Hepgul, N., et al. (2016). Psychometric properties of a generic, patient-centred palliative care outcome measure of symptom burden for people with progressive long term neurological conditions. *PLoS One*, *11*(10), e0165379. https://doi.org/10.1371/journal.pone.0165379.

Gerlach, O. H., Winogrodzka, A., & Weber, W. E. (2011). Clinical problems in the hospitalized Parkinson's disease patient: Systematic review. *Movement Disorders*, *26*(2), 197–208. https://doi.org/10.1002/mds.23449.

Hainque, E., & Grabli, D. (2020). Rapid worsening in Parkinson's disease may hide COVID-19 infection. *Parkinsonism & Related Disorders*, *75*, 126–127. https://doi.org/10.1016/j.parkreldis.2020.05.008.

Hearn, J., & Higginson, I. J. (1999). Development and validation of a core outcome measure for palliative care: The palliative care outcome scale. Palliative care Core audit project advisory group. *Quality in Health Care*, *8*(4), 219–227. https://doi.org/10.1136/qshc.8.4.219.

Hudson, P., & Aranda, S. (2014). The Melbourne family support program: Evidence-based strategies that prepare family caregivers for supporting palliative care patients. *BMJ Supportive & Palliative Care*, *4*(3), 231–237. https://doi.org/10.1136/bmjspcare-2013-000500.

Jenkinson, C., Fitzpatrick, R., Peto, V., Greenhall, R., & Hyman, N. (1997). The Parkinson's disease questionnaire (PDQ-39): Development and validation of a Parkinson's disease summary index score. *Age and Ageing*, *26*(5), 353–357. https://doi.org/10.1093/ageing/26.5.353.

Kataoka-Yahiro, M. R., McFarlane, S., Koijane, J., & Li, D. (2017). Culturally competent palliative and hospice care training for ethnically diverse staff in long-term care facilities. *The American Journal of Hospice & Palliative Care*, *34*(4), 335–346. https://doi.org/10.1177/1049909116638347.

Katz, M., Goto, Y., Kluger, B. M., Galifianakis, N. B., Miyasaki, J. M., Kutner, J. S., et al. (2018). Top ten tips palliative care clinicians should know about Parkinson's disease and related disorders. *Journal of Palliative Medicine*, *21*(10), 1507–1517. https://doi.org/10.1089/jpm.2018.0390.

Kennedy, M., Helfand, B. K. I., Gou, R. Y., Gartaganis, S. L., Webb, M., Moccia, J. M., et al. (2020). Delirium in older patients with COVID-19 presenting to the emergency department. *JAMA Network Open*, *3*(11), e2029540. https://doi.org/10.1001/jamanetworkopen.2020.29540.

Kluger, B. M., Miyasaki, J., Katz, M., Galifianakis, N., Hall, K., Pantilat, S., et al. (2020). Comparison of integrated outpatient palliative care with standard care in patients with Parkinson disease and related disorders: A randomized clinical trial. *JAMA Neurology*, 77(5), 551–560. https://doi.org/10.1001/jamaneurol.2019.4992.

Kluger, B. M., & Quill, T. E. (2021). Advancing neuropalliative care. *Lancet Neurology*, *20*(11), 885. https://doi.org/10.1016/S1474-4422(21)00326-4.

Koljack, C. E., Miyasaki, J., Prizer, L. P., Katz, M., Galifianakis, N., Sillau, S. H., et al. (2021). Predictors of spiritual well-being in family caregivers for individuals with Parkinson's disease. *Journal of Palliative Medicine*. https://doi.org/10.1089/jpm.2020.0797.

Kurpershoek, E., Hillen, M. A., Medendorp, N. M., de Bie, R. M. A., de Visser, M., & Dijk, J. M. (2021). Advanced care planning in Parkinson's disease: In-depth interviews with patients on experiences and needs. *Frontiers in Neurology*, *12*, 683094. https://doi.org/10.3389/fneur.2021.683094.

Lazcano-Ocampo, C., Wan, Y. M., van Wamelen, D. J., Batzu, L., Boura, I., Titova, N., et al. (2020). Identifying and responding to fatigue and apathy in Parkinson's disease: A review of current practice. *Expert Review of Neurotherapeutics*, *20*(5), 477–495. https://doi.org/10.1080/14737175.2020.1752669.

Lee, C. (2009). "race" and "ethnicity" in biomedical research: How do scientists construct and explain differences in health? *Social Science & Medicine*, *68*(6), 1183–1190. https://doi.org/10.1016/j.socscimed.2008.12.036.

Leta, V., Rodriguez-Violante, M., Abundes, A., Rukavina, K., Teo, J. T., Falup-Pecurariu, C., et al. (2021). Parkinson's disease and post-COVID-19 syndrome: The Parkinson's long-COVID Spectrum. *Movement Disorders*, *36*(6), 1287–1289. https://doi.org/10.1002/mds.28622.

Luis-Martinez, R., Di Marco, R., Weis, L., Cianci, V., Pistonesi, F., Baba, A., et al. (2021). Impact of social and mobility restrictions in Parkinson's disease during COVID-19 lockdown. *BMC Neurology*, *21*(1), 332. https://doi.org/10.1186/s12883-021-02364-9.

Macchi, Z. A., Koljack, C. E., Miyasaki, J. M., Katz, M., Galifianakis, N., Prizer, L. P., et al. (2020). Patient and caregiver characteristics associated with caregiver burden in Parkinson's disease: A palliative care approach. *Annals of Palliative Medicine*, *9*(Suppl 1), S24–S33. https://doi.org/10.21037/apm.2019.10.01.

McLean, G., Hindle, J. V., Guthrie, B., & Mercer, S. W. (2017). Co-morbidity and polypharmacy in Parkinson's disease: Insights from a large Scottish primary care database. *BMC Neurology*, *17*(1), 126. https://doi.org/10.1186/s12883-017-0904-4.

Miyasaki, J. M., Lim, S. Y., Chaudhuri, K. R., Antonini, A., Piemonte, M., Richfield, E., et al. (2021). Access and attitudes toward palliative care among movement disorders clinicians. *Movement Disorders*. https://doi.org/10.1002/mds.28773.

Miyasaki, J. M., & Robinson, M. T. (2020). Editorial: Neuropalliative care for movement disorders in the time of COVID-19. *Parkinsonism & Related Disorders*, *80*, 201–202. https://doi.org/10.1016/j.parkreldis.2020.10.045.

Oliver, D., Borasio, G. D., Veronese, S., Voltz, R., Lorenzl, S., & Hepgul, N. (2020). Current collaboration between palliative care and neurology: A survey of clinicians in Europe. *BMJ Supportive & Palliative Care*. https://doi.org/10.1136/bmjspcare-2020-002322.

Otu, A., Ahinkorah, B. O., Ameyaw, E. K., Seidu, A. A., & Yaya, S. (2020). One country, two crises: What Covid-19 reveals about health inequalities among BAME communities in the United Kingdom and the sustainability of its health system? *International Journal for Equity in Health*, *19*(1), 189. https://doi.org/10.1186/s12939-020-01307-z.

Palma, A., Rojas, V., Ihl, F., Avila, C., Plaza-Parrochia, F., Estuardo, N., et al. (2021). Implementation of a palliative hospital-centered spiritual and psychological telehealth system during COVID-19 pandemic. *Journal of Pain and Symptom Management*, *62*(5), 1015–1019. https://doi.org/10.1016/j.jpainsymman.2021.04.016.

Parihar, R., Ferastraoaru, V., Galanopoulou, A. S., Geyer, H. L., & Kaufman, D. M. (2021). Outcome of hospitalized Parkinson's disease patients with and without COVID-19. *Movement Disorders Clinical Practice*. https://doi.org/10.1002/mdc3.13231.

Prizer, L. P., Kluger, B. M., Sillau, S., Katz, M., Galifianakis, N. B., & Miyasaki, J. M. (2020). The presence of a caregiver is associated with patient outcomes in patients with Parkinson's disease and atypical parkinsonisms. *Parkinsonism & Related Disorders*, *78*, 61–65. https://doi.org/10.1016/j.parkreldis.2020.07.003.

Quill, T. E., & Abernethy, A. P. (2013). Generalist plus specialist palliative care—Creating a more sustainable model. *The New England Journal of Medicine*, *368*(13), 1173–1175. https://doi.org/10.1056/NEJMp1215620.

Rukavina, K., Batzu, L., Boogers, A., Abundes-Corona, A., Bruno, V., & Chaudhuri, K. R. (2021). Non-motor complications in late stage Parkinson's disease: Recognition, management and unmet needs. *Expert Review of Neurotherapeutics*, *21*(3), 335–352. https://doi.org/10.1080/14737175.2021.1883428.

Rukavina, K., Leta, V., Sportelli, C., Buhidma, Y., Duty, S., Malcangio, M., et al. (2019). Pain in Parkinson's disease: New concepts in pathogenesis and treatment. *Current Opinion in Neurology*, *32*(4), 579–588. https://doi.org/10.1097/WCO.0000000000000711.

Rukavina, K., Ocloo, J., Skoric, M. K., Sauerbier, A., Thomas, O., Staunton, J., et al. (2022). Ethnic disparities in treatment of chronic pain in individuals with Parkinson's disease living in the United Kingdom. *Movement Disorders Clinical Practice*, *9*(3), 369–374. https://doi.org/10.1002/mdc3.13430. PMID: 35392300; PMCID: PMC8974878.

Sainz-Amo, R., Baena-Alvarez, B., Parees, I., Sanchez-Diez, G., Perez-Torre, P., Lopez-Sendon, J. L., et al. (2020). COVID-19 in Parkinson's disease: What holds the key? *Journal of Neurology*. https://doi.org/10.1007/s00415-020-10272-0.

Salari, M., Zali, A., Ashrafi, F., Etemadifar, M., Sharma, S., Hajizadeh, N., et al. (2020). Incidence of anxiety in Parkinson's disease during the coronavirus disease (COVID-19) pandemic. *Movement Disorders*, *35*(7), 1095–1096. https://doi.org/10.1002/mds.28116.

Sauerbier, A., Schrag, A., Brown, R., Martinez-Martin, P., Aarsland, D., Mulholland, N., et al. (2021). Clinical non-motor phenotyping of black and Asian minority ethnic compared to white individuals with Parkinson's disease living in the United Kingdom. *Journal of Parkinson's Disease*, *11*(1), 299–307. https://doi.org/10.3233/JPD-202218.

Schmotz, C., Richinger, C., & Lorenzl, S. (2017). High burden and depression among late-stage idiopathic Parkinson disease and progressive Supranuclear palsy caregivers. *Journal of Geriatric Psychiatry and Neurology*, *30*(5), 267–272. https://doi.org/10.1177/0891988717720300.

Seppi, K., Ray Chaudhuri, K., Coelho, M., Fox, S. H., Katzenschlager, R., Perez Lloret, S., et al. (2019). Update on treatments for nonmotor symptoms of Parkinson's disease-an evidence-based medicine review. *Movement Disorders*, *34*(2), 180–198. https://doi.org/10.1002/mds.27602.

Shalash, A., Roushdy, T., Essam, M., Fathy, M., Dawood, N. L., Abushady, E. M., et al. (2020). Mental health, physical activity, and quality of life in Parkinson's disease during COVID-19 pandemic. *Movement Disorders*, *35*(7), 1097–1099. https://doi.org/10.1002/mds.28134.

Skelly, R., Brown, L., Fakis, A., Kimber, L., Downes, C., Lindop, F., et al. (2014). Does a specialist unit improve outcomes for hospitalized patients with Parkinson's disease? *Parkinsonism & Related Disorders*, *20*(11), 1242–1247. https://doi.org/10.1016/j.parkreldis.2014.09.015.

Straw, S., McGinlay, M., Drozd, M., Slater, T. A., Cowley, A., Kamalathasan, S., et al. (2021). Advanced care planning during the COVID-19 pandemic: Ceiling of care decisions and their implications for observational data. *BMC Palliative Care*, *20*(1), 10. https://doi.org/10.1186/s12904-021-00711-8.

Subramanian, I., Hinkle, J. T., Chaudhuri, K. R., Mari, Z., Fernandez, H., & Pontone, G. M. (2021). Mind the gap: Inequalities in mental health care and lack of social support in Parkinson disease. *Parkinsonism & Related Disorders*, *93*, 97–102. https://doi.org/10.1016/j.parkreldis.2021.11.015.

Subramanian, I., & Vaughan, L. C. (2020). Hoping for the best, planning for the worst: Palliative care approach to Parkinson disease during the COVID-19 pandemic. *Parkinsonism & Related Disorders*, *80*, 203–205. https://doi.org/10.1016/j.parkreldis.2020.09.042.

Suzuki, K., Numao, A., Komagamine, T., Haruyama, Y., Kawasaki, A., Funakoshi, K., et al. (2021). Impact of the COVID-19 pandemic on the quality of life of patients with Parkinson's disease and their caregivers: A single-center survey in Tochigi prefecture. *Journal of Parkinson's Disease*, *11*(3), 1047–1056. https://doi.org/10.3233/JPD-212560.

van den Bergh, R., Bloem, B. R., Meinders, M. J., & Evers, L. J. W. (2021). The state of telemedicine for persons with Parkinson's disease. *Current Opinion in Neurology*, *34*(4), 589–597. https://doi.org/10.1097/WCO.0000000000000953.

Vardy, E. R., Teodorczuk, A., & Yarnall, A. J. (2015). Review of delirium in patients with Parkinson's disease. *Journal of Neurology*, *262*(11), 2401–2410. https://doi.org/10.1007/s00415-015-7760-1.

Walter, H. A. W., Seeber, A. A., Willems, D. L., & de Visser, M. (2018). The role of palliative Care in Chronic Progressive Neurological Diseases-a Survey Amongst Neurologists in the Netherlands. *Frontiers in Neurology*, *9*, 1157. https://doi.org/10.3389/fneur.2018.01157.

Wang, D., Hu, B., Hu, C., Zhu, F., Liu, X., Zhang, J., et al. (2020). Clinical characteristics of 138 hospitalized patients with 2019 novel coronavirus-infected pneumonia in Wuhan, China. *JAMA*, *323*(11), 1061–1069. https://doi.org/10.1001/jama.2020.1585.

Xia, Y., Kou, L., Zhang, G., Han, C., Hu, J., Wan, F., et al. (2020). Investigation on sleep and mental health of patients with Parkinson's disease during the coronavirus disease 2019 pandemic. *Sleep Medicine*, *75*, 428–433. https://doi.org/10.1016/j.sleep.2020.09.011.

CHAPTER ELEVEN

COVID-19: The cynosure of rise of Parkinson's disease

Prashanth Lingappa Kukkle*
Parkinson's Disease and Movement Disorders Clinic, Bangalore, India
Center for Parkinson's Disease and Movement Disorders, Manipal Hospital, Miller's Road, Bangalore, India
*Corresponding author: e-mail address: drprashanth.lk@gmail.com

Contents

Abstract

Parkinson's disease (PD) is one of the most common age-related disorders globally. The pathophysiological mechanisms and precipitating factors underlying PD manifestations, including genetic and environmental parameters, inflammation/stress and ageing, remain elusive. Speculations about whether the Coronavirus Disease 2019 (Covid-19) pandemic could be a pivotal factor in affecting the prevalence and severity of PD or triggering a wave of new-onset parkinsonism in both the near and distant future have recently become very popular, with researchers wondering if there is a changing trend in current parkinsonism cases. Could the current understanding of the Covid-19 pathophysiology provide clues for an impending rise of parkinsonism cases in the future? Are there any lessons to learn from previous pandemics? Our aim was to look into these questions and available current literature in order to investigate if Covid-19 could constitute a cardinal event affecting the parkinsonism landscape.

1. Introduction

The 20th century has been one of the greatest timelines in human evolution. The advancements in science have been unparalleled to any of the previous parts of human history. The conquest of various infectious disorders has been leading directly to one of society's greatest achievements: "increase in life expectancy." Indeed, we are at a cusp of science where every

International Review of Neurobiology, Volume 165
ISSN 0074-7742
https://doi.org/10.1016/bs.irn.2022.06.007
Copyright © 2022 Elsevier Inc.
All rights reserved.

effort is made to increase life expectancy and quality of life. More specifically, life expectancy in the first half of the 20th century was less than 50 years, while now it has ascended beyond 70 years of age (The World Bank.IBRD.IDA, 2019). The leading causes of death have been shifting from infectious disorders to non-communicable and chronic conditions (World Health Organization, 2020), with the ever enigma of elixir of life hanging on winning over these age-related degenerative disorders, among which Alzheimer's dementia (AD) and Parkinson's disease (PD) are at the helm. There have been ongoing efforts to understand these disorders and factors affecting them, along with discovering early interventions to move past these hurdles.

Whether the Coronavirus Disease 2019 (Covid-19) could be a major spoke of the wheel in this race over degenerative disorders, can only be a soothsayer's predication. Can we only be retrospectively enlightened, or should we be taught by the past before facing the future? As George Santayana, an 19th century philosopher, quotes "Those who cannot remember the past are condemned to repeat it." There have been ample numbers of examples showing how prior medical outbreaks had an impact on life expectancy and chronic illnesses (Huremović, 2019). We will be covering on the lessons from the past to current available evidence to show what might be in hold for the future of new-onset parkinsonism due to the Covid-19 pandemic.

2. Lessons from the past

The past major epidemic and pandemic outbreaks have significantly contributed to the understanding of various disorders and have promoted scientific progress. Secondary parkinsonism developing in the context of or shortly after viral infections constitutes a well-established condition with a plethora of viruses acknowledged to be involved, including the West Nile virus (WNV), Herpes viruses, the Influenza A virus, the Human immunodeficiency virus (HIV) and others (Limphaibool, Iwanowski, Holstad, Kobylarek, & Kozubski, 2019; Valerio, Whitehouse, Menon, & Newcombe, 2021).

A peak into past pandemics, including outbreaks of the Coronaviruses (CoV) species, does give insight into potential future effects of the Severe Acute Respiratory Syndrome Coronavirus-2 (SARS-CoV-2). The pandemic of 1918 (Spanish Flu), 2003 (SARS-CoV-1) and 2012 (Middle East Respiratory Syndrome, MERS-CoV) have yielded a load of neurological issues, some of which came to the forefronts after months or even years, in some cases up to seven decades, following the exposure to the pathogen (Weir, 2020).

Encephalitis lethargica (EL), also known as von Economo encephalitis, in 1916–17, has been one of the greatest mysteries in medical science, while its role in PD and parkinsonism cases has not been fully understood even nowadays (Reid, McCall, Henry, & Taubenberger, 2001). The acute phase of EL was characterized by intractable somnolence, oculomotor palsies and, occasionally, by hyperkinetic or hypokinetic movement disorders (Hoffman & Vilensky, 2017). Those who survived the acute phase were often left with long term sequelae, including parkinsonian-like signs such as severe bradykinesia, rigidity, tremor, and hypomimia (post-encephalitic parkinsonism). Such phenomena, along with associated lesions in the substantia nigra of these patients, have led to a deeper understanding of PD or parkinsonism pathology. However, this link of parkinsonism and EL was more widely accepted by the late 1930s, almost two decades from the initial outbreak of EL (Lutters, Foley, & Koehler, 2018), giving an insight that many times the acknowledgement of a potential association by a wider peer group might take decades from the occurrence of the relevant events and might remain under question for an extended time period (Estupinan, Nathoo, & Okun, 2013; Vilensky, Gilman, & McCall, 2010).

EL was soon followed by the Spanish flu outbreak (1918–20), which was triggered by the H1N1 influenza strain and affected more than 500 million people worldwide (Taubenberger & Morens, 2006). Prenatal exposure to this strain was associated with an increased risk of more than 20% in cardiovascular diseases in the age group of 60–80 years (Mazumder, Almond, Park, Crimmins, & Finch, 2010). These people would also experience growth retardations and were found to have a lower level of education and economic productivity over their lifetime. Interestingly, such associations imply that affected individuals would continue to suffer from long term implications during their life span due to exposure to an infectious agent in their antenatal lives.

The 2003 SARS-CoV-1 outbreak, which originated in China, affected 8422 people and caused 916 deaths (Institute of Medicine Forum on Microbial, 2004; O'Sullivan, 2021). A 15-year prospective study of 80 healthcare workers who were exposed to SARS-CoV-1, has revealed residual radiological lesions in the subjects' lungs with corresponding compromise of the respiratory function (Zhang, Li, et al., 2020). In a different study, 22 individuals previously affected by SARS-CoV-1 were found to exhibit neurological symptoms in the long term, such as fatigue, myalgia, depression or poor sleep, leading the researchers to suggest a post-SARS-CoV-1 syndrome similar to fibromyalgia, with chronic fatigue and somnolence (Moldofsky & Patcai, 2011). The MERS-CoV, which emerged in Jeddah, Saudi Arabia in 2012 and infected about 2400 people worldwide (Donnelly, Malik, Elkholy, Cauchemez, & Van Kerkhove, 2019), has also left survivors who were diagnosed with chronic fatigue syndrome, depression and post-traumatic stress disorder (PTSD) (Elkholy et al., 2020; O'Sullivan, 2021). Given that some of these long-term effects, including neurological sequelae, were related to CoV species, it would not come as a surprise that any potential long-term complications of Covid-19 would be of mammoth proportions, taking into account the sheer volume of population affected by SARS-CoV-2 during the last 2 years.

3. Insights from the present

Covid-19 has been predominantly associated with a hyperinflammatory and hypercoagulable state (Fearon, Mikulis, & Lang, 2021; Nalbandian et al., 2021). The pathophysiologic mechanisms underlying SARS-CoV-2 effects include the following:

1. Direct viral toxicity.
2. Endothelial damage and microvascular injury.
3. Immune system dysregulation and stimulation of hyperinflammatory state.
4. Hypercoagulability with resultant in situ thrombosis and macrothrombosis.
5. Maladaptation of angiotensin converting enzyme 2 (ACE2) pathway.
6. Hypoxic injury.

These mechanisms might lead to acute and chronic sequelae, which may be broadly classified, based upon the prominence of underlying organs affected during the acute phase, as follows: pulmonary, hematologic, cardiovascular, neuropsychiatric, renal, endocrine, gastrointestinal, dermatologic, and multisystem inflammatory syndrome (in children) (Nalbandian et al., 2021).

84–88% of Covid-19 patients have been thought to exhibit some kind of neurological symptoms, including headache, dizziness, hyposmia, hypogeusia, encephalopathy/encephalitis (Helms et al., 2020; Mao et al., 2020), but also seizures, polyneuropathy, Guillain-Barre syndrome, and vascular events (arterial and venous) (Frontera et al., 2021; Paterson et al., 2020; Peterson, Sarangi, & Bangash, 2021).

There have been various reports of exacerbation of pre-existing neurological symptoms, including those of PD. Cilia and colleagues have reviewed the effects of Covid-19 on 141 people with PD (PwP) and noted a significant worsening of both motor and non-motor parkinsonian features (Cilia et al., 2020). The postulation of this worsening has been attributed to the infection *per se*, along with possible drug interactions during the Covid-19 management. In another series of 27 PwP an aggravation of motor symptoms was noted in 51.9% of the patients, while an increased levodopa equivalent daily dose was found to be required in 48.2% of them (Leta et al., 2021). Antonini and colleagues have reviewed the effects of Covid-19 on 10 PwP, noting that individuals with advanced age and longer duration of PD were particularly susceptible to higher mortality rates (Antonini, Leta, Teo, & Chaudhuri, 2020). In a recent systematic review of the effects of Covid-19 on pre-existing neurological disorders ($n = 2278$, 26 articles), it was reported that about 59% of PwP experienced an exacerbation of their symptoms (Kubota & Kuroda, 2021).

Méndez-Guerrero and colleagues have reported the first case of a subject with Covid-19 who developed parkinsonism in the form of an asymmetric, hypokinetic—rigid syndrome with mild resting and postural tremor, vertical oculomotor abnormalities and hyposmia (Méndez-Guerrero et al., 2020). The dopamine transporter single-photon emission computerized tomography (SPECT) imaging with ioflupane I-123 injection (DaTscan) depicted an asymmetric loss of dopamine in the nigrostriatal pathway. Various case reports of new-onset parkinsonism in the context of Covid-19 have been reported since then, with some of them having a fairly good response to dopaminergic therapies (Boura & Chaudhuri, 2022). In most of these cases, the assumption of Covid-19 acting as a precipitating factor for parkinsonism has been based on the temporal association of the two conditions (parkinsonism manifesting during or shortly after Covid-19), the fact that patients belonged to a relatively younger age group without any family history of parkinsonism or pre-existing prodromal parkinsonian features, such as rapid eye movement (REM) sleep behavior disorder (RBD), hyposmia or gastrointestinal issues, or genetic evidence of known PD-related mutations in screened subjects, although there were exceptions and the heterogeneity of these cases is highlighted (Boura & Chaudhuri, 2022). Furthermore, case reports of parkinsonism following Covid-19 have shown functional imaging evidence of defects in the dopaminergic pathway and a clinical benefit of dopaminergic therapies, mimicking typical PD (Cohen et al., 2020; Faber et al., 2020; Makhoul & Jankovic, 2021).

The possibility of Covid-19 causing parkinsonism could be based upon:

1. *The role of inflammatory mediators:* The pro-inflammatory cytokines may stimulate neuronal expression of alpha synuclein (α-syn) leading to cascading events with a final culmination to α-syn aggregation (Cheng, Fransson, & Mani, 2022). α-syn is the main protein component of the Lewy bodies and neuritis, which are considered the pathological trademark of PD (Poewe et al., 2017). Recently published data suggests that the innate expression of α-syn might confine viral transmission in the central nervous system (CNS) (Beatman et al., 2015; Massey & Beckham, 2016). However, whether a compromise of the above α-syn function might lead to a virus-induced over-production of intracellular α-syn, predisposing to PD, remains to be investigated. It has also been shown that a H5N1 infection might induce an aggregation of α-syn in various brain areas of rodents and a significant loss of dopaminergic neurons in their substantia nigra, thus, predisposing to synucleinopathies, including PD (Jang, Boltz, Webster, & Smeyne, 2009; Marreiros et al., 2020). More specifically, Marreiros and colleagues have reported that the infection of dopaminergic neurons with the H1N1 influenza virus resulted specifically in the formation of α-syn aggregates, indicating a highly selective process to develop parkinsonism.

   It is also of interest that there have been some case reports of post-Covid parkinsonism in the clinical practice where symptoms exhibited a good response to the administration of immunomodulatory/immunosuppressive therapies, like intravenous immunoglobulin (IVIg) and plasmapheresis, further highlighting the role of immune mechanisms in Covid-induced parkinsonism (Akilli & Yosunkaya, 2021; Tiraboschi et al., 2021).

2. *A direct neurotrophic invasion:* It has been shown that intranasal injections of either SARS-CoV-1 or MERS-CoV in animal models resulted in penetration of these viruses into the brain through the olfactory nerves, leading to an extensive spread of the virus in the CNS, including the thalamus and brainstem (Li et al., 2016; Netland, Meyerholz, Moore, Cassell, & Perlman, 2008). Song and colleagues have demonstrated a neuro-replicative potential and lethal consequences of SARS-CoV-2 infection in transgenic mice models expressing human ACE2 (Song et al., 2020). These preceding evidence, along with a high level of

genetic similarity between SARS-CoV-1 and SARS-CoV-2 (Williams et al., 2021), suggest a neuro-invasive potential of the latter in the CNS. SARS-CoV-2 has been thought to infect cells by interacting with the spike glycoprotein and ACE2 (Zhang, Penninger, Li, Zhong, & Slutsky, 2020). Cells with a high expression of ACE2, including airway epithelia, lung parenchyma, vascular endothelia, kidney cells and small intestinal cells, are believed to be highly susceptible to SARS-CoV-2 (Williams et al., 2021). In the brain, ACE2 are believed to be found in the striatum, the substantia nigra and the posterior hypothalamic area (Pavel, Murray, & Stoessl, 2020; Wan et al., 2021), although this has not been clarified.

It is also of interest that L-Dopa decarboxylase (DDC), an essential enzyme in the biosynthesis of both dopamine and serotonin, is the most significantly co-expressed and co-regulated gene with ACE2 in non-neuronal cell types, significantly affecting the dopamine blood levels (Nataf, 2020). A SARS-CoV-2 infection of monkey cell lines was found to induce downregulation of DDC, an effect which was also noticed with dengue and hepatitis C infections (Mpekoulis et al., 2021). These pathogens have also been associated with parkinsonism (Bopeththa & Ralapanawa, 2017; Tsai et al., 2016). Researchers showed that DDC levels rose in asymptomatic or mild severity Covid-19 patients, while an inverse relationship was noted between SARS-CoV-2 RNA levels and DDC expression (Mpekoulis et al., 2021). Whether a severe Covid-19 infection might lead to a dopamine depletion needs to be further investigated. Moreover, a dopamine D1 receptor agonist was found to suppress endoxin-induced pulmonary inflammation in mice, suggesting that a potential protective role of dopamine in inflammation needs to be further explored (Bone, Liu, Pittet, & Zmijewski, 2017).

In an autopsy study of 43 Covid-19 patients Matschke and colleagues confirmed the presence of SARS-CoV-2 in 53% of the subjects' brain (Matschke et al., 2020). It was also noted that microglia activation and cytotoxic T-cell infiltration was more pronounced in the brainstem and cerebellum. These preceding findings could be interpreted as indications that SARS-CoV-2 has the potential to directly invade the CNS.

3. *A part of multi-pathophysiological processes:* SARS-CoV-2 could also play a critical role by being part of multipronged pathophysiological processes. Up to now, many possible subjacent mechanisms have been speculated

to mediate the development of PD. Some possible hypotheses that might connect the dots between Covid-19 and parkinsonism would be the following: (1) the dual hit hypothesis (Hawkes, Del Tredici, & Braak, 2007; Klingelhoefer & Reichmann, 2015); (2) the multiple hit hypothesis (Meng, Shen, & Ji, 2019; Sulzer, 2007); and (3) the clustering of PD theory (Tsui, Calne, Wang, Schulzer, & Marion, 1999). Braak and colleagues have proposed that PD originates in the gut/ nasal cavity before affecting the brain (Rietdijk, Perez-Pardo, Garssen, van Wezel, & Kraneveld, 2017). This was further supported by loss of smell and gastrointestinal manifestations, appearing as prodromal symptoms in many PwP. Similar physiological processes might be replicated in the Covid-19 setting with many patients complaining of loss of smell and gastrointestinal issues; some of these symptoms might even persist in the long term (long COVID syndrome) (Leta et al., 2021).

4. The post Covid-19 future

The post-Covid-19 scenario varies from immediate changes noted in the near future to possible long term effects seen in Covid-19 patients, but also in antenatally exposed individuals, similarly to the Spanish flu. Some of these sequelae might be recognized early, while many might only be acknowledged decades later, as with EL. We need to learn from past experiences and undertake systematic follow-up assessments in the future in order to understand potential long term implications. The current pathophysiological understanding and research have given us some hints that Covid-19 might be a trigger for a parkinsonism wave in the next few decades, so vigilance is advised.

References

Akilli, N. B., & Yosunkaya, A. (2021). Part of the Covid19 puzzle: Acute parkinsonism. *The American Journal of Emergency Medicine*, *47*, 333.e331–333.e333. https://doi.org/10.1016/j.ajem.2021.02.050.

Antonini, A., Leta, V., Teo, J., & Chaudhuri, K. R. (2020). Outcome of Parkinson's disease patients affected by COVID-19. *Movement Disorders: Official Journal of the Movement Disorder Society*, *35*(6), 905–908. https://doi.org/10.1002/mds.28104.

Beatman, E. L., Massey, A., Shives, K. D., Burrack, K. S., Chamanian, M., Morrison, T. E., et al. (2015). Alpha-synuclein expression restricts RNA viral infections in the brain. *Journal of Virology*, *90*(6), 2767–2782. https://doi.org/10.1128/jvi.02949-15.

Bone, N. B., Liu, Z., Pittet, J.-F., & Zmijewski, J. W. (2017). Frontline science: D1 dopaminergic receptor signaling activates the AMPK-bioenergetic pathway in macrophages and alveolar epithelial cells and reduces endotoxin-induced ALI. *Journal of Leukocyte Biology*, *101*(2), 357–365. https://doi.org/10.1189/jlb.3HI0216-068RR.

Bopeththa, B., & Ralapanawa, U. (2017). Post encephalitic parkinsonism following dengue viral infection. *BMC Research Notes*, *10*(1), 655. https://doi.org/10.1186/s13104-017-2954-5.

Boura, I., & Chaudhuri, K. R. (2022). Coronavirus disease 2019 and related parkinsonism: The clinical evidence thus far. In *Movement disorders clinical practice* John Wiley & Sons, Ltd.

Cheng, F., Fransson, L., & Mani, K. (2022). Complex modulation of cytokine-induced α-synuclein aggregation by glypican-1-derived heparan sulfate in neural cells. *Glycobiology*, *32*(4), 333–342. https://doi.org/10.1093/glycob/cwab126.

Cilia, R., Bonvegna, S., Straccia, G., Andreasi, N. G., Elia, A. E., Romito, L. M., et al. (2020). Effects of COVID-19 on Parkinson's disease clinical features: A community-based case-control study. *Movement Disorders*, *35*(8), 1287–1292. https://doi.org/10.1002/mds.28170.

Cohen, M. E., Eichel, R., Steiner-Birmanns, B., Janah, A., Ioshpa, M., Bar-Shalom, R., et al. (2020). A case of probable Parkinson's disease after SARS-CoV-2 infection. *Lancet Neurology*, *19*(10), 804–805. https://doi.org/10.1016/s1474-4422(20)30305-7.

Donnelly, C. A., Malik, M. R., Elkholy, A., Cauchemez, S., & Van Kerkhove, M. D. (2019). Worldwide reduction in MERS cases and deaths since 2016. *Emerging Infectious Diseases*, *25*(9), 1758–1760. https://doi.org/10.3201/eid2509.190143.

Elkholy, A. A., Grant, R., Assiri, A., Elhakim, M., Malik, M. R., & Van Kerkhove, M. D. (2020). MERS-CoV infection among healthcare workers and risk factors for death: Retrospective analysis of all laboratory-confirmed cases reported to WHO from 2012 to 2 June 2018. *Journal of Infection and Public Health*, *13*(3), 418–422. https://doi.org/10.1016/j.jiph.2019.04.011.

Estupinan, D., Nathoo, S., & Okun, M. S. (2013). The demise of Poskanzer and Schwab's influenza theory on the pathogenesis of Parkinson's disease. *Parkinsons Disease*, *2013*, 167843. https://doi.org/10.1155/2013/167843.

Faber, I., Brandão, P. R. P., Menegatti, F., de Carvalho Bispo, D. D., Maluf, F. B., & Cardoso, F. (2020). Coronavirus disease 2019 and parkinsonism: A non-post-encephalitic case. *Movement Disorders*, *35*(10), 1721–1722. https://doi.org/10.1002/mds.28277.

Fearon, C., Mikulis, D. J., & Lang, A. E. (2021). Parkinsonism as a sequela of SARS-CoV-2 infection: Pure hypoxic injury or additional COVID-19-related response? *Movement Disorders*, *36*(7), 1483–1484. https://doi.org/10.1002/mds.28656.

Frontera, J. A., Yang, D., Lewis, A., Patel, P., Medicherla, C., Arena, V., et al. (2021). A prospective study of long-term outcomes among hospitalized COVID-19 patients with and without neurological complications. *Journal of the Neurological Sciences*, *426*, 117486. https://doi.org/10.1016/j.jns.2021.117486.

Hawkes, C. H., Del Tredici, K., & Braak, H. (2007). Parkinson's disease: A dual-hit hypothesis. *Neuropathology and Applied Neurobiology*, *33*(6), 599–614. https://doi.org/10.1111/j.1365-2990.2007.00874.x.

Helms, J., Kremer, S., Merdji, H., Clere-Jehl, R., Schenck, M., Kummerlen, C., et al. (2020). Neurologic features in severe SARS-CoV-2 infection. *New England Journal of Medicine*, *382*(23), 2268–2270. https://doi.org/10.1056/NEJMc2008597.

Hoffman, L. A., & Vilensky, J. A. (2017). Encephalitis lethargica: 100 years after the epidemic. *Brain*, *140*(8), 2246–2251. https://doi.org/10.1093/brain/awx177.

Huremović, D. (2019). Brief history of pandemics (pandemics throughout history). In *Psychiatry of pandemics: a mental health response to infection outbreak* (pp. 7–35). Springer, Cham.

Institute of Medicine Forum on Microbial, T. (2004). The National Academies Collection: Reports funded by National Institutes of Health. In S. Knobler, A. Mahmoud, S. Lemon, A. Mack, L. Sivitz, & K. Oberholtzer (Eds.), *Learning from SARS: Preparing for the next disease outbreak: Workshop summary*. Washington (DC), USA: National Academies Press. Copyright © 2004, National Academy of Sciences.

Jang, H., Boltz, D. A., Webster, R. G., & Smeyne, R. J. (2009). Viral parkinsonism. *Biochimica et Biophysica Acta, 1792*(7), 714–721. https://doi.org/10.1016/j.bbadis.2008.08.001.

Klingelhoefer, L., & Reichmann, H. (2015). Pathogenesis of Parkinson disease—the gut-brain axis and environmental factors. *Nature Reviews. Neurology, 11*(11), 625–636. https://doi.org/10.1038/nrneurol.2015.197.

Kubota, T., & Kuroda, N. (2021). Exacerbation of neurological symptoms and COVID-19 severity in patients with preexisting neurological disorders and COVID-19: A systematic review. *Clinical Neurology and Neurosurgery, 200*, 106349. https://doi.org/10.1016/j.clineuro.2020.106349.

Leta, V., Rodríguez-Violante, M., Abundes, A., Rukavina, K., Teo, J. T., Falup-Pecurariu, C., et al. (2021). Parkinson's disease and post-COVID-19 syndrome: The Parkinson's long-COVID spectrum. *Movement Disorders, 36*(6), 1287–1289. https://doi.org/10.1002/mds.28622.

Li, K., Wohlford-Lenane, C., Perlman, S., Zhao, J., Jewell, A. K., Reznikov, L. R., et al. (2016). Middle east respiratory syndrome coronavirus causes multiple organ damage and lethal disease in mice transgenic for human dipeptidyl peptidase 4. *The Journal of Infectious Diseases, 213*(5), 712–722. https://doi.org/10.1093/infdis/jiv499.

Limphaibool, N., Iwanowski, P., Holstad, M. J. V., Kobylarek, D., & Kozubski, W. (2019). Infectious etiologies of parkinsonism: Pathomechanisms and clinical implications. *Frontiers in Neurology, 10*, 652. https://doi.org/10.3389/fneur.2019.00652.

Lutters, B., Foley, P., & Koehler, P. J. (2018). The centennial lesson of encephalitis lethargica. *Neurology, 90*(12), 563–567. https://doi.org/10.1212/wnl.0000000000005176.

Makhoul, K., & Jankovic, J. (2021). Parkinson's disease after COVID-19. *Journal of the Neurological Sciences, 422*, 117331. https://doi.org/10.1016/j.jns.2021.117331.

Mao, L., Jin, H., Wang, M., Hu, Y., Chen, S., He, Q., et al. (2020). Neurologic manifestations of hospitalized patients with coronavirus disease 2019 in Wuhan, Cina. *JAMA Neurology*, 77(6), 683–690. https://doi.org/10.1001/jamaneurol.2020.1127.

Marreiros, R., Müller-Schiffmann, A., Trossbach, S. V., Prikulis, I., Hänsch, S., Weidtkamp-Peters, S., et al. (2020). Disruption of cellular proteostasis by H1N1 influenza A virus causes α-synuclein aggregation. *Proceedings of the National Academy of Sciences of the United States of America, 117*(12), 6741–6751. https://doi.org/10.1073/pnas.1906466117.

Massey, A. R., & Beckham, J. D. (2016). Alpha-synuclein, a novel viral restriction factor hiding in plain sight. *DNA and Cell Biology, 35*(11), 643–645. https://doi.org/10.1089/dna.2016.3488.

Matschke, J., Lütgehetmann, M., Hagel, C., Sperhake, J. P., Schröder, A. S., Edler, C., et al. (2020). Neuropathology of patients with COVID-19 in Germany: A post-mortem case series. *The Lancet. Neurology, 19*(11), 919–929. https://doi.org/10.1016/S1474-4422(20)30308-2.

Mazumder, B., Almond, D., Park, K., Crimmins, E. M., & Finch, C. E. (2010). Lingering prenatal effects of the 1918 influenza pandemic on cardiovascular disease. *Journal of Developmental Origins of Health and Disease, 1*(1), 26–34. https://doi.org/10.1017/S2040174409990031.

Méndez-Guerrero, A., Laespada-García, M. I., Gómez-Grande, A., Ruiz-Ortiz, M., Blanco-Palmero, V. A., Azcarate-Diaz, F. J., et al. (2020). Acute hypokinetic-rigid syndrome following SARS-CoV-2 infection. *Neurology, 95*(15), e2109–e2118. https://doi.org/10.1212/wnl.0000000000010282.

Meng, L., Shen, L., & Ji, H. F. (2019). Impact of infection on risk of Parkinson's disease: A quantitative assessment of case-control and cohort studies. *Journal of Neurovirology, 25*(2), 221–228. https://doi.org/10.1007/s13365-018-0707-4.

Moldofsky, H., & Patcai, J. (2011). Chronic widespread musculoskeletal pain, fatigue, depression and disordered sleep in chronic post-SARS syndrome; a case-controlled study. *BMC Neurology*, *11*, 37. https://doi.org/10.1186/1471-2377-11-37.

Mpekoulis, G., Frakolaki, E., Taka, S., Ioannidis, A., Vassiliou, A. G., Kalliampakou, K. I., et al. (2021). Alteration of L-Dopa decarboxylase expression in SARS-CoV-2 infection and its association with the interferon-inducible ACE2 isoform. *PLoS One*, *16*(6), e0253458. https://doi.org/10.1371/journal.pone.0253458.

Nalbandian, A., Sehgal, K., Gupta, A., Madhavan, M. V., McGroder, C., Stevens, J. S., et al. (2021). Post-acute COVID-19 syndrome. *Nature Medicine*, *27*(4), 601–615. https://doi.org/10.1038/s41591-021-01283-z.

Nataf, S. (2020). An alteration of the dopamine synthetic pathway is possibly involved in the pathophysiology of COVID-19. *Journal of Medical Virology*, *92*(10), 1743–1744. https://doi.org/10.1002/jmv.25826.

Netland, J., Meyerholz, D. K., Moore, S., Cassell, M., & Perlman, S. (2008). Severe acute respiratory syndrome coronavirus infection causes neuronal death in the absence of encephalitis in mice transgenic for human ACE2. *Journal of Virology*, *82*(15), 7264–7275. https://doi.org/10.1128/jvi.00737-08.

O'Sullivan, O. (2021). Long-term sequelae following previous coronavirus epidemics. *Clinical Medicine (London, England)*, *21*(1), e68–e70. https://doi.org/10.7861/clinmed.2020-0204.

Paterson, R. W., Brown, R. L., Benjamin, L., Nortley, R., Wiethoff, S., Bharucha, T., et al. (2020). The emerging spectrum of COVID-19 neurology: Clinical, radiological and laboratory findings. *Brain*, *143*(10), 3104–3120. https://doi.org/10.1093/brain/awaa240.

Pavel, A., Murray, D. K., & Stoessl, A. J. (2020). COVID-19 and selective vulnerability to Parkinson's disease. *The Lancet. Neurology*, *19*(9), 719. https://doi.org/10.1016/S1474-4422(20)30269-6.

Peterson, C. J., Sarangi, A., & Bangash, F. (2021). Neurological sequelae of COVID-19: A review. *The Egyptian Journal of Neurology, Psychiatry and Neurosurgery*, *57*(1), 122–129. https://doi.org/10.1186/s41983-021-00379-0.

Poewe, W., Seppi, K., Tanner, C. M., Halliday, G. M., Brundin, P., Volkmann, J., et al. (2017). Parkinson disease. *Nature Reviews. Disease Primers*, *3*(1), 17013. https://doi.org/10.1038/nrdp.2017.13.

Reid, A. H., McCall, S., Henry, J. M., & Taubenberger, J. K. (2001). Experimenting on the past: The enigma of von Economo's encephalitis lethargica. *Journal of Neuropathology and Experimental Neurology*, *60*(7), 663–670. https://doi.org/10.1093/jnen/60.7.663.

Rietdijk, C. D., Perez-Pardo, P., Garssen, J., van Wezel, R. J. A., & Kraneveld, A. D. (2017). Exploring Braak's hypothesis of Parkinson's disease. *Frontiers in Neurology*, *8*, 37–45. https://doi.org/10.3389/fneur.2017.00037.

Song, E., Zhang, C., Israelow, B., Lu-Culligan, A., Prado, A. V., Skriabine, S., et al. (2020). Neuroinvasion of SARS-CoV-2 in human and mouse brain. *bioRxiv*, *218*(3), e20202135. https://doi.org/10.1101/2020.06.25.169946.

Sulzer, D. (2007). Multiple hit hypotheses for dopamine neuron loss in Parkinson's disease. *Trends in Neurosciences*, *30*(5), 244–250. https://doi.org/10.1016/j.tins.2007.03.009.

Taubenberger, J. K., & Morens, D. M. (2006). 1918 Influenza: The mother of all pandemics. *Emerging Infectious Diseases*, *12*(1), 15–22. https://doi.org/10.3201/eid1201.050979.

The World Bank.IBRD.IDA. (2019). Life expectancy at birth, total (years). Retrieved from https://data.worldbank.org/indicator/SP.DYN.LE00.IN.

Tiraboschi, P., Xhani, R., Zerbi, S. M., Corso, A., Martinelli, I., Fusi, L., et al. (2021). Postinfectious neurologic complications in COVID-19: A complex case report. *Journal of Nuclear Medicine*, *62*(8), 1171–1176. https://doi.org/10.2967/jnumed.120.256099.

Tsai, H. H., Liou, H. H., Muo, C. H., Lee, C. Z., Yen, R. F., & Kao, C. H. (2016). Hepatitis C virus infection as a risk factor for Parkinson disease: A nationwide cohort study. *Neurology, 86*(9), 840–846. https://doi.org/10.1212/wnl.0000000000002307.

Tsui, J. K., Calne, D. B., Wang, Y., Schulzer, M., & Marion, S. A. (1999). Occupational risk factors in Parkinson's disease. *Canadian Journal of Public Health. Revue Canadienne de Sante Publique, 90*(5), 334–337. https://doi.org/10.1007/BF03404523.

Valerio, F., Whitehouse, D. P., Menon, D. K., & Newcombe, V. F. J. (2021). The neurological sequelae of pandemics and epidemics. *Journal of Neurology, 268*(8), 2629–2655. https://doi.org/10.1007/s00415-020-10261-3.

Vilensky, J. A., Gilman, S., & McCall, S. (2010). A historical analysis of the relationship between encephalitis lethargica and postencephalitic parkinsonism: A complex rather than a direct relationship. *Movement Disorders, 25*(9), 1116–1123. https://doi.org/10.1002/mds.22908.

Wan, D., Du, T., Hong, W., Chen, L., Que, H., Lu, S., et al. (2021). Neurological complications and infection mechanism of SARS-COV-2. *Signal Transduction and Targeted Therapy, 6*(1), 406–421. https://doi.org/10.1038/s41392-021-00818-7.

Weir, K. (2020). How COVID-19 attacks the brain. *Monitor on Psychology, 51*(8), 20. Retrieved from https://www.apa.org/monitor/2020/11/attacks-brain.

Williams, A., Branscome, H., Khatkar, P., Mensah, G. A., Al Sharif, S., Pinto, D. O., et al. (2021). A comprehensive review of COVID-19 biology, diagnostics, therapeutics, and disease impacting the central nervous system. *Journal of Neurovirology, 27*(5), 667–690. https://doi.org/10.1007/s13365-021-00998-6.

World Health Organization. (2020). The top 10 causes of death. Retrieved from https://www.who.int/news-room/fact-sheets/detail/the-top-10-causes-of-death.

Zhang, P., Li, J., Liu, H., Han, N., Ju, J., Kou, Y., et al. (2020). Long-term bone and lung consequences associated with hospital-acquired severe acute respiratory syndrome: A 15-year follow-up from a prospective cohort study. *Bone Research, 8*, 8. https://doi.org/10.1038/s41413-020-0084-5.

Zhang, H., Penninger, J. M., Li, Y., Zhong, N., & Slutsky, A. S. (2020). Angiotensin-converting enzyme 2 (ACE2) as a SARS-CoV-2 receptor: Molecular mechanisms and potential therapeutic target. *Intensive Care Medicine, 46*(4), 586–590. https://doi.org/10.1007/s00134-020-05985-9.

Parkinson's disease and Covid-19: The effect and use of telemedicine

Aleksandra M. Podlewska[a,b,c,*] **and Daniel J. van Wamelen**[a,b,c,d]

[a]Department of Neurosciences, Institute of Psychiatry, Psychology & Neuroscience, King's College London, London, United Kingdom
[b]Department of Basic and Clinical Neuroscience, Division of Neuroscience, King's College London, London, United Kingdom
[c]Parkinson's Foundation Centre of Excellence, King's College Hospital NHS Foundation Trust, London, United Kingdom
[d]Department of Neurology, Centre of Expertise for Parkinson & Movement Disorders, Donders Institute for Brain, Cognition and Behaviour, Radboud University Medical Center, Nijmegen, The Netherlands
*Corresponding author: e-mail address: aleksandra.podlewska@kcl.ac.uk

Contents

Abstract

As a result of the Coronavirus Disease 2019 (Covid-19) pandemic the use of telemedicine and remote assessments for patients has increased exponentially, enabling healthcare professionals to reduce the need for in-person clinical visits and, consequently, reduce the exposure to the Severe Acute Respiratory Syndrome Coronavirus-2 (SARS-CoV-2). This development has been aided by increased guidance on digital health technologies and cybersecurity measures, as well as reimbursement options within healthcare systems. Having been able to continue to connect with people with Parkinson's Disease (PwP, PD) has been crucial, since many saw their symptoms worsen over the pandemic. Inspite of the success of telemedicine, sometimes even enabling delivery of treatment and research, further validation and a unified framework are necessary to measure the true benefit to both clinical outcomes and health economics. Moreover, the use of telemedicine seems to have been biased towards people from a white background, those with higher

International Review of Neurobiology, Volume 165
ISSN 0074-7742
https://doi.org/10.1016/bs.irn.2022.04.002
Copyright © 2022 Elsevier Inc.
All rights reserved.

education, and reliable internet connections. As such, efforts should be pursued by being inclusive of all PwP, regardless of geographical area and ethnic background. In this chapter, we describe the effect he Covid-19 pandemic has had on the use of telemedicine for care and research in people with PD, the limiting factors for further rollout, and how telemedicine might develop further.

1. Introduction

As a result of the Coronavirus Disease 2019 (Covid-19) pandemic the use of telemedicine and remote assessments for patients has increased exponentially, including the introduction and use of interdisciplinary telehealth services (Ben-Pazi et al., 2018; Cubo, Hassan, Bloem, & Mari, 2020). Telemedicine has enabled the reduction in the need for in-person clinical visits and has, consequently, reduced exposure to the Severe Acute Respiratory Syndrome Coronavirus-2 (SARS-CoV-2). This development was further aided by the increased guidance on digital health technologies (AAN, 2020; Society, 2020) and cybersecurity measures, as well as expanded reimbursement options within healthcare systems (Hassan et al., 2020; Rockwell & Gilroy, 2020). In addition, over the course of the pandemic the advice related to data security management and data protection for video conferencing has become more clear (John & Wellmann, 2020). Moving to remote assessments, care delivery, and clinical appointments was crucial, since lockdowns, as imposed in many countries during the pandemic, had severely detrimental effects on people with Parkinson's Disease (PwP, PD). In a large study with over 2500 PwP, over 40% reported a worsening of symptoms during lockdown, with especially tremor, pain, and rigidity, but also psychic state worsening (Fabbri et al., 2021).

Telemedicine may continue to be necessary, possibly also in relation to future pandemics or other global public health problems, given the option of rapid mobilization of telemedicine (Hassan et al., 2020). Different definitions exist of telemedicine, often also referred to as telehealth or digital health (eHealth). In this respect, the Royal College of Nursing in the United Kingdom provides a useful definition and defines eHealth as technology-based means to empower, promote and facilitate health and wellbeing of individuals, their families and communities (Royal College of Nursing, 2022), offering an alternative to classical face-to-face consultations. Nonetheless, not all patients seem keen on continuing with remote assessments beyond the Covid-19 pandemic, with numbers ranging between 26% and 48% (de Rus Jacquet et al., 2021), necessitating further improvements to the delivery of remote care. In this respect also the delivery of remote care needs to be taken into account, as, e.g., during telephone consultation medication changes are less likely to take place than compared to in-person settings (Ahmad & Postuma, 2021). However, with pressure on hospitals and staff and the eventual almost complete suspension of all outpatient face-to-face appointments due to Covid-19, implementation of telemedicine as the primary care management method has seen a rapid up-scale (Kichloo et al., 2020; Peine et al., 2020).

In this book chapter, we review the evidence for telemedicine and remote assessments for PD and how the Covid-19 pandemic has highlighted the need and underlined the feasibility of such assessments. The use of technology in healthcare can be broadly categorized into two entities: patient-facing and professional services. The former concerns the means to deliver safe, secure, effective and high-quality care to the patients, while the latter focuses on the use of technology to educate staff and streamline or improve the care delivered to the patients.

2. Patient-facing services

With some literature suggesting a wide array of benefits to home-based patient consultations, the proposal for the use of technology to deliver patient care in PD has been made in the years preceding the outbreak of the pandemic (Dorsey et al., 2016; Espay et al., 2016). The call for home-based care has identified six essential features for consideration to deliver a truly patient-centered care: location, driving, mobility, cognition, disease course, and caregivers (Dorsey et al., 2016). It has been highlighted that access to specialist services is limited to those with the ability to physically attend appointments, which further deepens the disparity in the quality of care provided to the patients. However, with most non-urgent outpatient services being moved to telemedicine, patients in remote locations were able to access the multidisciplinary services at their doorstep, which they would not have otherwise had a chance to utilize. For example, specialist referrals and attending multiple appointments far away from home had no longer been considered an issue, as the appointments were now conducted remotely. In addition, the availability of virtual appointments, irrespective of lockdowns and recommendations regarding shielding, ensured regular follow-ups and, therefore, accurate and adequate management. Currently, models of care for remote assessments are being proposed to ensure a uniform basic structure of the consultations, and that both the motor and non-motor concerns are addressed in PwP (Larson, Schneider, & Simuni, 2021; van Munster et al., 2021).

The increased implementation and availability of telehealth allowed healthcare professionals to provide their services to patients who otherwise would not be able to receive the care, worsened by lockdowns limitations or even eliminating in person contact. Especially care and nursing homes have been reported to have been particularly affected given the vulnerability of their populations (Solis, Franco-Paredes, Henao-Martínez, Krsak, & Zimmer, 2020), with some forced to go into full lockdowns with staff having to stay on-site, as the risk of viral transmission remained high. Here, remote services, conducted via means of telehealth, played a vital role in ensuring that these patients were not without review and treatment advice with some preliminary findings showing that in particular, psychiatric, geriatric and palliative care can be provided effectively through telemedicine (Groom, McCarthy, Stimpfel, & Brody, 2021). As such, we have moved away from the rapid and improvised transition from in-person visits to telemedicine at the start of the pandemic to a global increase in the use of telemedicine in movement disorders and PD (Hassan et al., 2020), with some consensus for the delivery format. Currently, policies are varied and flexible, and e.g., in the United States of America (USA) flexibility in platform selection exists (Larson, Schneider, & Simuni, 2021). As a consequence, telemedicine use among PwP increased from 9.7% prior to the pandemic to 63.5% during the pandemic (Feeney et al., 2021). Even management and evaluation of eligibility criteria for device-aided therapies has shifted to telemedicine (Fasano et al., 2020; York et al., 2021).

Other outcomes of the transition to telehealth have been observed in participation of "hard-to-reach" populations. Substantial differences in healthcare access and utilization existed in the USA prior to the Covid-19 pandemic. Here, one study found that Hispanic patients were 40% less likely and Black patients 30% less likely than white patients to see an outpatient neurologist (Saadi, Himmelstein, Woolhandler, & Mejia, 2017), along with a lower referral rate to PD specialists compared to white men (Willis, Schootman, Evanoff, Perlmutter, & Racette, 2011). Moreover, prepandemic newly diagnosed Black PwP were up to four times less likely than white patients to be initiated on treatment (12% and 38%, respectively) (Dahodwala, Xie, Noll, Siderowf, & Mandell, 2009). The pandemic seems to have exacerbated disparities in care access, as telemedicine use was highest among those with higher incomes, higher levels of education, and white PwP (Feeney et al., 2021). Similarly, another study has found that the percentage of telemedicine visits compared to face-to-face visits increased for white PwP and decreased for Black PwP (Esper et al., 2021).

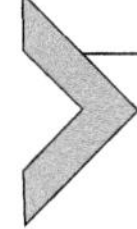

3. Videoconferencing and assessments via video systems

The evidence and options behind videoconferencing software for assessment of movement disorder patients have been recently reviewed by Cubo et al. (2021). The authors were able to identify 26 videoconferencing software programs that were available for remote assessment. Of these, half were specifically designed for general healthcare, of which 23.0% were compliant with European and US regulations (including General Data Protection Regulation, GDPR), and 19.2% provided information regarding security (Table 1). Limiting factors to videoconferencing and video assessments include slow internet speed, and in some parts of the world, the limited availability of smart phones, but also reduced dexterity impacting on device control (Garg et al., 2021).

A field related to telemedicine is that of home videos. Although such home videos appear common place in the field of epilepsy, for movement disorders and PD this appears to be less pronounced. In one small study looking at the value of home videos for movement disorders, it was shown that out of 20 home videos, 50% were determined to be of additional clinical value and in 62.4% of evaluations it was possible to identify the phenomenology from the home videos and were found to be consistent with the final diagnosis. In this respect, it is also important to note that video quality was a key determinant for identifying the correct phenomenology; videos rated as 'poor quality' had significantly lower odds of leading to a correct phenomenology (Billnitzer & Jankovic, 2021).

4. Additional telemedicine assessments

In a cohort study of 116 PD outpatients using the telemedicine tool "myParkinsoncoach," involving periodic monitoring, feedback, knowledge modules, and text message functionality. By using this tool, a 29% reduction in need for outpatient visits was observed in the year after introducing the tool compared to the year before introduction, coinciding with a 39% reduction in overall PD-related healthcare costs. The time needed to be dedicated to the tool monitoring and follow-up activities was on average of 15.5 min per patient per month (Wijers, Hochstenbach, & Tissingh, 2021). Similarly, in another video-based study in 22 PwP who underwent three non-motor assessments and a modified version of the Movement

Table 1 Applications for videoconferencing.

	Features					Security					
Application	**Chat**	**Call type**	**File share**	**Screen share**	**Healthcare based**	**Pricing**	**Supported OS and platforms**	**Communications protection**	**Extra security**	**Security in group meetings**	**Security standards compliance**
Facebook messenger	√	Video	√	X	X	Free	Windows, MacOs, iOS, Android	E2EE	2FA	Invitation, Admin control	SOC2, GDPR
FaceTime	√	Video	X	X	X	Free	MacOs. iOs	E2EE	2FA, Face ID and iPhone security	X	?
Google Duo	X	Video	X	X	X	Free	Movil based: Android, iOs, Web browser-based	E2EE	2FA, Google Account security	Invitation and user block option	HIPAA - BAA, GDPR
Google Hangouts	√	Video	√	√	√	Contact sales	Android, iOs, Web browser-based	IETF, SRTP and DTLS client-Server	2FA, APP, SSO and Google's MFA	Invitation, admin control PIN	HIPAA, HITRUST, SOC2, GDPR
Jitsi Meet	√	Video	√	√	X	Free and License	Windows, Linux, MacOs, iOS, Android	E2EE DTLS-SRTP	X	Password Admin control	?
Line	√	Video	√	√	X	Free	Windows, MacOs, iOS, Android	E2EE	X	Invitation	?

Signal	√	Video	√	X	X	Free	Windows, Linux, MacOs, iOS, Android	E2EE	Screen lock	X	X
Skype Business	√	Video	√	√	X	License	Windows, Linux, MacOs, iOS, Android	E2EE	2FA	Invitation, Admin control	GDPR, HIPAA, HITRUST, HITECH, CCPA.
Telegram	√	Video	X	X	X	Free	Windows, Linux, MacOs, iOS, Android	E2EE	2FA, block code, secret chats, and active sessions	X	GDPR
WeChat	√	Video	√	X	X	Free and License	Windows, Web browser-based, MacOs, iOS, Android	TLS client-Server	X	X	EEA
WhatsApp	√	Video	√	X	X	Free	Windows, Web browser-based, MacOs, iOS, Android	E2EE	2FA	Invitation	GDPR, EEA

Continued

Table 1 Applications for videoconferencing.—cont'd

	Features					Security					
Application	Chat	Call type	File share	Screen share	Healthcare based	Pricing	Supported OS and platforms	Communications protection	Extra security	Security in group meetings	Security standards compliance
Zoom	√	Video	√	√	√	License	Windows, Linux, MacOs, iOS, Android	E2EE, DTLS	2FA, SSO	Invitation, Password, Admin control	HIPAA - BAA, PHIPA/ PIPEDA, SOC2
Teams	√	Video	√	√	√	License	Windows, Linux, MacOs, iOS, Android, Web browser-based	E2EE	2FA	Invitation, Password, Admin control	HIPAA, HITECH, SOC2, HITRUST, GDPR

Abbreviations: √, available; X, not available; ?, unknown; 2FA, two factor authentication; APP, advanced protection program; BAA, business associate agreement; CCPA, California Consumer Privacy Act; DTLS, datagram transport layer security; E2EE, end-to-end encryption; EEA, European Economic Area; GDPR, general data protection regulation; HIPAA, hardware as a service program; HITECH, Health Information Technology for Economic and Clinical Health Act; HITRUST, prescriptive set of controls that meet the requirements of multiple regulations and standards; IETF, internet engineering task force; MFA, multi-factor authentication; PHIPA, Personal Health Information Protection Act; PIPEDA, Personal Information Protection and Electronic Documents Act; SRTP, secure real-time transport protocol; SOC, system and organization control; SSO, single sign-on; TLS, transport layer security.

Adapted from Cubo, E., Arnaiz-Rodriguez, A., Arnaiz-González, Á., Díez-Pastor, J. F., Spindler, M., Cardozo, A., … Bloem, B. R. (2021). Videoconferencing software options for telemedicine: A review for movement disorder neurologists. *Frontiers in Neurology*, *12*, 745917. doi:10.3389/fneur.2021.745917.

Disorder Society Unified Parkinson's Disease Rating Scale part III (MDS-UPDRS III), showed high satisfaction (100%) among patients, even though the duration to complete the visits was longer (41.3 min for video assessment and 17.5 min for non-motor assessment via telephone). Moreover, in some studies there was an increase in recruitment numbers following the switch to a virtual environment (Quinn, Macpherson, Long, & Shah, 2020).

Interestingly, the estimated costs for the remote assessments were 20-fold lower than for in-person visits (Xu et al., 2021). The latter seems to be backed by other studies (Delgoshaei et al., 2017), although this is not consistent with some studies only reporting increased patient satisfaction without a significant cost reduction (Snoswell et al., 2020). Assessments which are not possible to perform over video, such as rigidity and postural stability, can be complemented with the use of wearable sensors. For example, Safarpour and colleagues showed that, using three inertial sensors measuring over the period of a week, number of walking bouts and turns were significantly associated with the rigidity subscore of the MDS-UPDRS and that the number of turns, foot pitch angle, and sway area while standing were associated with the postural stability and gait subscore (Safarpour et al., 2021).

Questions might arise in relation to the reliability and reproducibility of assessments over video, although the same could also be said for in-person visits. Nevertheless, motor examination, including finger tapping and pronation/supination, showed a high rate of rater agreement and correlations between findings made by experienced examiners were high with low variability (Luiz, Marques, Folador, & Andrade, 2021).

5. Remote delivery of treatment

In relation to delivery of treatment, and more generally to research delivery, Myers and colleagues have recently shown that recruitment of large, geographically dispersed remote cohorts from a single location was feasible, with an average study recruitment of 4.9 participants per week over 1 year across three large PD studies (Myers et al., 2021). An example of remote delivery is the "ParkinDANCE" program, where researchers adapted the in-person visits to online delivery during the Covid-19 pandemic and participants completed eight 1 h sessions of online therapeutic dancing. During the sessions, each participant was assigned their own dance teacher and selected music for the classes, resulting in high attendance (100%) with participants able to quickly adapt to online delivery with support and resources (Morris et al., 2021). The feasibility of this study was echoed in a recent systematic review of 15 studies looking at the effect of telerehabilitation on gait and balance, dexterity of the upper limbs, and speech disorders, but also quality of life and patient satisfaction, where these symptoms could be improved through the use of telemedicine (Vellata et al., 2021). Importantly, patients seem to be satisfied with remote delivery of care and research, as exemplified by the results of the Fox Insight study. It demonstrated that 97.9% of participants were satisfied with the study taking place virtually, 98.5% were willing to participate in a future observational study with virtual visits, and 76.1% were willing to participate in an interventional trial with virtual visits (Myers et al., 2021).

6. Wearable technologies

As also mentioned above, one of the drawbacks of remote outpatient visits compared to classical face-to-face appointments is the limited opportunity to assess motor and non-motor symptoms, although even in a face-to-face setting the risk of reporting bias and symptom recall may not be representative of the patient's real functioning at home (Bot et al., 2016; Ossig et al., 2016). A call for the use of wearable technology to further inform clinical assessments has therefore been made pre-pandemic, though large-scale implementation of such devices had proven difficult operationally. Fueled by the Covid-19 pandemic, and inspite of barriers to engagement, some countries have started adopting objective remote measures as tools to inform clinical assessments. Wearable sensors, such as the Parkinson's KinetiGraph™, which have been validated for use to measure some motor symptoms of PD, such as bradykinesia and dyskinesia (Chen et al., 2020; Pahwa, Bergquist, Horne, & Minshall, 2020), could play an important role in extending the clinical impression based on a telephone or a video consultation.

While wearable sensor systems may be considered costly, also taking into account that such technology might not be available in every country, they could nonetheless play an important role in clinical decisions making. Healthcare systems in countries such as Germany and Sweden now provide full funding of the remote wearable readings, and some other countries such as the UK are currently reviewing evidence to adapt the tool for use in general practice for PD. Moreover, some UK-based centers have demonstrated a good feasibility of such equipment in virtual settings during the Covid-19 pandemic (Evans, Mohamed, & Thomas, 2020). Further evidence is, however, necessary to establish cost-effectiveness, although the preliminary reports suggest that, overall, remote care utilizing wearable technology can lead to a more personalized and effective healthcare delivery (James et al., 2021).

7. Patient education

Education is an important aspect of empowering patients in their care. With shielding having been advised for PwP during the pandemic, most educational seminars and meetings had been moved online to avoid a gap in knowledge delivery, especially for individuals newly diagnosed with PD. This had a significant positive impact on the patients, given that the resources which once were only available locally, could be accessed from anywhere. For example, the King's Parkinson's Educational Evening Clinics, organized by Prof K. Ray Chaudhuri at the Parkinson's Foundation Centre of Excellence at King's College Hospital in London have seen an increase in attendance, from a smaller number of patients who could make their way to the hospital site physically, to a larger number, as more people were able to join the online meeting. In a similar manner, educational meetings for specific aspects of living with PD, such as consideration of advanced therapies, like the levodopa-carbidopa intestinal gel (LCIG) infusion therapy, had been organized and widely publicized, and given its accessibility online, well attended by both patients and clinicians who wished to learn more about the therapies. Associations of PwP, such as the American Parkinson Disease Association (APDA) or the European Parkinson Disease Association (EPDA), have not only created event calendars outlining all the educational online events, but have also started blog and vlog entries and have organized structured meetings based on the needs expressed by patients online. The UK-based Parkinson's Disease Nurse Specialist Association has contributed to a multitude of free accessible online meetings organized for PwP, sharing their knowledge and expertise on the topics relevant to PD during the pandemic.

In addition, in a recent qualitative study, Zipprich and colleagues demonstrated that 30% of the participants had insufficient knowledge and understanding of Covid-19 regulations and preventative measures (Zipprich, Teschner, Witte, Schönenberg, & Prell, 2020), highlighting the need for robust education about the current issues outside of the direct management of PD as a condition. To that end, a recent systematic review recommended a proactive approach to education of PwP regarding Covid-19, including infection prevention (Kubota & Kuroda, 2021). It is therefore clear that technology plays a vital role in the patient education, which is an integral part of effective care management.

Lastly, celebratory events such as the World Parkinson's day, which always provide an opportunity for education, have also moved to online platforms, therefore, becoming more available globally, rather than just locally. Anecdotal feedback demonstrates that PwP, as well as their carers and families, have found the accessibility of online educational events helpful in times of uncertainty and have reported that the role of PD specialists, including the PD Nurse Specialists (PNS), has been instrumental in providing the freely available resources and ensuring that these are accessible.

8. Communication between professionals

Electronic patient records and referrals have been a part of standard practice in many countries, though data sharing and privacy concerns have impacted the extent of implementation of services nationally in many countries. With Microsoft Teams deemed a secure platform for sharing of confidential data, multidisciplinary meetings could continue in the era of social distancing, therefore enabling PNS to effectively engage all the specialists required to provide personalized care, essential to an effective management of PwP (Titova & Chaudhuri, 2017).

9. Limitations and future of telehealth in Parkinson's disease

Despite clear progress that has been made with the different aspects of telehealth in PwP (Fig. 1) over the course of the Covid-19 pandemic, it is not without its drawbacks. One of the major limitations is the access to the appropriate technology. For example, in many suburban and remote locations, access to internet remains limited and patients may not always own mobile devices with adequate system requirements and features, such as camera, to fully engage in a video consultation with a clinician. Other factors include slow internet speed, and in certain areas, the limited availability of smart phones, but also patient-related factors, such as reduced dexterity and age, impacting on device control (Garg et al., 2021; Scott Kruse et al., 2018). Moreover, cognitive decline, a prominent non-motor symptom present in the majority of PwP across the disease pan (Aarsland et al., 2017, 2021), may be a major barrier to implementation of remote care in PD. Finally, the disparities related to healthcare access by PwP from ethnic

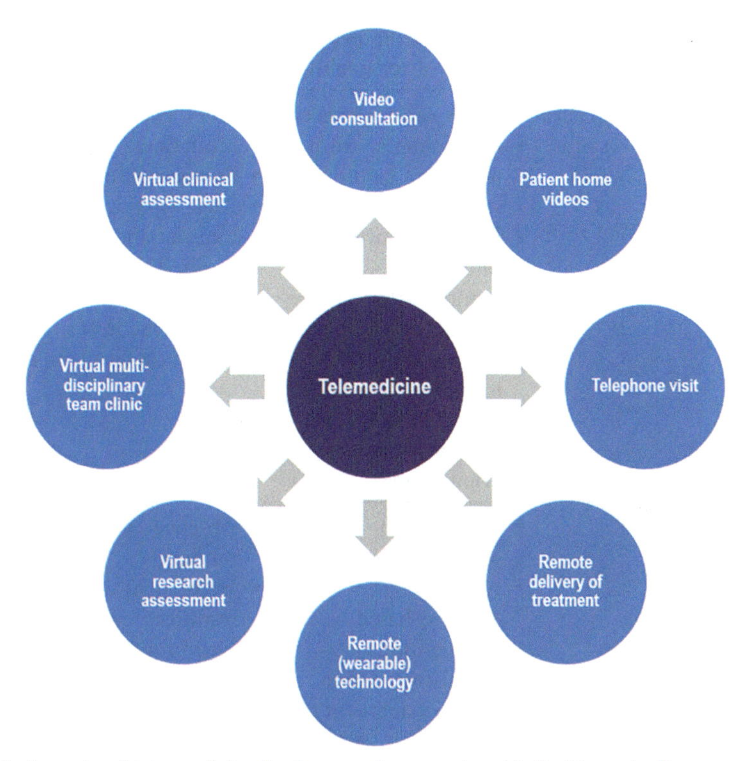

Fig. 1 Aspects of telemedicine in the care for people with Parkinson's disease.

minority backgrounds need to be taken into account (Mantri & Mitchell, 2021). These factors (Table 2) together can, therefore, further drive the disparity between patients and ultimately lead to a difference in the care provided to individuals in different economic situations, thus, creating a digital divide, which could leave the most vulnerable patients at risk (Eyrich, Andino, & Fessell, 2021).

Further areas of improvement include social connection and physical activity in overall well-being, helping PwP to stay connected (Macchi et al., 2021). One strategy to achieve this might be social prescribing, involving the connection of PwP with non-medical services (Alderwick, Gottlieb, Fichtenberg, & Adler, 2018), exemplified by the Togetherness Program at CareMore and the Veteran's Administration Compassionate Contact Corps Program in the USA, which include proactive phone calls and visits to isolated patients (Subramanian, Farahnik, & Mischley, 2020). Also virtual support groups may help to improve social connection of PwP and carers, including programs to help strengthen coping strategies and manage

Table 2 Benefits and barriers of telehealth in Parkinson's disease.

	Benefits of telehealth	Barriers
Integration of telehealth into healthcare systems	• Improved care access • Convenience from joining in home setting • Risk reduction for infection • Continuation of care • Reduction in impact of geographical location • Potential to reach underserved populations	• Software and license limitations • Disparities in access • Lack of widespread reliable internet and connection • Limited digital literacy • Limited ability for management of device-aided therapies
Support for PwP	• Connection with social services • Access to exercise-based interventions • Social connection • Self-management skills	• Absence of appropriate funding • Insufficient community resources • Need for research on effective models
Support for carers	• Improvement of quality of life for carers • Reduction in carer burden	• Lack of financial support • Caregiver burnout • Insufficient community resources

Abbreviations: PwP, people with Parkinson's disease.
Modified from Valdovinos, B. Y., Modica, J. S., & Schneider, R. B. (2022). Moving forward from the COVID-19 pandemic: Needed changes in movement disorders care and research. *Current Neurology and Neuroscience Reports*, 1–10. doi:10.1007/s11910-022-01178-7.

anticipatory grief (Chaudhuri et al., 2021), lifestyle changes, and enhancement of medication adherence (Yogev-Seligmann & Kafri, 2021). Finally, targeted education for caregivers may improve self-management and self-efficacy (Duits, Boots, Mulders, Moonen, & Vugt, 2021).

10. Conclusions

Telemedicine has demonstrated its success in providing access for patients with movement disorders to clinic services and even treatment during the time of the Covid-19 pandemic (Ben-Pazi et al., 2018). Nonetheless, further validation and a unified framework are necessary to measure the true benefit to both clinical outcomes and health economics. The focus should be on creating more robust healthcare systems by optimizing interdisciplinary care models incorporating telehealth, developing novel ways to remotely manage and initiate device-aided therapies, and developing clinical programs to encourage self-management skills and provide access to social support, including that of carers of PwP. These efforts should be pursued by being inclusive of all PwP, regardless of geographical area and ethnic and financial background.

References

AAN. (2020). https://www.aan.com/practice/telemedicine.

Aarsland, D., Batzu, L., Halliday, G. M., Geurtsen, G. J., Ballard, C., Ray Chaudhuri, K., et al. (2021). Parkinson disease-associated cognitive impairment. *Nature Reviews. Disease Primers*, 7(1), 47. https://doi.org/10.1038/s41572-021-00280-3.

Aarsland, D., Creese, B., Politis, M., Chaudhuri, K. R., Ffytche, D. H., Weintraub, D., et al. (2017). Cognitive decline in Parkinson disease. *Nature Reviews. Neurology*, *13*(4), 217–231. https://doi.org/10.1038/nrneurol.2017.27.

Ahmad, F. A., & Postuma, R. B. (2021). Telephone visit efficacy for Parkinson's disease during the COVID-19 pandemic. *Clinical Parkinsonism & Related Disorders*, *5*, 100107. https://doi.org/10.1016/j.prdoa.2021.100107.

Alderwick, H. A. J., Gottlieb, L. M., Fichtenberg, C. M., & Adler, N. E. (2018). Social prescribing in the U.S. and England: Emerging interventions to address patients' social needs. *American Journal of Preventive Medicine*, *54*(5), 715–718. https://doi.org/10.1016/j.amepre.2018.01.039.

Ben-Pazi, H., Browne, P., Chan, P., Cubo, E., Guttman, M., Hassan, A., et al. (2018). The promise of telemedicine for movement disorders: An interdisciplinary approach. *Current Neurology and Neuroscience Reports*, *18*(5), 26. https://doi.org/10.1007/s11910-018-0834-6.

Billnitzer, A., & Jankovic, J. (2021). The clinical value of patient home videos in movement disorders. *Tremor and Other Hyperkinetic Movement (New York, N.Y.)*, *11*, 37. https://doi.org/10.5334/tohm.651.

Bot, B. M., Suver, C., Neto, E. C., Kellen, M., Klein, A., Bare, C., et al. (2016). The mPower study, Parkinson disease mobile data collected using ResearchKit. *Scientific Data*, *3*, 160011. https://doi.org/10.1038/sdata.2016.11.

Chaudhuri, K. R., Rukavina, K., McConvey, V., Antonini, A., Lorenzl, S., Bhidayasiri, R., et al. (2021). The impact of COVID-19 on palliative care for people with Parkinson's and response to future pandemics. *Expert Review of Neurotherapeutics*, *21*(6), 615–623. https://doi.org/10.1080/14737175.2021.1923480.

Chen, L., Cai, G., Weng, H., Yu, J., Yang, Y., Huang, X., et al. (2020). More sensitive identification for bradykinesia compared to tremors in Parkinson's disease based on Parkinson's KinetiGraph (PKG). *Frontiers in Aging Neuroscience*, *12*. https://doi.org/10.3389/fnagi.2020.594701.

Cubo, E., Arnaiz-Rodriguez, A., Arnaiz-González, Á., Díez-Pastor, J. F., Spindler, M., Cardozo, A., et al. (2021). Videoconferencing software options for telemedicine: A review for movement disorder neurologists. *Frontiers in Neurology*, *12*, 745917. https://doi.org/10.3389/fneur.2021.745917.

Cubo, E., Hassan, A., Bloem, B. R., & Mari, Z. (2020). Implementation of telemedicine for urgent and ongoing healthcare for patients with Parkinson's disease during the COVID-19 pandemic: New expectations for the future. *Journal of Parkinson's Disease*, *10*(3), 911–913. https://doi.org/10.3233/jpd-202108.

Dahodwala, N., Xie, M., Noll, E., Siderowf, A., & Mandell, D. S. (2009). Treatment disparities in Parkinson's disease. *Annals of Neurology*, *66*(2), 142–145. https://doi.org/10.1002/ana.21774.

de Rus Jacquet, A., Bogard, S., Normandeau, C. P., Degroot, C., Postuma, R. B., Dupré, N., et al. (2021). Clinical perception and management of Parkinson's disease during the COVID-19 pandemic: A Canadian experience. *Parkinsonism & Related Disorders*, *91*, 66–76. https://doi.org/10.1016/j.parkreldis.2021.08.018.

Delgoshaei, B., Mobinizadeh, M., Mojdekar, R., Afzal, E., Arabloo, J., & Mohamadi, E. (2017). Telemedicine: A systematic review of economic evaluations. *Medical Journal of the Islamic Republic of Iran*, *31*, 113. https://doi.org/10.14196/mjiri.31.113.

Dorsey, E. R., Vlaanderen, F. P., Engelen, L. J., Kieburtz, K., Zhu, W., Biglan, K. M., et al. (2016). Moving Parkinson care to the home. *Movement Disorders*, *31*(9), 1258–1262. https://doi.org/10.1002/mds.26744.

Duits, A. A., Boots, L. M. M., Mulders, A. E. P., Moonen, A. J. H., & Vugt, M. E. (2021). Covid proof self-management training for caregivers of patients with Parkinson's disease. *Movement Disorders*, *36*(3), 529–530. https://doi.org/10.1002/mds.28457.

Espay, A. J., Bonato, P., Nahab, F. B., Maetzler, W., Dean, J. M., Klucken, J., et al. (2016). Technology in Parkinson's disease: Challenges and opportunities. *Movement Disorders*, *31*(9), 1272–1282. https://doi.org/10.1002/mds.26642.

Esper, C. D., Scorr, L., Papazian, S., Bartholomew, D., Esper, G. J., & Factor, S. A. (2021). Telemedicine in an academic movement disorders center during COVID-19. *Journal of Movement Disorders*, *14*(2), 119–125. https://doi.org/10.14802/jmd.20099.

Evans, L., Mohamed, B., & Thomas, E. C. (2020). Using telemedicine and wearable technology to establish a virtual clinic for people with Parkinson's disease. *BMJ Open Quality*, *9*(3). https://doi.org/10.1136/bmjoq-2020-001000.

Eyrich, N. W., Andino, J. J., & Fessell, D. P. (2021). Bridging the digital divide to avoid leaving the most vulnerable behind. *JAMA Surgery*, *156*(8), 703–704. https://doi.org/10.1001/jamasurg.2021.1143.

Fabbri, M., Leung, C., Baille, G., Béreau, M., Brefel Courbon, C., Castelnovo, G., et al. (2021). A French survey on the lockdown consequences of COVID-19 pandemic in Parkinson's disease. The ERCOPARK study. *Parkinsonism & Related Disorders*, *89*, 128–133. https://doi.org/10.1016/j.parkreldis.2021.07.013.

Fasano, A., Antonini, A., Katzenschlager, R., Krack, P., Odin, P., Evans, A. H., et al. (2020). Management of advanced therapies in Parkinson's disease patients in times of humanitarian crisis: The COVID-19 experience. *Movement Disorders Clinical Practice*, 7(4), 361–372. https://doi.org/10.1002/mdc3.12965.

Feeney, M. P., Xu, Y., Surface, M., Shah, H., Vanegas-Arroyave, N., Chan, A. K., et al. (2021). The impact of COVID-19 and social distancing on people with Parkinson's disease: A survey study. *NPJ Parkinson's Disease*, 7(1), 10. https://doi.org/10.1038/s41531-020-00153-8.

Garg, D., Majumdar, R., Chauhan, S., Preenja, R., Parihar, J., Saluja, A., et al. (2021). Teleneurorehabilitation among person with Parkinson's disease in India: The initial experience and barriers to implementation. *Annals of Indian Academy of Neurology*, *24*(4), 536–541. https://doi.org/10.4103/aian.AIAN_127_21.

Groom, L. L., McCarthy, M. M., Stimpfel, A. W., & Brody, A. A. (2021). Telemedicine and telehealth in nursing homes: An integrative review. *Journal of the American Medical Directors Association*, *22*(9), 1784–1801.e1787. https://doi.org/10.1016/j.jamda.2021.02.037.

Hassan, A., Mari, Z., Gatto, E. M., Cardozo, A., Youn, J., Okubadejo, N., et al. (2020). Global survey on telemedicine utilization for movement disorders during the COVID-19 pandemic. *Movement Disorders*, *35*(10), 1701–1711. https://doi.org/10.1002/mds.28284.

James, S., Ashley, C., Williams, A., Desborough, J., McInnes, S., Calma, K., et al. (2021). Experiences of Australian primary healthcare nurses in using telehealth during COVID-19: A qualitative study. *BMJ Open*, *11*(8), e049095. https://doi.org/10.1136/bmjopen-2021-049095.

John, N., & Wellmann, M. (2020). Data security management and data protection for video conferencing software. *International Cybersecurity Law Review*, *1*(1), 39–50.

Kichloo, A., Albosta, M., Dettloff, K., Wani, F., El-Amir, Z., Singh, J., et al. (2020). Telemedicine, the current COVID-19 pandemic and the future: A narrative review and perspectives moving forward in the USA. *Family Medicine and Community Health*, *8*(3). https://doi.org/10.1136/fmch-2020-000530.

Kubota, T., & Kuroda, N. (2021). Exacerbation of neurological symptoms and COVID-19 severity in patients with preexisting neurological disorders and COVID-19: A systematic review. *Clinical Neurology and Neurosurgery*, *200*, 106349. https://doi.org/10.1016/j.clineuro.2020.106349.

Larson, D. N., Schneider, R. B., & Simuni, T. (2021). A new era: The growth of video-based visits for remote management of persons with Parkinson's disease. *Journal of Parkinson's Disease*, *11*(s1), S27–s34. https://doi.org/10.3233/jpd-202381.

Luiz, L. M. D., Marques, I. A., Folador, J. P., & Andrade, A. O. (2021). Intra and inter-rater remote assessment of bradykinesia in Parkinson's disease. *Neurologia (Engl Ed)*. https://doi.org/10.1016/j.nrl.2021.08.005.

Macchi, Z. A., Ayele, R., Dini, M., Lamira, J., Katz, M., Pantilat, S. Z., et al. (2021). Lessons from the COVID-19 pandemic for improving outpatient neuropalliative care: A qualitative study of patient and caregiver perspectives. *Palliative Medicine*, *35*(7), 1258–1266. https://doi.org/10.1177/02692163211017383.

Mantri, S., & Mitchell, K. T. (2021). Falling down the digital divide: A cautionary tale. *Parkinsonism & Related Disorders*, *93*, 33–34. https://doi.org/10.1016/j.parkreldis.2021.10.032.

Morris, M. E., Slade, S. C., Wittwer, J. E., Blackberry, I., Haines, S., Hackney, M. E., et al. (2021). Online dance therapy for people with Parkinson's disease: Feasibility and impact on consumer engagement. *Neurorehabilitation and Neural Repair*. https://doi.org/10.1177/15459683211046254.

Movement Disorder Society. (2020). https://www.movementdisorders.org/MDS/About/Committees- -Other-Groups/Telemedicine-in-Your-Movement-Disorders-Practice-A-Step-by-Step-Guide.htm.

Myers, T. L., Augustine, E. F., Baloga, E., Daeschler, M., Cannon, P., Rowbotham, H., et al. (2021). Recruitment for remote decentralized studies in Parkinson's disease. *Journal of Parkinson's Disease*. https://doi.org/10.3233/jpd-212935.

Myers, T. L., Tarolli, C. G., Adams, J. L., Barbano, R., Cristina Gil-Díaz, M., Spear, K. L., et al. (2021). Video-based Parkinson's disease assessments in a nationwide cohort of Fox Insight participants. *Clinical Parkinsonism & Related Disorders*, *4*, 100094. https://doi.org/10.1016/j.prdoa.2021.100094.

Ossig, C., Antonini, A., Buhmann, C., Classen, J., Csoti, I., Falkenburger, B., et al. (2016). Wearable sensor-based objective assessment of motor symptoms in Parkinson's disease. *Journal of Neural Transmission (Vienna)*, *123*(1), 57–64. https://doi.org/10.1007/s00702-015-1439-8.

Pahwa, R., Bergquist, F., Horne, M., & Minshall, M. E. (2020). Objective measurement in Parkinson's disease: A descriptive analysis of Parkinson's symptom scores from a large population of patients across the world using the personal KinetiGraph®. *Journal of Clinical Movement Disorders*, 7(1), 5. https://doi.org/10.1186/s40734-020-00087-6.

Peine, A., Paffenholz, P., Martin, L., Dohmen, S., Marx, G., & Loosen, S. H. (2020). Telemedicine in Germany during the COVID-19 pandemic: Multi-professional national survey. *Journal of Medical Internet Research*, *22*(8), e19745. https://doi.org/10.2196/19745.

Quinn, L., Macpherson, C., Long, K., & Shah, H. (2020). Promoting physical activity via telehealth in people with Parkinson disease: The path forward after the COVID-19 pandemic? *Physical Therapy*, *100*(10), 1730–1736. https://doi.org/10.1093/ptj/pzaa128.

Rockwell, K. L., & Gilroy, A. S. (2020). Incorporating telemedicine as part of COVID-19 outbreak response systems. *The American Journal of Managed Care*, *26*(4), 147–148. https://doi.org/10.37765/ajmc.2020.42784.

Royal College of Nursing. (2022). *What is e-health?*. Retrieved from https://www.rcn.org.uk/clinical-topics/ehealth.

Saadi, A., Himmelstein, D. U., Woolhandler, S., & Mejia, N. I. (2017). Racial disparities in neurologic health care access and utilization in the United States. *Neurology*, *88*(24), 2268–2275. https://doi.org/10.1212/wnl.0000000000004025.

Safarpour, D., Dale, M. L., Shah, V. V., Talman, L., Carlson-Kuhta, P., Horak, F. B., et al. (2021). Surrogates for rigidity and PIGD MDS-UPDRS subscores using wearable sensors. *Gait & Posture*, *91*, 186–191. https://doi.org/10.1016/j.gaitpost.2021.10.029.

Scott Kruse, C., Karem, P., Shifflett, K., Vegi, L., Ravi, K., & Brooks, M. (2018). Evaluating barriers to adopting telemedicine worldwide: A systematic review. *Journal of Telemedicine and Telecare*, *24*(1), 4–12. https://doi.org/10.1177/1357633X16674087.

Snoswell, C. L., Taylor, M. L., Comans, T. A., Smith, A. C., Gray, L. C., & Caffery, L. J. (2020). Determining if telehealth can reduce health system costs: Scoping review. *Journal of Medical Internet Research*, *22*(10), e17298. https://doi.org/10.2196/17298.

Solis, J., Franco-Paredes, C., Henao-Martínez, A. F., Krsak, M., & Zimmer, S. M. (2020). Structural vulnerability in the U.S. revealed in three waves of COVID-19. *The American Journal of Tropical Medicine and Hygiene*, *103*(1), 25–27. https://doi.org/10.4269/ajtmh.20-0391.

Subramanian, I., Farahnik, J., & Mischley, L. K. (2020). Synergy of pandemics-social isolation is associated with worsened Parkinson severity and quality of life. *NPJ Parkinson's Disease*, *6*(1), 28. https://doi.org/10.1038/s41531-020-00128-9.

Titova, N., & Chaudhuri, K. R. (2017). Personalized medicine in Parkinson's disease: Time to be precise. *Movement Disorders*, *32*(8), 1147–1154.

van Munster, M., Stümpel, J., Thieken, F., Pedrosa, D. J., Antonini, A., Côté, D., et al. (2021). Moving towards integrated and personalized care in Parkinson's disease: A framework proposal for training Parkinson nurses. *Journal of Personalized Medicine*, *11*(7), 623. https://doi.org/10.3390/jpm11070623.

Vellata, C., Belli, S., Balsamo, F., Giordano, A., Colombo, R., & Maggioni, G. (2021). Effectiveness of telerehabilitation on motor impairments, non-motor symptoms and compliance in patients with Parkinson's disease: A systematic review. *Frontiers in Neurology*, *12*, 627999. https://doi.org/10.3389/fneur.2021.627999.

Wijers, A., Hochstenbach, L., & Tissingh, G. (2021). Telemonitoring via questionnaires reduces outpatient healthcare consumption in Parkinson's disease. *Movement Disorders Clinical Practice*, *8*(7), 1075–1082. https://doi.org/10.1002/mdc3.13280.

Willis, A. W., Schootman, M., Evanoff, B. A., Perlmutter, J. S., & Racette, B. A. (2011). Neurologist care in Parkinson disease: A utilization, outcomes, and survival study. *Neurology*, 77(9), 851–857. https://doi.org/10.1212/WNL.0b013e31822c9123.

Xu, X., Zeng, Z., Qi, Y., Ren, K., Zhang, C., Sun, B., et al. (2021). Remote video-based outcome measures of patients with Parkinson's disease after deep brain stimulation using smartphones: A pilot study. *Neurosurgical Focus*, *51*(5), E2. https://doi.org/10.3171/2021.8.Focus21383.

Yogev-Seligmann, G., & Kafri, M. (2021). COVID-19 social distancing: Negative effects on people with Parkinson disease and their associations with confidence for self-management. *BMC Neurology*, *21*(1), 284. https://doi.org/10.1186/s12883-021-02313-6.

York, M. K., Farace, E., Pollak, L., Floden, D., Lin, G., Wyman-Chick, K., et al. (2021). The global pandemic has permanently changed the state of practice for pre-DBS neuropsychological evaluations. *Parkinsonism & Related Disorders*, *86*, 135–138. https://doi.org/10.1016/j.parkreldis.2021.04.029.

Zipprich, H. M., Teschner, U., Witte, O. W., Schönenberg, A., & Prell, T. (2020). Knowledge, attitudes, practices, and burden during the COVID-19 pandemic in people with Parkinson's disease in Germany. *Journal of Clinical Medicine*, *9*(6), 1643. https://doi.org/10.3390/jcm9061643.

CHAPTER THIRTEEN

Impact of Covid-19 on research and training in Parkinson's disease

Yi-Min Wan[a,b,c,*], Daniel J. van Wamelen[a,c,d], Yue Hui Lau[a,c], Silvia Rota[a,c], and Eng-King Tan[e]

[a]Parkinson's Foundation Centre of Excellence, King's College Hospital NHS Foundation Trust, London, United Kingdom
[b]Department of Psychiatry, Ng Teng Fong General Hospital, Singapore, Singapore
[c]Department of Basic and Clinical Neuroscience, Institute of Psychiatry, Psychology & Neuroscience, King's College London, London, United Kingdom
[d]Department of Neurology, Centre of Expertise for Parkinson & Movement Disorders, Donders Institute for Brain, Cognition and Behaviour, Radboud University Medical Center, Nijmegen, the Netherlands
[e]Duke-NUS Medical School, National Neuroscience Institute, Singapore, Singapore
*Corresponding author: e-mail address: yi_min.wan@kcl.ac.uk

Contents

Abstract

The Coronavirus Disease 2019 (Covid-19) pandemic and the consequent restrictions imposed worldwide have posed an unprecedented challenge to research and training in Parkinson's disease (PD). The pandemic has caused loss of productivity, reduced access to funding, an oft-acute switch to digital platforms, and changes in daily work protocols, or even redeployment. Frequently, clinical and research appointments were suspended or changed as a solution to limit the risk of Severe Acute Respiratory Syndrome Coronavirus-2 (SARS-CoV-2) spread and infection, but since the care and research in the field of movement disorders had traditionally been performed at in-person settings, the repercussions of the pandemic have even been more keenly felt in these areas. In this chapter, we review the implications of this impact on neurological research and training, with an emphasis on PD, as well as highlight lessons that can be learnt from how the Covid-19 pandemic has been managed in terms of restrictions in these crucial aspects of the neurosciences. One of the solutions brought to the fore has

International Review of Neurobiology, Volume 165
ISSN 0074-7742
https://doi.org/10.1016/bs.irn.2022.04.003
Copyright © 2022 Elsevier Inc.
All rights reserved.

been to replace the traditional way of performing research and training with remote, and therefore socially distanced, alternatives. However, this has introduced fresh challenges in international collaboration, contingency planning, study prioritization, safety precautions, artificial intelligence, and various forms of digital technology. Nonetheless, in the long-term, these strategies will allow us to mitigate the adverse impact on PD research and training in future crises.

1. Introduction

In response to the Coronavirus Disease 2019 (Covid-19) pandemic, not only have many hospitals canceled elective procedures, but other clinical and research appointments have been changed to limit the risk of Severe Acute Respiratory Syndrome Coronavirus-2 (SARS-CoV-2) spread and infection to patients, as well as to clinical and research staff. Since medical care and research in movement disorders and Parkinson's disease (PD) have been traditionally performed face-to-face (Cohen, Busis, & Ciccarelli, 2020), a clear impact of the pandemic was felt in clinical and research settings. One of the solutions that have been brought forward was to replace the traditional way of performing research and training with remote, and therefore socially-distanced, alternatives. Although such methods had been active across some centers even before the Covid-19 pandemic, their use has not been widespread.

In this chapter, we review the impact of the Covid-19 pandemic, directly or indirectly linked to the virus itself, as well as of the associated changes effected in response on research and training in the field of neurology with a special focus on PD. Accordingly, we attempt to make recommendations for the post-pandemic world of PD research and training.

2. Research activities

2.1 Impact

The Covid-19 pandemic has transformed the way global research is approached (Lau, Lau, & Ibrahim, 2021), with research projects around the world having to face tremendous challenges at multiple levels (Fig. 1), while some have been forced to halt completely. Those that continued have had to significantly adapt to a new reality (Tan et al., 2021). The term "research resilience" is appropriate in response to a world trying to manage this research crisis (Rahman, Tuckerman, Vorley, & Gherhes, 2021; Tan et al., 2021). It is, indeed, important to proactively incorporate resilience into the research process, as well as to

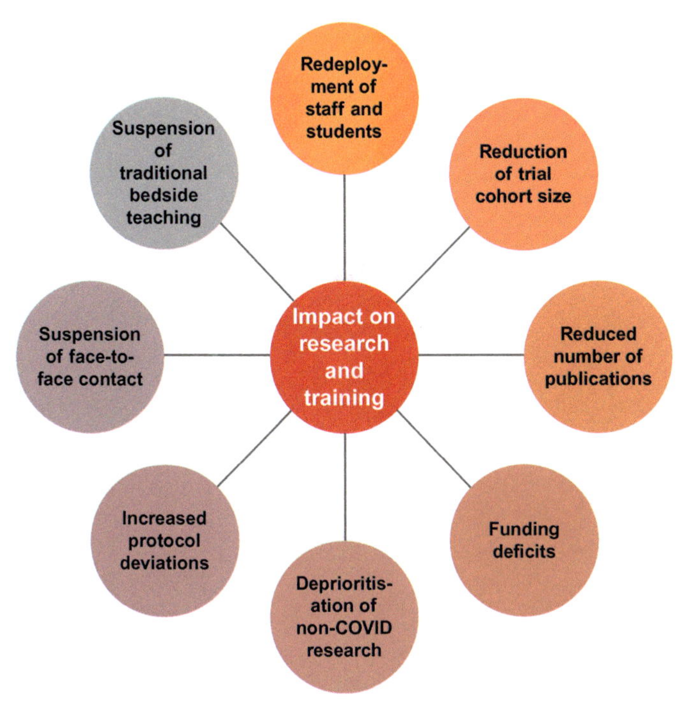

Fig. 1 The impact of Covid-19 pandemic and related restrictions on Parkinson's disease research and training.

provide lessons and considerations on adaptability and flexibility in the research field to obviate challenges faced (Rahman et al., 2021) (Fig. 2).

A striking example of how research has changed in response to the pandemic was the call for Urgent Public Health (UPH) studies in the United Kingdom (UK) (National Institute for Health Research, 2021), where trials related to Covid-19 were prioritized at the expense of non-Covid-related research, such as the ones related to PD. Because of this call, 90% of non-commercial UK government funded National Institute for Healthcare Research (NIHR) studies and over 50% of the commercial studies were temporarily halted (Iacobucci, 2020). Similar actions have been taken in other countries, such as the United States of America (US) (Kardas-Nelson, 2020). Reasons, which have affected the conduct of clinical trials during the crisis, included patient hesitancy or inability to continue interventions due to self-isolation or quarantine, as well as limited access to medical institutions, such as hospitals. Interruption of supply chains and study monitoring has also caused scarcity in research resources (AlNaamani, AlSinani, & Barkun, 2020).

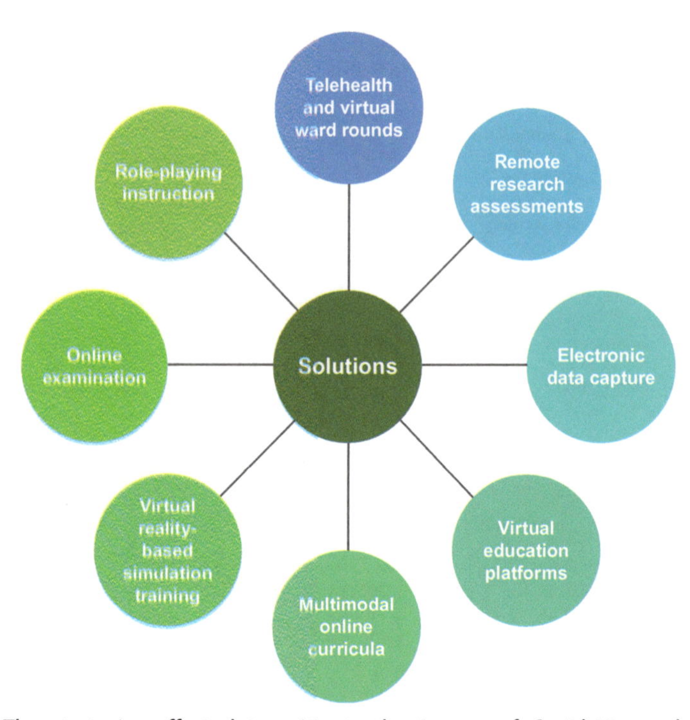

Fig. 2 The strategies effected to mitigate the impact of Covid-19 pandemic on Parkinson's Disease research and training.

As a consequence, researchers had to shift their work from a live modality to a remote one and change their priorities, focusing more on data analysis, drafting papers, and writing grant proposals (Chan & Tan, 2020). According to some authors, there have been about 6–12 months of lost productivity in research with a reduction of trial size by 30% due to patients' drop-outs (Kardas-Nelson, 2020). This has also been reflected by the drop in the number of published studies since the beginning of the pandemic (Fig. 3). Additionally, Covid-19 has caused a fundraising deficit with an average of 38% reduction in donations in medical research charities (Parkinson's UK, 2020). When looking specifically at PD research, a survey run among the Parkinson Study Group (PSG) clinician members has reported that most of the researchers stated a 75–100% reduction of their activities, while many of them had to perform protocol deviations (38.2%) or exceptions (25.5%), and some even had to change their work profile because of layoffs (16.8%) (Shivkumar et al., 2021).

Other areas where the impact of Covid-19 has been felt, were that of disease symptomatology. Data interpretation in PD clinical studies needs

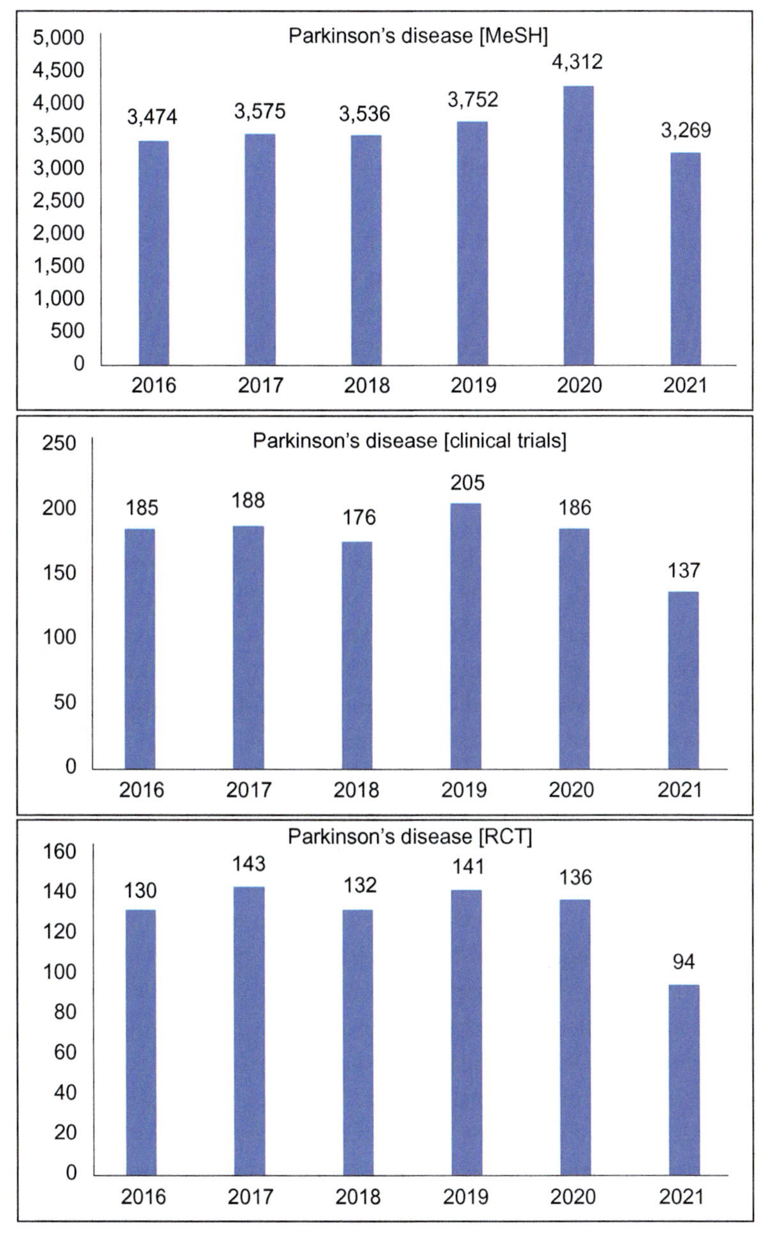

Fig. 3 The decline in the number of studies in people with Parkinson's disease over the year 2021 compared to the preceding 5 years; note the even greater decline in the number of clinical trials and randomized clinical trials (RCT). *Source: PubMed.*

to be handled cautiously, as neurological symptoms of Covid-19 may overlap with PD (and related disorders), with long-term effects on PD remaining unclear (Tan et al., 2021). Methods, which have facilitated more complete data collection during the Covid-19 pandemic, are also crucial in improving the validity of assessments, efficacy, and safety (Council., U. N. R, 2010; Fleming, 2011).

Recognizing the potential effect of pandemics as a source of bias is important to allow a balanced interpretation of results. The Covid-19 pandemic constitutes a potential source of unintentional bias and researchers should be careful not to overgeneralize the results and be more vigilant in the inspection of the data for type I/II errors (Simundic, 2013). The pandemic has also disrupted research participation among people with PD (PwP); one study has demonstrated that 40% of PwP (without Covid-19), who had been actively participating in research prior to the pandemic, had to cancel, and 35% were forced to postpone in-person research visits, although the remaining 25% were able to conduct research visits via other means (Brown et al., 2020). On the whole, while 11% felt that the pandemic has made them less likely to participate in research, 6.3% felt that the pandemic has actually made them more likely to participate (Brown et al., 2020). However, compensatory activities, such as telemedicine and digital research platforms, may exclude those lacking technological resources or competency.

Nonetheless, some of the effects of the pandemic on research seem to have been positive, providing prospects for growth for individual researchers and the scientific community. For instance, many research conferences have pivoted to a virtual format, drastically reducing the costs of registration and attendance, which has allowed for greater and more inclusive participation in global networking. Scientists have also taken greater advantage of alternative platforms for disseminating research findings, such as PsyArXiv, and of online data collection systems, which have responded by increasing research capacity. Online modalities have the capability of reaching a wider and more diverse population than traditional, center-based research, while targeted recruitment will allow future studies to clarify the extent of and plan for disparities in healthcare access during and after the pandemic (Brown et al., 2020). Strategies that have been used to reduce the impact of the Covid-19 pandemic are outlined below.

2.2 Strategies

The common limitations imposed by the pandemic which may have affected the integrity of PD (and other) clinical studies included delays in study enrolment, staffing changes, limited clinic space, remote interactions due to social distancing measures, manpower shortages due to staff redeployment to Covid-19 frontline, lack of training to handle emergent infectious disease protocols, and inadequate financial support (Tan et al., 2021). As such, research and clinical teams have increasingly switched to telehealth modalities and the adoption of flexible research protocols and clinical trials to maintain study and data integrity (Fleming, Labriola, & Wittes, 2020; Medicines and Healthcare Products Regulatory Agency, 2020; Rose, 2020). In fact, during the pandemic, telehealth use has increased from 9.7% to 63.5%, and almost half of the patients (46%) preferred to continue using it, especially patients with a higher income and higher education, or those where technical support could be provided by a carer (Feeney et al., 2021). Nevertheless, considering the determinants of its use among patients, telehealth was able to reach only a limited, probably non-representative part of the population, a fact that could have important implications, especially in medical research.

Other crucial factors influencing trial continuation and resumption were Covid-19 disease burden and community spread (Tan et al., 2021), necessitating the prioritization of preventive, diagnostic, and interventional measures to further reduce the risk of exposure. Part of this, for all ongoing and new projects, included ranking of studies in terms of importance in the order of therapeutics and/or vaccines, diagnostics, and epidemiology, mainly impacting those studies evaluating interventions aimed at preventing or treating diseases other than Covid-19 (Tan et al., 2021). Clinical trials investigating potential disease-modifying therapies in progressive neurodegenerative diseases (such as PD), or high-impact novel symptomatic therapies were to be minimized (Papa et al., 2020). Over the course of the pandemic, clinical trials related to non-Covid-19 conditions, such as life-saving interventions for acute strokes and neurology trials for diseases with rapid progression, like amyotrophic lateral sclerosis and multiple sclerosis, resumed or continued, while other studies involving human subjects restarted gradually (Tan et al., 2020).

Additional solutions included classifying the different types of PD studies, weighing the merits of each individual study in deciding which ones to resume first. For instance, priority was given to clinical studies with no patient contact, such as telephone surveys assessing quality of life or drug-related side effects in PD, studies which could be conducted in non-clinical areas, such as an imaging study in a dedicated research scanner, and clinical trials in their passive phases (AlNaamani et al., 2020).

Potential benefits from a particular trial must be balanced against the possible risks of exposing patients and researchers to Covid-19 (Tan et al., 2020). For any study with face-to-face contact, stringent efforts should be undertaken to comply with new safety protocols regarding social distancing, protective personal equipment, and minimizing the number of researchers physically present on-site during visits (Tan et al., 2020). To maintain physical distancing between laboratories and trial units, reopening was meant to start at reduced capacity and all staff on-site mandated to wear a mask to prevent the risk of SARS-CoV-2 transmission, with allocation of select staff to work from home. At the same time, negotiations should be done with sponsors to extend intervals for investigations, schedule study patients at the same time as regular appointments, and to defer non-essential biochemical tests. Other steps that could be taken are to sanction remote consent taking with subjects, as well as virtual investigator training, and to allow research coordinators to cross-cover different studies and work on staggered schedules (Tan et al., 2020).

An alternative was to delay initiation of enrolment in trials that had not yet started or to pause enrolment in ongoing trials, perhaps on a site-specific basis, until the SARS-CoV-2 burden in that setting is markedly reduced. In trials that were relatively near completion when severe disruption began, the study team (and not the data monitoring committee) could decide to terminate the trial, thus, sacrificing a small degree of statistical power in exchange for more interpretable inferences (Fleming et al., 2020). Valid statistical approaches should guide the presentation of results of clinical trials for which the conduct has been meaningfully influenced by the pandemic, taking into account missing data and protocol deviations (Fleming et al., 2020). Also, potential medico-legal implications must be considered in situations where trial participants may be subjected to additional risk from accidental exposure to Covid-19 while conducting clinical trials (Tan et al., 2020). Population screening (PCR and/or antigen testing) of all asymptomatic individuals, including study subjects and research staff for Covid-19, could be impractical, although testing in healthcare workers in the frontline has been frequently carried out in some places (Tan et al., 2021).

For those studies where virtual assessments could be an option, other factors need to be taken into consideration. Virtual consent-taking, remote prescription with delivery during enforced quarantine, and possibly institution-backed legal recourse have been among the available options (Tan et al., 2021). These measures have enabled research staff to work effectively from home by telecommunications (Tan et al., 2020). Also, supplemental approaches, such as electronic data capture implemented at home by the patient or caregiver, telemedicine, or telephone interviews (McDermott & Newman, 2020) have proven to be effective.

Further developments in telemedicine could take the form of artificial intelligence with an integrated-omics approach, aiding in better defining disease models and discovering new therapeutic targets in the field of PD (Adly, Adly, & Adly, 2020; Welton & Tan, 2021). However, despite the positive emergence of telemedicine (Bloem, Dorsey, & Okun, 2020) and the ingenious use of virtual reality for certain types of PD therapy (Chen, Gao, He, & Bian, 2020), compliance will be difficult to implement fully on the ground (Tan et al., 2021). Furthermore, validated clinical scales to assess PD and related complications through remote evaluations are currently not available and, hence, require further investigations (Tan et al., 2021).

The repercussions on research funding and policies during this pandemic also depend significantly on the resources available in that specific jurisdiction and the prevalence of SARS-CoV-2 infection of the study locality. Complications, such as unfunded extensions of grants, can lead to unfinished research or delayed initiation of studies. For instance, research funded by existing grants that had been halted, recommenced during transition period between lockdown and re-opening in the UK, whereas overall research funding has not been much affected in Singapore, with surplus funds provided for Covid-19 research (Tan et al., 2020). In Japan, telemedicine has been included in universal insurance systems since 2018, but reimbursement for it has not proven sufficient during the pandemic (Suzuki et al., 2021).

Taking all the above into consideration, the ultimate goal for researchers in a priori research planning should be to incorporate more reflection and flexibility into their research preparation, and contemplate integrating alternative data collection methods (Rahman et al., 2021). In the case of grant-funded projects, such considerations can be applied into research contingency plans (Rahman et al., 2021). There is also an extensive arsenal of studies which explore the topic of digital research ethics, encompassing issues such as informed consent, confidentiality, data security, privacy, anonymity, data storage and processing, and ethical decision making while using digital methods (Rahman et al., 2021). By applying a critical reflection framework to research practices, we have been able to provide empirical insights and lessons into research continuity in the current global pandemic context (White, Fook, & Gardner, 2006). This will allow research output obtained during the Covid-19 pandemic into making crucial contributions to the literature (Rahman et al., 2021).

Due to the limited time available to enroll large patient cohorts required for many high-quality clinical trials, as well as the resources required during such a crisis, an early coordinated effort is imperative to assimilate data and produce research findings with more meaningful scientific rigor (AlNaamani et al., 2020). International research collaborations would be particularly essential in this regard (Fearon & Fasano, 2021), preferably to the ends of developing consensus guidelines in the conduct of PD research (AlNaamani et al., 2020).

Collaborative work with multi-center participation from different countries will enable more cohesive diverse participation and allow sufficient support of those with limited resources, permitting enrolment of larger sample sizes (Adly et al., 2020) toward favoring the broad generalizability of results, as well as adequately powered subgroup analyses.

Planning for future emergencies is a coordinated global effort in establishing worldwide protocols for the rapid implementation of strategic research prioritization and redeployment of research infrastructure and capacity (Wyatt, Faulkner-Gurstein, Cowan, & Wolfe, 2021). Dedicated databases of PD cohorts with and without Covid-19 should be assembled for future studies (Fearon & Fasano, 2021).

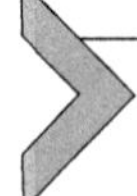

3. Medical and specialty training

3.1 Impact

The Covid-19 pandemic has not only directly impacted patient care, where many hospitals had to upscale their capacity and care delivery, but has also effected an unprecedented influence over healthcare professional education (Fig. 1), including the training of nursing staff and medical students (Tan et al., 2020; van der Meulen, Kleineberg, Schreier, Garcia-Azorin, & Di Lorenzo, 2020). This is exemplified by a study from Italy, showing that across 36 Medical Schools, all lessons, seminars, and conferences had been suspended, with 92% of centers rescheduling these on virtual platforms and seminars in the other centers (Di Lorenzo et al., 2021).

The above has been supported by another study showing similar rates of disruption (Abati & Costamagna, 2020). The impact has also been felt in teaching hospitals where traditional bedside clinical teaching often had to be suspended, affecting higher grade training, such as for speciality registrars and residents. Overall, the majority of those affected by the restrictions have been feeling that these would have major repercussions on their training. In fact, 79% of respondents in a European Academy of Neurology survey have indicated that the pandemic will probably have a serious impact on their training and career (Cuffaro et al., 2021). In this section, we have tried to limit ourselves to evidence in relation to the impact of the Covid-19 pandemic and related restrictions on neurological (specifically movement disorders) training, as well as stipulate some of the solutions to overcome these problems.

Clinical instructional experiences and exposure to real patients have traditionally been considered the cornerstones of nursing and medical education, an indispensable part of the healthcare professionals' development (AlThiga, Mohidin, Park, & Tekian, 2017; Mandan, Sidhu, & Mahmood, 2016). As a result of the Covid-19 pandemic, concerns have arisen that interruption of the clinical practicum could lead to a reduced quality of medical education. Thus, faculty had to urgently address this matter and ensure the quality of care delivered by future healthcare workers.

3.2 Strategies

As a learning experience for medical and nursing students, some medical schools have encouraged their students to work on the frontline during the Covid-19 pandemic (Lee, Park, & Seo, 2020; Leigh et al., 2020). However, shortage of personal protective equipment (PPE), restricted availability of Covid-19 testing, and infection control schemes have been some of the limiting factors (Wang, Deng, & Tsui, 2020). Therefore, digital education without the risks associated with in-person education could drive the way forward (Park, Park, Lim, Rhim, & Lee, 2020), and, indeed, have been applied by many Medical Schools with the adoption of virtual platforms for remotely delivered lectures or digital banks of resources, especially in the pre-clinical years. However, some concerns have been raised regarding the impact of online teaching during the clinical years, with arguments regarding the ability of students to develop clinical competence (Dost, Hossain, Shehab, Abdelwahed, & Al-Nusair, 2020; Huddart et al., 2020).

As such, these developments should ideally be standardized, by evaluating teaching methods and telemedicine, reinforcing wellbeing, and promoting international educational collaborations, which have been suggested to improve neurology training during and after the pandemic (Sandrone et al., 2021).

A large part of the available knowledge regarding the repercussion of the Covid-19 pandemic on medical healthcare professional training has recently been summarized by Hao and colleagues, who performed a systematic review of studies on this topic available up to April 2021 (Hao et al., 2022). They were able to identify 16 studies, spanning the world (United States (*n*=7), China (*n*=3), UK (*n*=1), Japan (n=1), Korea (n=1), Italy (n=1), Saudi Arabia (n = 1), and Israel (n = 1)) and a total of 1174 participants (457 were undergraduate nursing students and 717 medical students), looking at the effect of the Covid-19 pandemic on nursing and medical training. Based on their results, it seems that universities and colleges resorted to four different types of education replacing classical training:

(1) virtual reality-based simulation training;
(2) teleconsultation and virtual rounds;
(3) web-based specialized skills learning; and.
(4) multimodal online curricula.

Virtual reality-based stimulation training seemed to be a good alternative to in-person training, whereby Assessment Technologies Institute (ATI) scores, the perceived quality of this training modality, has been comparable between students participating in this course compared to previous cohorts who were trained in-person (Weston & Zauche, 2021). Some results have also suggested that more than half of students preferred simulation training with virtual platforms rather than online formal teaching (De Ponti et al., 2020). This has also been confirmed by other studies showing that online teaching has been perceived as having increased convenience, enhanced quality, a sense of comfort and safety (Pokryszko-Dragan, Marschollek, Nowakowska-Kotas, & Aitken, 2021). Nonetheless, the promoting effect of simulation training on overall clinical performance was found to be slightly inferior (Kang, Kim, Lee, Kim, & Kim, 2020).

For surgical education relevant to the PD field (deep brain stimulation), most students believed new web-based surgical instruction was as difficult/easy as conventional teaching, supported by another study looking at reviewing radiology exams, which has shown that remote training has been perceived as more active and entertaining than in-person sessions (Alpert, Young, Lala, & McGuinness, 2021; Co & Chu, 2020). Similarly, telemedicine, including teleconsultation and virtual rounds, has shown benefit for students. For example, telemedicine appointments using video calls or video conferencing to interact with patients during virtual rounds have improved clinical ability and professional confidence in most of the students (Gummerson et al., 2021; Weber et al., 2021). Some studies have postulated that the overall ability to teach tele-instructors in virtual rounds has been better than even in-person rounds (Bala et al., 2021; Sukumar et al., 2021; Weber et al., 2021).

Perhaps the best studied area of remote training is that of multimodal online curricula, where Hao and colleagues were able to identify eight studies looking at this topic (Hao et al., 2022). Such curricula included online videos, massive open online courses, discussion posts, virtual conferences, impromptu role playing and lectures, with most interested students opting for interactive discussion learning patterns and not passive teaching resources (Coffey, MacDonald, Shahrvini, Baxter, & Lander, 2020; He et al., 2021; Michener, Fessler, Gonzalez, & Miller, 2020; Weber et al., 2020). Other examples of useful resources included role-playing instruction, where a faculty member would play the role of a patient and a student acted as a physician (Kaliyadan, ElZorkany, & Al Wadani, 2020; Kasai et al., 2021). Interestingly, some programs have even allowed for students to participate in supervised remote direct patient care and support frontline healthcare workers by performing remote clinical tasks, which have been perceived as overall positive by the students (satisfaction scores 3.33–4.57 out of 5) (Safdieh et al., 2021).

For higher degree training, such as specialty registrars/residents, surprisingly less evidence is available, at least in the field of neurology and movement disorders. Perhaps the Covid-19 pandemic has caused the matter of training and research to be less pressing, and has, therefore, received less attention. This seems to be supported by the notion that for most registrars/residents, training and research have not been perceived as "most stressful" during the pandemic. Rather, safety, violence and aggression, and family matters have been perceived as the main stressors (Wu et al., 2021). Moreover, when virtual methods were used to solve some of the surrounding issues, like patient rounds on wards as an educational part of training, registrars/residents did not seem as positive. In the study by Kolikonda and colleagues, it has been shown that in "virtual rounding", consisting of patient presentation and discussion in the morning in on-line virtual team format, followed by in-person patient rounds in small groups on a stroke ward, the majority of neurology residents found telemedicine applications not useful compared to other healthcare staff providers. The authors attributed this to lack of prior exposure and preparedness (Kolikonda et al., 2022).

On the other hand, some studies have shown a more positive attitude towards virtual ward rounds (Gros et al., 2021; Zeinali, Almasi-Doghaee, & Haghi-Ashtiani, 2020) with a sharp increase in the number of hours per month spent on virtual educational meetings after the start of the pandemic (35 vs. 16h per month) (Zeinali et al., 2020). Some services have already tried to address this by considering contingency planning, maintenance of education, sustainability of team members, and promotion of the safe delivery of neurological care, for example in the University of Toronto Adult Neurology Residency Program (Muir et al., 2021).

Similar strategies have also been deployed elsewhere in the form of a virtual intensive care unit (tele-ICU) rotation for medical students to support the care of patients diagnosed with Covid-19 in the ICU, consisting of clinical engagement, a multiple-choice pre-test, faculty-supervised, student-led case and topic presentations, faculty-led debriefing sessions, evidence-based-medicine discussion forums, a multiple-choice post-test, and a final reflection (Ho et al., 2021).

Other strategies that have been proposed to improve teaching during the pandemic, included remote supervision of neurology fellows. Suarez-Cedeno and colleagues have shown that such an approach has been considered as successful by consultants, mainly due to increased patient access and a decrease in scheduling barriers, despite some technical difficulties and lack of portions of the examination, such as tone, reflexes, and sensory testing (Suarez-Cedeno et al., 2021). In relation to practical teaching, including electroencephalogram (EEG) teaching, more evidence regarding solutions is becoming available. Yadala and colleagues have shown that virtual EEG training via zoom could be effective, with clear improvement in post-test compared to pre-test scores and high satisfaction rates among registrars/residents (100% felt more confident after the training in scoring EEGs). Moreover, when compared to traditional EEG reading, 100% have agreed that the virtual sessions were more accessible, 72.7% have agreed that they were more interactive, 81.9% have felt engaged, and 90.9% have felt they were able to attend more sessions (Yadala et al., 2020).

Finally, other obstacles in relation to teaching, such as final exams, have been the topic of a recent study. Rajan and colleagues have shown that it was possible to perform a neurology exit examination in a hybrid virtual format. They have created a case repository with history and clinical examination findings, followed by structured questions for case discussions, where external examiners would assess candidates virtually through a video conferencing platform (Rajan et al., 2020).

For postgraduate and other healthcare professional teaching, the pandemic has necessitated changes to the conventional way these used to be delivered. Examples include Continuing Medical Education (CME) programs where a general solution would be to switch to virtual webinars. Although such webinars have comprised a major avenue for education during the pandemic and initial satisfaction among physicians was high (around 75% have reported to be satisfied or very satisfied), the majority of physicians (over 75%) have felt overwhelmed with the number and frequency of webinars as the pandemic continued, resulting in increasingly lower attendance rates, as was recently shown by Ismail, Abdelkarim, and Al-Hashel (2021). The authors have suggested that webinars should be viewed as complementing traditional in-person meetings, rather than replacing them.

Overall, although the Covid-19 pandemic and the imposed restrictions have had a major impact on teaching and training across the whole medical field, ranging from students to postgraduates, an increasing number of solutions have been recommended. After the initial cessation of in-person teaching and training sessions, many universities and hospital have now resorted to online courses, lectures, and curricula, supported by in-person sessions whenever possible. Available evidence suggests that this approach has been largely successful, achieving high rates of satisfaction with a demonstrable and adequate transfer of knowledge using virtual and other remote methods. The acquired knowledge would be important for future worldwide medical emergencies and pandemics, so that continuity in high-level quality of training and teaching can be assured.

4. Mental health in research and training

The Covid-19 pandemic has had a negative impact on the mental health of PwP. Among the PD non-motor symptoms, anxiety and depression have significantly increased during the pandemic (Dommershuijsen et al., 2021). The disruption of healthcare services, loss of usual activities and supports, and reductions in physical activity are some of the possible causes of this phenomenon (Brooks, Weston, & Greenberg, 2021). A recent study exploring the effect of the pandemic on the mental health of PwP has demonstrated a stronger association of Covid-19 stressors with mental health in women, highly educated people, individuals with advanced PD, and those prone to social distancing or seeking social support (Dommershuijsen et al., 2021).

A review published in March 2020, only a year after the outbreak, reported how the Covid-19 outbreak has led to the emergence of mental health issues, such as anxiety, insomnia, and denial, both in the SARS-CoV-2 infected and non-infected individuals, with an increase of the prevalence of post-traumatic stress disorder and depression by almost 40% and 7% respectively in the general population (Torales, O'Higgins, Castaldelli-Maia, & Ventriglio, 2020). Lockdown measures and the consequent isolation have been acknowledged as some of the possible factors underlying these mental health issues, similarly to past outbreaks, such as the Middle East Respiratory Syndrome Coronavirus (MERS-CoV) in

Korea in 2015 (Kim et al., 2019). Female sex, lower socioeconomic status or frequent social media use have been shown to be possible risks factor for the development of mental health issues (Li & Wang, 2020; Mazza et al., 2020; Pierce et al., 2020).

Although it might still be early to appreciate the overall long-term impact of the Covid-19 pandemic on academia, the immediate effect on researchers' productivity and mental health are already visible. A recent survey involving more than 2000 European and American researchers has shown a decrease of 5% and 36% in the hours spent at work and in the number of new non-Covid-19-related projects initiated in the past year respectively (Gao, Yin, Myers, Lakhani, & Wang, 2021). Indeed, a search through the major database of PubMed has revealed a stark decline in the quantity of PD studies compared to the pre-pandemic era (Fig. 3).

A gender gap has been found in the reduction of time dedicated to work, with female scientists and especially mothers with young children being most affected (Myers et al., 2020). This tendency was shown to be particularly true for black female mother researchers (Staniscuaski et al., 2021). Another study conducted in the UK has shown how the Covid-19 pandemic has affected researchers' mental approach to their work, with the lowest scores observed in questions related to the energy levels and feelings of relaxation, optimism and cheerfulness, with one in four responders showing signs of burnout, such as the daily feeling of being emotionally drained from work (Wray & Kinman, 2022).

In the above context, students have been largely affected, with an increase of mental health concerns related to school closures, loss of routine, and reduced social interactions (YoungMinds, 2020). The online learning format adopted by tertiary education institutions, despite being considered a good alternative to traditional learning (Abou El-Seoud et al., 2014) and offering some major advantages, such as flexibility (Dhawan, 2020), has been viewed as another additional stressor by some authors. Obstacles, such as technical difficulties, can reduce students' motivation and put them under increased pressure to learn autonomously, leading to potentially higher rates of dropout (Coman, Țîru, Meseșan-Schmitz, Stanciu, & Bularca, 2020; Grubic, Badovinac, & Johri, 2020).

As already mentioned, several factors have limited the interaction among medical students and patients, including the reduction of in-person assessments, the possibility of being potential vectors of SARS-CoV-2, or shortages in PPE (especially during the beginning of the pandemic)

(Ahmed, Allaf, & Elghazaly, 2020; Roberts, 2020; Rose, 2020), leading to recently emerged alarming data on medical students' mental health. A study conducted at six Jordanian Medical Schools, assessing the mental wellbeing status using Kessler's psychological stress scale (K10), has demonstrated that about half of the participants had severe mental disorders with high psychological distress (Seetan et al., 2021). The negative effect of isolation on physical exercise, the lack of training, and the reduced social interactions have been noted to be among the most frequent worries. Additionally, this study has shown higher scores in younger students compared to those being at their last years of studies, a finding which has also been confirmed in a Brazilian study, reporting a high level of mental illness burden (measured with the Hospital Anxiety and Depression Scale, the Self-Reporting Questionnaire, the Interpersonal Reactivity Index, and the Mindful Attention Awareness Scale) in medical students, particularly among freshmen (Perissotto et al., 2021).

5. Conclusions

Face-to-face patient assessments, conduct of research visits, collaboration between researchers, and the redeployment of researchers to clinical care to alleviate hospitals from the clinical burden of the pandemic have been recognized as key research areas, significantly affected by the Covid-19 crisis. These modifications have resulted in new ways to conduct research, providing more resilient long-term options for future pandemics and worldwide events, including remote visits, digital data collection and electronic capture of outcomes, as well as the establishment of virtual platforms for research and education.

In general, the pandemic has created an unprecedented opportunity for researchers and healthcare professionals to show resilience and innovation and adapt to a new norm, leading to a paradigm shift in PD research and training strategies globally in the post-Covid-19 era. While much ambiguity remains, what is certain is that future neurological research and education landscape will never be the same compared to pre-pandemic times. Many institutions have now enforced this 'new normal' and productively embedded the strategies sparked by the pandemic as part of their standardized guidelines. Hopefully, the evolving changes will enhance the resilience of the medical fraternity in improving the provided services and prepare for a potential next pandemic.

References

Abati, E., & Costamagna, G. (2020). Education research: Effect of the COVID-19 pandemic on neurology trainees in Italy: A resident-driven survey. *Neurology, 95*(23), 1061–1066. https://doi.org/10.1212/WNL.0000000000010878.

Abou El-Seoud, S., Seddiek, N., Taj-Eddin, I., Ghenghesh, P., Nosseir, A., & El-Khouly, M. (2014). E-learning and students' motivation: A research study on the effect of E-learning on higher education. *International Journal of Emerging Technologies in Learning, 9*(4), 20–26. https://doi.org/10.3991/ijet.v9i4.3465.

Adly, A. S., Adly, A. S., & Adly, M. S. (2020). Approaches based on artificial intelligence and the internet of intelligent things to prevent the spread of COVID-19: Scoping review. *Journal of Medical Internet Research, 22*(8), e19104. https://doi.org/10.2196/19104.

Ahmed, H., Allaf, M., & Elghazaly, H. (2020). COVID-19 and medical education. *The Lancet Infectious Diseases, 20*(7), 777–778. https://doi.org/10.1016/S1473-3099(20)30226-7.

AlNaamani, K., AlSinani, S., & Barkun, A. N. (2020). Medical research during the COVID-19 pandemic. *World Journal of Clinical Cases, 8*(15), 3156–3163. https://doi.org/10.12998/wjcc.v8.i15.3156.

Alpert, J. B., Young, M. G., Lala, S. V., & McGuinness, G. (2021). Medical student engagement and educational value of a remote clinical radiology learning environment: Creation of virtual read-out sessions in response to the COVID-19 pandemic. *Academic Radiology, 28*(1), 112–118. https://doi.org/10.1016/j.acra.2020.09.011.

AlThiga, H., Mohidin, S., Park, Y. S., & Tekian, A. (2017). Preparing for practice: Nursing intern and faculty perceptions on clinical experiences. *Medical Teacher, 39*(sup1), S55–S62. https://doi.org/10.1080/0142159X.2016.1254739.

Bala, L., Kinross, J., Martin, G., Koizia, L. J., Kooner, A. S., Shimshon, G. J., et al. (2021). A remote access mixed reality teaching ward round. *The Clinical Teacher, 18*(4), 386–390. https://doi.org/10.1111/tct.13338.

Bloem, B. R., Dorsey, E. R., & Okun, M. S. (2020). The coronavirus disease 2019 crisis as catalyst for telemedicine for chronic neurological disorders. *JAMA Neurology*, 77(8), 927–928. https://doi.org/10.1001/jamaneurol.2020.1452.

Brooks, S. K., Weston, D., & Greenberg, N. (2021). Social and psychological impact of the COVID-19 pandemic on people with Parkinson's disease: a scoping review. *Public Health, 199*, 77–86. https://doi.org/10.1016/j.puhe.2021.08.014.

Brown, E. G., Chahine, L. M., Goldman, S. M., Korell, M., Mann, E., Kinel, D. R., et al. (2020). The effect of the COVID-19 pandemic on people with Parkinson's disease. *Journal of Parkinson's Disease, 10*(4), 1365–1377. https://doi.org/10.3233/JPD-202249.

Chan, C. H., & Tan, E. K. (2020). Safeguarding non-COVID-19 research: Looking up from ground zero. *Archives of Medical Research, 51*(7), 731–732. https://doi.org/10.1016/j.arcmed.2020.05.023.

Chen, Y., Gao, Q., He, C. Q., & Bian, R. (2020). Effect of virtual reality on balance in individuals with Parkinson disease: A systematic review and Meta-analysis of randomized controlled trials. *Physical Therapy, 100*(6), 933–945. https://doi.org/10.1093/ptj/pzaa042.

Co, M., & Chu, K. M. (2020). Distant surgical teaching during COVID-19—A pilot study on final year medical students. *Surgical Practice*. https://doi.org/10.1111/1744-1633.12436.

Coffey, C. S., MacDonald, B. V., Shahrvini, B., Baxter, S. L., & Lander, L. (2020). Student perspectives on remote medical education in clinical core clerkships during the COVID-19 pandemic. *Medical Science Educator*, 1–8. https://doi.org/10.1007/s40670-020-01114-9.

Cohen, B. H., Busis, N. A., & Ciccarelli, L. (2020). Coding in the world of COVID-19: Non-face-to-face evaluation and management care. *Continuum (Minneap Minn), 26*(3), 785–798. https://doi.org/10.1212/CON.0000000000000874.

Coman, C., Țîru, L. G., Meseșan-Schmitz, L., Stanciu, C., & Bularca, M. C. (2020). Online teaching and learning in higher education during the coronavirus pandemic: Students' perspective. *Sustainability*, *12*(24), 10367. https://doi.org/10.3390/su122410367.

Council., U. N. R. (2010). The prevention and treatment of missing data in clinical trials, D. o. B. a. S. S. a. E. In *Panel on Handling Missing Data in Clinical Trials. Committee on National Statistics*, National Academies Press.

Cuffaro, L., Carvalho, V., Di Liberto, G., Klinglehoefer, L., Sauerbier, A., Garcia-Azorin, D., et al. (2021). Neurology training and research in the COVID-19 pandemic: a survey of the resident and research fellow section of the European academy of neurology. *European Journal of Neurology*, *28*(10), 3437–3442. https://doi.org/10.1111/ene.14696.

De Ponti, R., Marazzato, J., Maresca, A. M., Rovera, F., Carcano, G., & Ferrario, M. M. (2020). Pre-graduation medical training including virtual reality during COVID-19 pandemic: A report on students' perception. *BMC Medical Education*, *20*(1), 332. https://doi.org/10.1186/s12909-020-02245-8.

Dhawan, S. (2020). Online learning: A panacea in the time of COVID-19 crisis. *Journal of Educational Technology Systems*, *49*(1), 5–22. https://doi.org/10.1177/0047239520934018.

Di Lorenzo, F., Ercoli, T., Cuffaro, L., Barbato, F., Iodice, F., Tedeschi, G., et al. (2021). COVID-19 impact on neurology training program in Italy. *Neurological Sciences*, *42*(3), 817–823. https://doi.org/10.1007/s10072-020-04991-5.

Dommershuijsen, L. J., Van der Heide, A., Van den Berg, E. M., Labrecque, J. A., Ikram, M. K., Ikram, M. A., et al. (2021). Mental health in people with Parkinson's disease during the COVID-19 pandemic: Potential for targeted interventions? *NPJ Parkinsons Disease*, 7(1), 95. https://doi.org/10.1038/s41531-021-00238-y.

Dost, S., Hossain, A., Shehab, M., Abdelwahed, A., & Al-Nusair, L. (2020). Perceptions of medical students towards online teaching during the COVID-19 pandemic: A national cross-sectional survey of 2721 UK medical students. *BMJ Open*, *10*(11), e042378. https://doi.org/10.1136/bmjopen-2020-042378.

Fearon, C., & Fasano, A. (2021). Parkinson's disease and the COVID-19 pandemic. *Journal of Parkinson's Disease*, *11*(2), 431–444. https://doi.org/10.3233/JPD-202320.

Feeney, M. P., Xu, Y., Surface, M., Shah, H., Vanegas-Arroyave, N., Chan, A. K., et al. (2021). The impact of COVID-19 and social distancing on people with Parkinson's disease: A survey study. *NPJ Parkinsons Disease*, 7(1), 10. https://doi.org/10.1038/s41531-020-00153-8.

Fleming, T. R. (2011). Addressing missing data in clinical trials. *Annals of Internal Medicine*, *154*(2), 113–117. https://doi.org/10.7326/0003-4819-154-2-201101180-00010.

Fleming, T. R., Labriola, D., & Wittes, J. (2020). Conducting clinical research during the COVID-19 pandemic: Protecting scientific integrity. *JAMA*, *324*(1), 33–34. https://doi.org/10.1001/jama.2020.9286.

Gao, J., Yin, Y., Myers, K. R., Lakhani, K. R., & Wang, D. (2021). Potentially long-lasting effects of the pandemic on scientists. *Nature Communications*, *12*(1), 6188. https://doi.org/10.1038/s41467-021-26428-z.

Gros, P., Rotstein, D., Kinach, M., Chan, D. K., Montalban, X., Freedman, M., et al. (2021). Innovation in resident education—description of the neurology international residents videoconference and exchange (NIRVE) program. *Journal of the Neurological Sciences*, *420*, 117222. https://doi.org/10.1016/j.jns.2020.117222.

Grubic, N., Badovinac, S., & Johri, A. M. (2020). Student mental health in the midst of the COVID-19 pandemic: A call for further research and immediate solutions. *The International Journal of Social Psychiatry*, *66*(5), 517–518. https://doi.org/10.1177/0020764020925108.

Gummerson, C. E., Lo, B. D., Porosnicu Rodriguez, K. A., Cosner, Z. L., Hardenbergh, D., Bongiorno, D. M., et al. (2021). Broadening learning communities during COVID-19: Developing a curricular framework for telemedicine education in neurology. *BMC Medical Education*, *21*(1), 549. https://doi.org/10.1186/s12909-021-02979-z.

Hao, X., Peng, X., Ding, X., Qin, Y., Lv, M., Li, J., et al. (2022). Application of digital education in undergraduate nursing and medical interns during the COVID-19 pandemic: A systematic review. *Nurse Education Today, 108*, 105183. https://doi.org/10.1016/j.nedt.2021.105183.

He, M., Tang, X. Q., Zhang, H. N., Luo, Y. Y., Tang, Z. C., & Gao, S. G. (2021). Remote clinical training practice in the neurology internship during the COVID-19 pandemic. *Medical Education Online, 26*(1), 1899642. https://doi.org/10.1080/10872981.2021.1899642.

Ho, J., Susser, P., Christian, C., DeLisser, H., Scott, M. J., Pauls, L. A., et al. (2021). Developing the eMedical student (eMS)-a pilot project integrating medical students into the tele-ICU during the COVID-19 pandemic and beyond. *Healthcare (Basel), 9*(1). https://doi.org/10.3390/healthcare9010073.

Huddart, D., Hirniak, J., Sethi, R., Hayer, G., Dibblin, C., Meghna Rao, B., et al. (2020). #MedStudentCovid: How social media is supporting students during COVID-19. *Medical Education, 54*(10), 951–952. https://doi.org/10.1111/medu.14215.

Iacobucci, G. (2020). Covid-19 makes the future of UK clinical research uncertain. *BMJ, 369*, m1619. https://doi.org/10.1136/bmj.m1619.

Ismail, I. I., Abdelkarim, A., & Al-Hashel, J. Y. (2021). Physicians' attitude towards webinars and online education amid COVID-19 pandemic: When less is more. *PLoS One, 16*(4), e0250241. https://doi.org/10.1371/journal.pone.0250241.

Kaliyadan, F., ElZorkany, K., & Al Wadani, F. (2020). An online dermatology teaching module for undergraduate medical students amidst the COVID-19 pandemic: An experience and suggestions for the future. *Indian Dermatology Online Journal, 11*(6), 944–947. https://doi.org/10.4103/idoj.IDOJ_654_20.

Kang, K. A., Kim, S. J., Lee, M. N., Kim, M., & Kim, S. (2020). Comparison of learning effects of virtual reality simulation on nursing students caring for children with asthma. *International Journal of Environmental Research and Public Health, 17*(22). https://doi.org/10.3390/ijerph17228417.

Kardas-Nelson, M. (2020). Covid-19's impact on US medical research-shifting money, easing rules. *BMJ, 369*, m1744. https://doi.org/10.1136/bmj.m1744.

Kasai, H., Shikino, K., Saito, G., Tsukamoto, T., Takahashi, Y., Kuriyama, A., et al. (2021). Alternative approaches for clinical clerkship during the COVID-19 pandemic: Online simulated clinical practice for inpatients and outpatients-a mixed method. *BMC Medical Education, 21*(1), 149. https://doi.org/10.1186/s12909-021-02586-y.

Kim, Y. G., Moon, H., Kim, S. Y., Lee, Y. H., Jeong, D. W., Kim, K., et al. (2019). Inevitable isolation and the change of stress markers in hemodialysis patients during the 2015 MERS-CoV outbreak in Korea. *Scientific Reports, 9*(1), 5676. https://doi.org/10.1038/s41598-019-41964-x.

Kolikonda, M. K., Blaginykh, E., Brown, P., Kovi, S., Zhang, L. Q., & Uchino, K. (2022). Virtual rounding in stroke care and neurology education during the COVID-19 pandemic—A residency program survey. *Journal of Stroke and Cerebrovascular Diseases, 31*(1), 106177. https://doi.org/10.1016/j.jstrokecerebrovasdis.2021.106177.

Lau, Y. H., Lau, K. M., & Ibrahim, N. M. (2021). Management of Parkinson's disease in the COVID-19 pandemic and future perspectives in the era of vaccination. *Journal of Movement Disorder, 14*(3), 177–183. https://doi.org/10.14802/jmd.21034.

Lee, Y. M., Park, K. D., & Seo, J. H. (2020). New paradigm of pediatric clinical clerkship during the epidemic of COVID-19. *Journal of Korean Medical Science, 35*(38), e344. https://doi.org/10.3346/jkms.2020.35.e344.

Leigh, J., Bolton, M., Cain, K., Harrison, N., Bolton, N. Y., & Ratcliffe, S. (2020). Student experiences of nursing on the front line during the COVID-19 pandemic. *The British Journal of Nursing, 29*(13), 788–789. https://doi.org/10.12968/bjon.2020.29.13.788.

Li, L. Z., & Wang, S. (2020). Prevalence and predictors of general psychiatric disorders and loneliness during COVID-19 in the United Kingdom. *Psychiatry Research*, *291*, 113267. https://doi.org/10.1016/j.psychres.2020.113267.

Mandan, J., Sidhu, H. S., & Mahmood, A. (2016). Should a clinical rotation in hematology be mandatory for undergraduate medical students? *Advances in Medical Education and Practice*, 7, 519–521. https://doi.org/10.2147/AMEP.S112132.

Mazza, C., Ricci, E., Biondi, S., Colasanti, M., Ferracuti, S., Napoli, C., et al. (2020). A Nationwide survey of psychological distress among Italian people during the COVID-19 pandemic: Immediate psychological responses and associated factors. *International Journal of Environmental Research and Public Health*, *17*(9). https://doi.org/10.3390/ijerph17093165.

McDermott, M. M., & Newman, A. B. (2020). Preserving clinical trial integrity during the coronavirus pandemic. *JAMA*, *323*(21), 2135–2136. https://doi.org/10.1001/jama.2020.4689.

Medicines and Healthcare Products Regulatory Agency. (2020). *Guidance on minimising disruptions to the conduct and integrity of clinical trials of medicines during COVID-19.* Retrieved from https://www.gov.uk/guidance/guidance-on-minimising-disruptions-to-the-conduct-and-integrity-of-clinical-trials-of-medicines-during-covid-19.

Michener, A., Fessler, E., Gonzalez, M., & Miller, R. K. (2020). The 5 M's and more: A new geriatric medical student virtual curriculum during the COVID-19 pandemic. *Journal of the American Geriatrics Society*, *68*(11), E61–E63. https://doi.org/10.1111/jgs.16855.

Muir, R. T., Gros, P., Ure, R., Mitchell, S. B., Kassardjian, C. D., Izenberg, A., et al. (2021). Modification to neurology residency training: The Toronto neurology COVID-19 pandemic experience. *Neurology Clinical Practice*, *11*(2), e165–e169. https://doi.org/10.1212/CPJ.0000000000000894.

Myers, K. R., Tham, W. Y., Yin, Y., Cohodes, N., Thursby, J. G., Thursby, M. C., et al. (2020). Unequal effects of the COVID-19 pandemic on scientists. *Nature Human Behaviour*, *4*(9), 880–883. https://doi.org/10.1038/s41562-020-0921-y.

National Institute for Health Research. (2021). *Urgent public health COVID-19 studies.* Retrieved from https://www.nihr.ac.uk/covid-studies/.

Papa, S. M., Brundin, P., Fung, V. S. C., Kang, U. J., Burn, D. J., Colosimo, C., et al. (2020). Impact of the COVID-19 pandemic on Parkinson's disease and movement disorders. *Movement Disorder Clinical Practice*, 7(4), 357–360. https://doi.org/10.1002/mdc3.12953.

Park, J., Park, H., Lim, J. E., Rhim, H. C., & Lee, Y. M. (2020). Medical students' perspectives on recommencing clinical rotations during coronavirus disease 2019 at one institution in South Korea. *The Korean Journal of Medical Education*, *32*(3), 223–229. https://doi.org/10.3946/kjme.2020.170.

Parkinson's UK. (2020). *Protecting the future of research after coronavirus (COVID-19).* Retrieved from https://www.parkinsons.org.uk/news/protecting-future-research-after-coronavirus-covid-19.

Perissotto, T., Silva, T., Miskulin, F. P. C., Pereira, M. B., Neves, B. A., Almeida, B. C., et al. (2021). Mental health in medical students during COVID-19 quarantine: A comprehensive analysis across year-classes. *Clinics (São Paulo, Brazil)*, *76*, e3007. https://doi.org/10.6061/clinics/2021/e3007.

Pierce, M., Hope, H., Ford, T., Hatch, S., Hotopf, M., John, A., et al. (2020). Mental health before and during the COVID-19 pandemic: A longitudinal probability sample survey of the UK population. *Lancet Psychiatry*, 7(10), 883–892. https://doi.org/10.1016/S2215-0366(20)30308-4.

Pokryszko-Dragan, A., Marschollek, K., Nowakowska-Kotas, M., & Aitken, G. (2021). What can we learn from the online learning experiences of medical students in Poland during the SARS-CoV-2 pandemic? *BMC Medical Education*, *21*(1), 450. https://doi.org/10.1186/s12909-021-02884-5.

Rahman, S. A., Tuckerman, L., Vorley, T., & Gherhes, C. (2021). Resilient research in the field: Insights and lessons from adapting qualitative research projects during the COVID-19 pandemic. *International Journal of Qualitative Methods, 20*, 1–16. https://doi.org/10.1177/16094069211016106.

Rajan, R., Radhakrishnan, D. M., Srivastava, A. K., Vishnu, V. Y., Gupta, A., Shariff, A., et al. (2020). Conduct of virtual neurology DM final examination during COVID-19 pandemic. *Annals of Indian Academy of Neurology, 23*(4), 429–432. https://doi.org/10.4103/aian.AIAN_593_20.

Roberts, M. (2020). *Coronavirus: Has the NHS got enough PPE? [Online]*. BBC News.

Rose, S. (2020). Medical student education in the time of COVID-19. *JAMA, 323*(21), 2131–2132. https://doi.org/10.1001/jama.2020.5227.

Safdieh, J. E., Lee, J. I., Prasad, L., Mulcare, M., Eiss, B., & Kang, Y. (2021). Curricular response to COVID-19: Real-time interactive telehealth experience (RITE) program. *Medical Education Online, 26*(1), 1918609. https://doi.org/10.1080/10872981.2021.1918609.

Sandrone, S., Albert, D. V., Dunham, S. R., Kraker, J., Noviawaty, I., Palm, M., et al. (2021). Training in neurology: How lessons learned on teaching, well-being and telemedicine during the COVID-19 pandemic can shape the future of neurology education. *Neurology*. https://doi.org/10.1212/WNL.0000000000012010.

Seetan, K., Al-Zubi, M., Rubbai, Y., Athamneh, M., Khamees, A., & Radaideh, T. (2021). Impact of COVID-19 on medical students' mental wellbeing in Jordan. *PLoS One, 16*(6), e0253295. https://doi.org/10.1371/journal.pone.0253295.

Shivkumar, V., Subramanian, T., Agarwal, P., Mari, Z., Mestre, T. A., & Parkinson Study, G. (2021). Uptake of telehealth in Parkinson's disease clinical care and research during the COVID-19 pandemic. *Parkinsonism & Related Disorders, 86*, 97–100. https://doi.org/10.1016/j.parkreldis.2021.03.032.

Simundic, A. M. (2013). Bias in research. *Biochemica Medica (Zagreb), 23*(1), 12–15. https://doi.org/10.11613/bm.2013.003.

Staniscuaski, F., Kmetzsch, L., Soletti, R. C., Reichert, F., Zandona, E., Ludwig, Z. M. C., et al. (2021). Gender, race and parenthood impact academic productivity during the COVID-19 pandemic: From survey to action. *Frontiers in Psychology, 12*, 663252. https://doi.org/10.3389/fpsyg.2021.663252.

Suarez-Cedeno, G., Pantelyat, A., Mills, K. A., Murthy, M., Alshaikh, J. T., Rosenthal, L. S., et al. (2021). Movement disorders virtual fellowship training in times of coronavirus disease 2019: A single-center experience. *Telemedicine Journal and E-Health, 27*(10), 1160–1165. https://doi.org/10.1089/tmj.2020.0419.

Sukumar, S., Zakaria, A., Lai, C. J., Sakumoto, M., Khanna, R., & Choi, N. (2021). Designing and implementing a novel virtual rounds curriculum for medical students' internal medicine clerkship during the COVID-19 pandemic. *MedEdPORTAL, 17*, 11106. https://doi.org/10.15766/mep_2374-8265.11106.

Suzuki, K., Numao, A., Komagamine, T., Haruyama, Y., Kawasaki, A., Funakoshi, K., et al. (2021). Impact of the COVID-19 pandemic on the quality of life of patients with Parkinson's disease and their caregivers: A single-center survey in Tochigi prefecture. *Journal of Parkinson's Disease, 11*(3), 1047–1056. https://doi.org/10.3233/JPD-212560.

Tan, E. K., Albanese, A., Chaudhuri, K., Lim, S. Y., Oey, N. E., Shan Chan, C. H., et al. (2021). Adapting to post-COVID19 research in Parkinson's disease: Lessons from a multinational experience. *Parkinsonism & Related Disorders, 82*, 146–149. https://doi.org/10.1016/j.parkreldis.2020.10.009.

Tan, E. K., Albanese, A., Chaudhuri, K. R., Opal, P., Wu, Y. C., Chan, C. H., et al. (2020). Neurological research & training after the easing of lockdown in countries impacted by COVID-19. *Journal of the Neurological Sciences, 418*, 117105. https://doi.org/10.1016/j.jns.2020.117105.

Torales, J., O'Higgins, M., Castaldelli-Maia, J. M., & Ventriglio, A. (2020). The outbreak of COVID-19 coronavirus and its impact on global mental health. *The International Journal of Social Psychiatry*, *66*(4), 317–320. https://doi.org/10.1177/0020764020915212.

van der Meulen, M., Kleineberg, N. N., Schreier, D. R., Garcia-Azorin, D., & Di Lorenzo, F. (2020). COVID-19 and neurological training in Europe: From early challenges to future perspectives. *Neurological Sciences*, *41*(12), 3377–3379. https://doi.org/10.1007/s10072-020-04723-9.

Wang, J. J., Deng, A., & Tsui, B. C. H. (2020). COVID-19: Novel pandemic, novel generation of medical students. *British Journal of Anaesthesia*, *125*(3), e328–e330. https://doi.org/10.1016/j.bja.2020.05.025.

Weber, D. J., Albert, D. V. F., Aravamuthan, B. R., Bernson-Leung, M. E., Bhatti, D., & Milligan, T. A. (2020). Training in neurology: Rapid implementation of cross-institutional neurology resident education in the time of COVID-19. *Neurology*, *95*(19), 883–886. https://doi.org/10.1212/WNL.0000000000010753.

Weber, A. M., Dua, A., Chang, K., Jupalli, H., Rizwan, F., Chouthai, A., et al. (2021). An outpatient telehealth elective for displaced clinical learners during the COVID-19 pandemic. *BMC Medical Education*, *21*(1), 174. https://doi.org/10.1186/s12909-021-02604-z.

Welton, T., & Tan, E. K. (2021). Applying artificial intelligence to multi-omic data: New functional variants in Parkinson's disease. *Movement Disorders*, *36*(2), 347. https://doi.org/10.1002/mds.28481.

Weston, J., & Zauche, L. H. (2021). Comparison of virtual simulation to clinical practice for prelicensure nursing students in pediatrics. *Nurse Educator*, *46*(5), E95–E98. https://doi.org/10.1097/NNE.0000000000000946.

White, S., Fook, J., & Gardner, F. (2006). *Critical reflection in health and social care*. Open University Press, McGraw-Hill Education.

Wray, S., & Kinman, G. (2022). The challenges of COVID-19 for the well-being of academic staff. *Occupational Medicine (London)*, *72*(1), 2–3. https://doi.org/10.1093/occmed/kqab007.

Wu, A., Parris, R. S., Scarella, T. M., Tibbles, C. D., Torous, J., & Hill, K. P. (2021). What gets resident physicians stressed and how would they prefer to be supported? A best-worst scaling study. *Postgraduate Medical Journal*. https://doi.org/10.1136/postgradmedj-2021-140719.

Wyatt, D., Faulkner-Gurstein, R., Cowan, H., & Wolfe, C. D. A. (2021). Impacts of COVID-19 on clinical research in the UK: A multi-method qualitative case study. *PLoS One*, *16*(8), e0256871. https://doi.org/10.1371/journal.pone.0256871.

Yadala, S., Nalleballe, K., Sharma, R., Lotia, M., Kapoor, N., Veerapaneni, K. D., et al. (2020). Resident education during COVID-19 pandemic: Effectiveness of virtual electroencephalogram learning. *Cureus*, *12*(10), e11094. https://doi.org/10.7759/cureus.11094.

YoungMinds. (2020). *Coronavirus: Impact on young people with mental health needs*.

Zeinali, M., Almasi-Doghaee, M., & Haghi-Ashtiani, B. (2020). Facing COVID-19, jumping from in-person training to virtual learning: A review on educational and clinical activities in a neurology department. *Basic and Clinical Neuroscience*, *11*(2), 151–154. https://doi.org/10.32598/bcn.11.covid19.910.2.